Lecture Notes in Computer Science   6713

Commenced Publication in 1973
Founding and Former Series Editors:
Gerhard Goos, Juris Hartmanis, and Jan van Leeuwen

Carlo Sansone   Josef Kittler
Fabio Roli (Eds.)

Multiple Classifier Systems

10th International Workshop, MCS 2011
Naples, Italy, June 15-17, 2011
Proceedings

 Springer

Volume Editors

Carlo Sansone
Università di Napoli Federico II, Dipartimento di Informatica e Sistemistica
Via Claudio, 21, 80125 Napoli, Italy
E-mail: carlosan@unina.it

Josef Kittler
University of Surrey, Centre for Vision, Speech and Signal Processing
Guildford, Surrey, GU2 7XH, UK
E-mail: eesljk@ee.surrey.ac.uk

Fabio Roli
University of Cagliari, Department of Electrical and Electronic Engineering
Piazza d'Armi, 09123, Cagliari, Italy
E-mail: roli@diee.unica.it

ISSN 0302-9743   e-ISSN 1611-3349
ISBN 978-3-642-21556-8   e-ISBN 978-3-642-21557-5
DOI 10.1007/978-3-642-21557-5
Springer Heidelberg Dordrecht London New York

Library of Congress Control Number: 2011929083

CR Subject Classification (1998): I.4, H.3, I.5, F.1, J.3, H.4

LNCS Sublibrary: SL 6 – Image Processing, Computer Vision, Pattern Recognition, and Graphics

Typesetting: Camera-ready by author, data conversion by Scientific Publishing Services, Chennai, India

Printed on acid-free paper

Springer is part of Springer Science+Business Media (www.springer.com)

Preface

This volume contains the papers presented at the Multiple Classifier Systems Workshop, MCS 2011, held in Naples, Italy, during June 15–17, 2011. Being the tenth in a well-established series of meetings providing an international forum for discussion of issues in multiple classifier system design, the workshop achieved its objective of bringing together researchers from the diverse communities (neural networks, pattern recognition, machine learning and statistics) working on this research topic.

From more than 50 submissions, the Program Committee selected 36 papers to create an interesting scientific program. Papers were organized into sessions dealing with classifier combination and classifier selection, diversity, bagging and boosting, combination of multiple kernels, and applications, especially in the computer security field. The workshop program and this volume were enriched by two invited talks given by Shai Avidan (Tel Aviv University, Israel), and Nicolò Cesa-Bianchi (University of Milan, Italy).

This workshop would not have been possible without the help of many individuals and organizations. First of all, our thanks go to the members of the MCS 2011 Program Committee and to the reviewers, whose expertise was instrumental for selecting contributions that could characterize the progress made in the field over the last years and could aspire to chart its future research. The management of the papers, including the preparation of this proceedings volume, was done by the EasyChair conference management system.

A special thank goes to the members of the Local Organizing Committee, Francesco Gargiulo, Emanuela Marasco, Claudio Mazzariello and Vincenzo Paduano for their valuable contributions to the organization, and their availability to solve the practical problems arising during the preparation of MCS 2011. We also want to thank Giorgio Fumera who managed the MCS website.

This workshop was organized by the Department of Computer and Systems Engineering of the University of Naples Federico II, Italy, the Center for Vision, Speech and Signal Processing, University of Surrey, UK, and the Department of Electrical and Electronic Engineering of the University of Cagliari, Italy.

This MCS edition was included in the list of events celebrating the bicentenary of the School of Engineering of the University of Naples Federico II. We thank the International Association for Pattern Recognition (IAPR) and the Italian group of researchers affiliated to the IAPR (GIRPR) for endorsing MCS 2011.

We finally wish to express our appreciation to our financial sponsors: Nettuno Solutions, Ericsson Telecomunicazioni and the AIRobots European Project – 7FP.

June 2011

Carlo Sansone
Joseph Kittler
Fabio Roli

Organization

MCS 2011 was organized by the Department of Computer Engineering and Systems (DIS) of the University of Naples Federico II (Italy), in association with the Center for Vision, Speech and Signal Processing (CVSSP), University of Surrey, UK and the Department of Electrical and Electronic Engineering (DIEE) of the University of Cagliari, Italy.

Program Committee

Conference Chairs   Carlo Sansone (University of Naples Federico II, Italy)
Josef Kittler (University of Surrey, UK)
Fabio Roli (University of Cagliari, Italy)

Scientific Committee

J.A. Benediktsson (Iceland)
G. Brown (UK)
H. Bunke (Switzerland)
L.P. Cordella (Italy)
R.P.W. Duin (The Netherlands)
N. El-Gayar (Egypt)
G. Fumera (Italy)
C. Furlanello (Italy)
J. Ghosh (USA)
V. Govindaraju (USA)
M. Haindl (Czech Republic)
L. Hall (USA)
T.K. Ho (USA)
N. Intrator (Israel)
P. Kegelmeyer (USA)

K. Kryszczuk (Switzerland)
L.I. Kuncheva (UK)
V. Mottl (Russia)
K. Nandakumar (Singapore)
N. Oza (USA)
E. Pekalska (UK)
R. Polikar (USA)
J.J. Rodriguez (Spain)
A. Ross (USA)
A. Sharkey (UK)
F. Tortorella (Italy)
G. Valentini (Italy)
T. Windeatt (UK)
D. Windridge (UK)
Z.-H. Zhou (China)

Local Organizing Committee

Francesco Gargiulo
Emanuela Marasco
Claudio Mazzariello
Vincenzo Paduano

Endorsed by

IAPR - International Association for Pattern Recognition
GIRPR - Italian Chapter of the IAPR

Supported by

AIRobots European Project - 7FP
Ericsson Telecomunicazioni S.p.A.
Nettuno Solutions s.r.l.

Additional Reviewers

Acharyya, Sreangsu	Kramer, Kurt	Shoemaker, Larry
Basu, Satrajit	Lee, Wan-Jui	Simeone, Paolo
Fischer, Andreas	Li, Nan	Tulyakov, Sergey
Fontanella, Francesco	Marpu, Prashanth	Yan, Fei
Hruschka, Eduardo	Marrocco, Claudio	Zhan, De-Chuan
Indermühle, Emanuel	Re, Matteo	
Korecki, John	Ricamato, Maria Teresa	

Table of Contents

ECOC

Diversity

Clustering

Biometrics

Computer Security

Ensemble Methods for Tracking and Segmentation (Abstract)

Shai Avidan

School of Electrical Engineering,
Tel-Aviv University,
Tel-Aviv 69978, Israel
avidan@eng.tau.ac.il
http://www.www.tau.ac.il/avidan

Abstract. Ensemble methods offer an elegant way of training an ensemble of weak classifiers into a strong classifier through the use of the AdaBoost algorithm. In this abstract we discuss two extensions of AdaBoost and demonstrate them on two problems in the field of Computer Vision. The first, termed *Ensemble Tracking*, extends AdaBoost in the temporal domain and adapts it to the problem of tracking an object in a video sequence. The second, termed *SpatialBoost*, extends AdaBoost in the spatial domain and adapts it to the problem of interactive image segmentation.

In the case of *Ensemble Tracking*, we consider tracking as a binary classification problem, where an ensemble of weak classifiers is trained online to distinguish between the object and the background. But because of real-time constraints we can only train one weak classifier per frame and can not maintain an ensemble of more than a limited number of weak classifiers. We propose an online extension of AdaBoost, the *Ensemble Tracker*, that constantly, and efficiently, combines this stream of weak classifiers into a strong classifier, without going beyond the bound on the number of weak classifiers.

In the case of *SpatialBoost* we show how to extend AdaBoost to incorporate spatial reasoning and demonstrate it on the problem of interactive image segmentation. The user marks some of the pixels as positive and negative examples and then lets the algorithm label the rest of the pixels. Simply training AdaBoost on the appearance of the labeled pixels and using it to label the unlabeled pixels yields unsatisfactory results. This is because AdaBoost lacks spatial reasoning. But in fact, we know that nearby pixels should, quite often, have the same label. To this end we introduce spatial reasoning in the form of weak classifiers that attempt to infer pixel label from the pixel labels of surrounding pixels, after each boosting iteration. SpatialBoost combines these spatial weak classifiers with the appearance based weak classifier to give superior results.

Taken together, these extensions demonstrate the flexibility of ensemble methods, and the ways in which they can be modified to account for special properties of images and video. We conclude by sketching a number of possible extensions to this line of work.

C. Sansone, J. Kittler, and F. Roli (Eds.): MCS 2011, LNCS 6713, p. 1, 2011.
© Springer-Verlag Berlin Heidelberg 2011

Ensembles and Multiple Classifiers: A Game-Theoretic View

Nicolò Cesa-Bianchi

DSI, Università degli Studi di Milano, Italy

1 Aggregating Strategies

The study of multiple classifier systems is a fundamental topic in modern machine learning. However, early work on aggregation of predictors can be traced back to the Fifties, in the area of game theory. At that time, the pioneering work of James Hannan [11] and David Blackwell [2] laid down the foundations of repeated game theory. In a nutshell, a repeated game is the game-theoretic interpretation of learning. In games played once, lacking any information about the opponent, the best a player can do is to play the minimax strategy (the best strategy against the worst possible opponent). In repeated games, by examining the history of past opponent moves, the player acquires information about the opponent's behavior and can *adapt* to it, in order to achieve a better payoff than that guaranteed by the minimax strategy.

Years later, Volodya Vovk [18] and Tom Cover [7,8] in Information Theory, Nick Littlestone and Manfred Warmuth [15] in Computer Science, and others in different fields, re-discovered, and greatly extended, these results by viewing the repeated game as a problem of strategy aggregation with a changing payoff structure. This gave rise to the paradigm of *prediction with expert advice*, in which an aggregating strategy combines the predictions of a number of base strategies (the experts) in a sequential prediction game. The kind of guarantees one can prove are game-theoretic. For examples, there exists a (simple) randomized aggregating strategy that, on any sequential classification problem and for any set of N base classifiers, guarantees an average mistake rate that converges to that of the best base classifier at rate $\sqrt{T \ln N}$, where T is the number of prediction steps [5].

In the last fifteen years, researchers in learning theory started to recognize the game-theoretic nature of certain learning results, such as the Perceptron convergence theorem [16], and developed a research program devoted to the cross-fertilization between game theory and machine learning. In the game-theoretic analysis of Perceptron, one combines features (the experts, in the game-theoretic view) with the goal of achieving the performance of the best linear combination of features on the observed data sequence (whose labels correspond to the moves played by the opponent). This paradigm of *online linear learning* allowed to analyze many gradient descent algorithms for classification and regression, such as p-norm Perceptrons [10], Winnow [14], the Widrow-Hoff rule [6], Ridge Regression [1,17], and others.

C. Sansone, J. Kittler, and F. Roli (Eds.): MCS 2011, LNCS 6713, pp. 2–5, 2011.

Although online learning algorithms are not multiple classifier systems, they naturally generate an ensemble of classifiers, or predictors: the online ensemble.

2   Online Ensembles

Online ensembles connect online learning to statistical learning theory. Namely, these ensembles allow to operate and analyze online algorithms in statistical scenarios, where the learner uses a training set to construct a classifier or regressor with small statistical risk. When run on a training set of examples, an online linear algorithm considers one example at a time. At step t, the algorithm evaluates, on the t-th training example, the current linear model built on the previous $t - 1$ examples (call the resulting loss the *online loss*). Then, the linear model is possibly updated, reflecting the information acquired by observing the t-th example. The online ensemble is the collection of linear models generated by the online algorithm in a pass on the training set. Now, assuming the training set is a statistical sample, the theory of large deviations implies that with high probability the average risk on the online ensemble is close to the average online loss. Moreover, online analysis can be used to bound the average online loss in terms of the empirical risk of the best linear model, which is in turn close to its true risk. This implies that a typical element of the online ensemble has a risk that can be bounded in terms of the risk of the best linear model [4].

3   Multikernel and Multitask Online Learning

Besides considering the ensemble of models generated by online algorithms, one can investigate directly the possibilities offered by an ensemble of online learners. A first example is multikernel learning, where K instances of an online kernel-based classifier (e.g., a kernel Perceptron) are run simultaneously, each in its own reproducing kernel Hilbert space. The weights of the K instances are linearly combined to compute the aggregated prediction. Online learning theory (more precisely, mirror descent analysis) prescribes that the linear coefficients be the components of the gradient of a certain norm, or potential, applied to the sum of past loss gradients. Using a certain family of norms, $(2, p)$ group norms, it can be proven that the system has a good online performance (on any individual data sequence) whenever there exists a good fixed multikernel linear classifier with a sparse vector of linear coefficients [13,12].

A second example is when each instance in the multiple classifier system is trained on a possible different classification problem. Here the idea is to have these instances interact, in order to exploit possible similarities among the tasks being solved. This is done as follows: whenever a single instance is updated (because a new example for the associated task is observed), then the update is shared among other instances, where the degree of sharing is ruled by a graph representing a priori knowledge about potential task similarities. This shared update can be described as a kernel. Hence, the whole system can be analyzed as a kernel Perceptron, with a bound on the number of mistakes that holds for any individual data sequence and depends on the closeness in Euclidean space of the best linear classifiers for all tasks [3].

4   Conclusions

These are just a few examples of the many possibilities of applying online learning, and other game-theoretic techniques, to the field of multiple classifier systems. We hope to have enthused the reader into looking at them in more detail. As a final remark, we recall that one of the most celebrated ensemble method, AdaBoost, is exactly equivalent to a repeated game in the framework of prediction with expert advice [9]. This equivalence provides an elegant game-theoretic analysis of AdaBoost.

References

1. Azoury, K.S., Warmuth, M.K.: Relative loss bounds for on-line density estimation with the exponential family of distributions. Machine Learning 43(3), 211–246 (2001)
2. Blackwell, D.: An analog of the minimax theorem for vector payoffs. Pacific Journal of Mathematics 6, 1–8 (1956)
3. Cavallanti, G., Cesa-Bianchi, N., Gentile, C.: Linear algorithms for online multitask classification. Journal of Machine Learning Research 11, 2597–2630 (2010)
4. Cesa-Bianchi, N., Conconi, A., Gentile, C.: On the generalization ability of on-line learning algorithms. IEEE Transactions on Information Theory 50(9), 2050–2057 (2004)
5. Cesa-Bianchi, N., Freund, Y., Helmbold, D.P., Haussler, D., Schapire, R., Warmuth, M.K.: How to use expert advice. Journal of the ACM 44(3), 427–485 (1997)
6. Cesa-Bianchi, N., Long, P.M., Warmuth, M.K.: Worst-case quadratic loss bounds for a generalization of the Widrow-Hoff rule. In: Proceedings of the 6th Annual ACM Workshop on Computational Learning Theory, pp. 429–438. ACM Press, New York (1993)
7. Cover, T.: Behaviour of sequential predictors of binary sequences. In: Proceedings of the 4th Prague Conference on Information Theory, Statistical Decision Functions and Random Processes, pp. 263–272. Publishing house of the Czechoslovak Academy of Sciences (1965)
8. Cover, T.: Universal portfolios. Mathematical Finance 1, 1–29 (1991)
9. Freund, Y., Schapire, R.: Game theory, on-line prediction and boosting. In: Proceedings of the 9th Annual Conference on Computational Learning Theory. ACM Press, New York (1996)
10. Gentile, C.: The robustness of the p-norm algorithms. Machine Learning 53(3), 265–299 (2003)
11. Hannan, J.: Approximation to Bayes risk in repeated play. Contributions to the theory of games 3, 97–139 (1957)
12. Jie, L., Orabona, F., Fornoni, M., Caputo, B., Cesa-Bianchi, N.: OM-2: An online multi-class multi-kernel learning algorithm. In: Proceedings of the 4th IEEE Online Learning for Computer Vision Workshop. IEEE Press, Los Alamitos (2007)
13. Kakade, S., Shalev-Shwartz, S., Tewari, A.: On the duality of strong convexity and strong smoothness: Learning applications and matrix regularization (2009)

14. Littlestone, N.: Learning quickly when irrelevant attributes abound: a new linear-threshold algorithm. Machine Learning 2(4), 285–318 (1988)
15. Littlestone, N., Warmuth, M.K.: The weighted majority algorithm. Information and Computation 108, 212–261 (1994)
16. Novikoff, A.B.J.: On convergence proofs of Perceptrons. In: Proceedings of the Symposium on the Mathematical Theory of Automata vol. XII, pp. 615–622 (1962)
17. Vovk, V.: Competitive on-line statistics. International Statistical Review 69, 213–248 (2001)
18. Vovk, V.G.: Aggregating strategies. In: Proceedings of the 3rd Annual Workshop on Computational Learning Theory, pp. 372–383 (1990)

Anomaly Detection Using Ensembles

Larry Shoemaker and Lawrence O. Hall

Computer Science and Engineering, University of South Florida, Tampa, FL 33620-5399
Tel.: (813)974-3652, Fax: (813)974-5456
lwshoema@cse.usf.edu,
hall@cse.usf.edu

Abstract. We show that using random forests and distance-based outlier partitioning with ensemble voting methods for supervised learning of anomaly detection provide similar accuracy results when compared to the same methods
without partitioning. Further, distance-based outlier and one-class support vector machine partitioning and ensemble methods for semi-supervised learning of
anomaly detection also compare favorably to the corresponding non-ensemble
methods. Partitioning and ensemble methods would be required for very large
datasets that need distributed computing approaches. ROC curves often show significant improvement from increased true positives in the low false positive range
for ensemble methods used on several datasets.

Keywords: outliers, anomalies, random forests, data partitioning, ROC curves.

1 Introduction

Anomaly detection, also known as outlier detection, deals with finding patterns in data
that are unusual, abnormal, unexpected, and/or interesting [7]. Anomalies are important
because they translate to significant information that can lead to critical action in a wide
variety of application domains, such as credit card fraud detection, security intrusion
detection, insurance, health care, fault detection, and military surveillance. Some of
the challenges presented by anomaly detection include imprecise boundaries between
normal and anomalous behavior, malicious actions that make anomalies appear normal,
evolution of normal behavior, different application domains, lack of labeled data, and
noise. Researchers have applied concepts from statistics, machine learning, data mining,
information theory, and spectral theory to form anomaly detection techniques [7,19].

Accurate data labels that denote normal or anomalous behavior usually require manual effort by a human expert and can be too expensive to acquire for many applications.
In addition, labeling all possible types of anomalies that can arise in an evolving domain can be difficult. Based on the type of data labels available, there are three modes in
which anomaly detection techniques can operate. They are supervised, semi-supervised,
and unsupervised anomaly detection modes [7].

In supervised mode, a training dataset with labeled instances for the normal and
anomaly classes is assumed. Typically, a predictive model is built for normal *vs.*
anomaly classes using the training data. Then the model is used to predict the class
of unseen data. Since the anomaly class is almost always rare, the imbalanced data
distribution problem must be addressed.

C. Sansone, J. Kittler, and F. Roli (Eds.): MCS 2011, LNCS 6713, pp. 6–15, 2011.

In semi-supervised mode, it is assumed that the training data has labels only for the normal class. A model is built for the normal class, and the model is used to identify anomalies in the test dataset. In unsupervised mode, no labeled training data is required. The only assumption made is that normal instances appear much more frequently than anomaly instances in the test data [7]. If this assumption is not valid, a high false alarm rate results.

Ensemble methods improve class accuracy by combining the predictions of multiple classifiers. Requirements for improved accuracy include having independent (or only slightly correlated) base classifiers that perform better than random guessing [20]. Breiman's random forest (RF) algorithm [6] is an ensemble method specifically designed for decision tree classifiers. One popular random forest method initially uses bagging or bootstrap aggregating to repeatedly sample with replacement from a dataset using a uniform probability distribution [20]. Another method of injecting randomness into each base decision tree, grown using each bag of training data, is to randomly select only some of the features as tests at each node of the decision tree. There are different ways of determining the number of features selected, but one commonly used is $log_2 n + 1$ given n features. Random forests weighted (RFW) predictions are based on the percentage of trees that vote for a class. The motivation for using this ensemble technique stems from the inherent speed benefit of analyzing only a few possible attributes from which a test is selected at an internal tree node. The accuracy of random forests was evaluated in [3] and shown to be comparable with or better than other well-known ensemble generation techniques. It is more impervious to noise than AdaBoost, a commonly used boosting ensemble method [20].

A recent intrusion detection approach used feature subsets and a modular ensemble of one-class classifiers chosen for different groups of network services, with false alarm rates and detection rates tuned for best ensemble performance [12]. Good results were shown for the KDD Cup 1999 dataset. A multiple classifier system for accurate payload-based anomaly detection also used an ensemble of one-class classifiers and focused on ROC curve results at low false alarm rates [18]. Another one-class classification approach combined density and class probability estimates [14].

Ensemble approaches for anomaly detection are explored and compared to some existing anomaly detection approaches. Random forests and distance-based outlier ensemble methods for supervised learning of anomaly detection are compared to traditional methods on the same datasets. Further, one-class support vector machines and distance-based outlier ensemble methods for semi-supervised learning of anomaly detection are compared to non-ensemble methods.

2 Datasets

In order to compare our ensemble based anomaly detection approaches to other approaches on the same dataset, we selected the same modified KDD Cup 1999 data subset used in [16,1]. The unmodified dataset "includes a wide variety of intrusions simulated in a military network environment"[2]. The full dataset, a 10% subset of the dataset, a full test dataset, and three different 10% subsets of the full test dataset are available from the UCI KDD Archive [2]. The KDD Cup 1999 10% test dataset with corrected labels is the one modified and used as in [16,1]. This dataset with 311,029 data records and five classes was modified to include only the normal class and the U2R intrusion attack class, which was considered the outlier or anomaly class. The modified KDD U2R dataset has 60,593 normal and 246 intrusion (outlier) data records. Table 1 shows the modified dataset characteristics, as well as those of a second modified KDD dataset with 60,593 normal and 4166 probe intrusion attack class (outlier) data records for additional testing. We also used the same two modified ann-thyroid datasets, and the same five modified shuttle datasets used in [16,1].

Table 1. Dataset characteristics

Modifications made	Size	# of features		Outliers	
		continuous	discrete	# of	% of
KDD U2R vs. normal	60,839	32	9	246	0.40%
KDD probe vs. normal	64,759	32	9	4,166	6.43%
Ann-thyroid 1 vs. normal	3,251	6	15	73	2.25%
Ann-thyroid 2 vs. normal	3,355	6	15	177	5.28%
shuttle 2 vs. normal	11,491	9	0	13	0.11%
shuttle 3 vs. normal	11,517	9	0	39	0.34%
shuttle 5 vs. normal	12,287	9	0	809	6.58%
shuttle 6 vs. normal	11,482	9	0	4	0.03%
shuttle 7 vs. normal	11,480	9	0	2	0.02%

3 Experiments

Anomaly detection, also known as outlier detection, deals with finding patterns in data that are unusual, abnormal, unexpected, and/or interesting [7] (see Section 1). In these experiments the outlier examples are the selected minority class of the dataset as described below.

As discussed in Section 2, we used the same modified KDD U2R dataset, the two modified ann-thyroid datasets, and the five modified shuttle datasets used in [16,1], which include only the normal class and one other minority class, considered as the outlier or anomaly class. Each dataset was randomly split into two groups a total of 30 times, with each group used as a train/test set for 60 total trials as in [16,1]. In order to investigate ensemble approaches to outlier detection for these datasets, and to compare the results to those in [16,1], each group was also partitioned into 20 stratified partitions (each partition had about the same number of positive instances and about the same number of negative instances) for testing on the other group of the random data split. This basic process was repeated for 30 different random splits.

The first group of experiments included using supervised outlier detection methods on the data in one data group or on partitions of that group to determine outliers in the other data group. The outlier detection methods included random forests weighted (RFW) [6] and distance-based outlier (DBO) methods [4]. The accuracy of random forests was evaluated in [3] and shown to be comparable with or better than other well-known ensemble generation techniques. Here, the number of random features chosen at each decision tree node was $log_2 n + 1$ given n features. RFW predictions are based on the percentage of trees that vote for a class. First RFW with 250 trees was trained on all of the data (without partitioning) in one group and then used to test all of the data in the remaining group. Breiman used random forests with 100 trees [6], while a later study used 1000 trees [3]. We chose 250 to achieve more accurate results than the established number of 100 without incurring the additional computational costs of 1000 trees, which might not necessarily prove beneficial in our experiments.

The random split was intended to duplicate the methods used in [16,1], and did not limit each group to the same number of examples. RFW with 250 trees was trained on the data in each training partition and tested on all of the test data in the other group. This simulates distributed data that can't be shared; a more difficult problem. We also trained an ensemble of 250 random forests on the data in each of the 20 partitions. The weighted votes from each partition's RFW ensemble for each test example were averaged for the final vote. The RFW weighted votes (scores) were sorted with the most likely to be an outlier first, for determining the ROC AUC.

The above train/test experiments were repeated using the distance-based outlier (DBO) algorithm except that the training data in all cases was used as the reference set of neighbors for determining the average distance from each test instance to its five nearest neighbors [4]. DBO was chosen as the conventional outlier method for comparison with RFW and for investigating DBO's ensemble performance. DBO is a state of the art approach to outlier detection that is reasonably fast, efficient, and flexible, being recently modified for streaming data [17]. The software implementation of DBO was ORCA, which is a method for finding outliers in near linear time [4]. Continuous features were scaled to zero mean and unit standard deviation. The distance measure used for continuous features was the weighted Euclidean distance, and for discrete features was the weighted Hamming distance. The number of outliers was specified as the number of test instances, so that as many test instances as possible were given outlier prediction scores. Unlike RFW, the ground truth training labels were not used for training, only for the stratified partitioning. The DBO outlier prediction scores were used instead of the corresponding RFW scores above. These prediction scores rated each test instance's likelihood of being an outlier based on the average distance from its five nearest neighbors in the reference training set. Those instances with a higher average distance were given a higher score to reflect the increased likelihood of being outliers.

The second group of experiments included using semi-supervised outlier detection methods on the data in each group or in partitions of that group to determine outliers in the other group. This was done for the same random splits of data that were used for the supervised group of experiments. However, only examples of the normal class were used for training. The outlier detection methods included distance-based outlier (DBO) and one-class support vector machines [15]. DBO was chosen for these experiments for direct comparison to its performance in the first group of supervised learning experiments. Each test was repeated by swapping the training and the test group. First DBO used all of the data in the training group as a reference set of neighbors for determining the average distance from each test instance to its five nearest neighbors. Then DBO used the data in each test partition as the reference set to choose neighbors. The LIB-SVM software implementation of one-class support vector machines [8] was modified so that one-class decision values were output for use in determining ROC curves and AUC. The radial basis function was used with the default settings, including values of 0.5 for nu and 1 divided by the number of features for gamma.

4   Evaluation Metrics

Receiver operating characteristics (ROC) graphs are commonly used in machine learning and data mining research to organize and visualize the performance of classifiers, especially for domains with skewed class distribution and unequal classification costs [11]. For two class problems, a classifier is a mapping from instances to predicted classes, positive or negative. Some classifier models only predict the class of the instance, while others produce a continuous output or score, to which different thresholds may be applied. The true positive (TP) rate (detection or hit rate) and the false positive (FP) rate (false alarm rate) are determined below [20].

$$\text{TP rate} = \frac{\text{Positives correctly classified}}{\text{Total positives}} \qquad (1)$$

$$\text{FP rate} = \frac{\text{Negatives correctly classified}}{\text{Total negatives}} \qquad (2)$$

The TP rate is plotted on the Y axis and FP rate is plotted on the X axis in ROC graphs. Notable points in ROC space include the lower left point (0,0), which represents the classifier never predicting positive for any instance, and (1,1), which represents the classifier always predicting positive for any instance. Another point of interest is (0,1) which represents perfect classification, with no FPs or FNs. Average or random performance lies on the diagonal from (0,0) to (1,1). An ROC curve is constructed by sorting test instances by the classifier's scores from most likely to least likely to be a member of the positive class [11]. Each classifier score establishes a point on the ROC curve and a threshold that can be used to classify instances with scores that are above that threshold as positive, otherwise as negative. Since there may be cases of classifiers assigning equal scores to some test instances with possibly different ground truth class labels, all equal classifier scores are processed by an ROC algorithm before establishing a new ROC point. This reflects the expected performance of the classifier, independent of the order of equally scored instances that reflect the arbitrary order of test instances.

An ROC curve is a two-dimensional representation of expected classifier performance. If a single scalar value is desired to represent expected classifier performance, it can be determined as the area under the ROC curve, abbreviated AUC [13,5,11]. Since random guessing produces the diagonal from (0,0) to (1,1) and an AUC of 0.5, a realistic classifier should have an AUC greater than this. The AUC is equivalent to the probability that its classifier will rank a randomly selected positive instance higher than a randomly selected negative instance [11]. This feature is equivalent to the Wilcoxon test of ranks [13,11]. If an area of interest is less than that of the full curve, the areas under the parts of the curves of interest may be used instead for comparing classifier performance.

When considering an average ROC for multiple test runs, one could simply merge sort the instances together by their assigned scores into one large set, and run an ROC algorithm on the set. In order to make a valid classifier comparison using ROC curves, a measure of the variance of multiple test runs should be used [11]. One valid method is vertical averaging, which takes vertical samples for fixed, regularly spaced FP rates and averages the corresponding TP rates. For each ROC curve, the maximum plotted TP rate is chosen for each fixed FP rate. This point is used in individual ROC curves to calculate the AUC, and thus is also used for averaging multiple curve AUCs. If no TP rate has been plotted for the fixed FP rate, the TP rate is interpolated between the maximum TP rates at the FP rates immediately before and immediately after the fixed FP rate. The mean of the maximum plotted TP rates for all curves at each fixed FP rate is plotted and confidence intervals of each mean can be added by drawing the corresponding confidence interval error bars [11]. These are calculated as shown below [10,9].

$$\text{Standard deviation (SD)} = \sqrt{\frac{\Sigma(X - M)^2}{n - 1}} \tag{3}$$

$$\text{Standard Error (SE)} = \frac{\text{SD}}{\sqrt{n}} \tag{4}$$

$$\text{Confidence Interval (CI)} = M \pm t_{(n-1)} \cdot \text{SD} \tag{5}$$

where X in this case is a TP rate for the selected FP rate, M is the mean of the TP rates for the selected FP rate, n is the number of trials, and $t_{(n-1)}$ is a critical value of t.

5 Results

Table 2 shows the supervised ROC AUC results for random forests weighted (RFW) and for distance-based outlier (DBO) methods. The U2R RFW no partitioning mean AUC, shown in Table 2, is 0.9976 which is very close to 1.0 or perfect classification. The U2R RFW 20 of 20 partition mean AUC for is 0.9985 which is almost identical to the above result. The DBO outlier prediction scores were used instead of the corresponding RFW scores above. The lowest DBO mean KDD U2R AUC is 0.9692 for the DBO using all training data without partitioning as the reference set of neighbors. The DBO mean KDD U2R AUC for the 20 partition vote is 0.9850 which is only 0.0135 less than the RFW mean AUC. For comparison, the feature bagging method of outlier detection in the results of [16,1] yielded an AUC of 0.74 ($\pm$ 0.1) for the KDD U2R dataset, so our approach is a big improvement. Feature bagging results for the ann-thyroid 1 and ann-thyroid 2 datasets were 0.869 and 0.769 in [16], which were also lower than our RFW results.

Figure 1 shows the U2R modified KDD Cup 1999 ROC vertically averaged curves using DBO with unpartitioned and 20 partitions of training data. Confidence bars are shown for the 99% confidence region of the ROC mean. The curve is only shown for the FP rate of 0.0 to 0.2 by 0.01 steps since both curves are very close to a detection rate of 1.0 for a FP rate of 0.2 to 1.0. The mean AUC for DBO using 20 partitions is 0.9850,

Table 2. Average supervised results

Dataset	Method	Mean ROC AUC		
		Partitioning		
		none	20	20 − none
KDD U2R	RFW	0.9976	0.9985	0.0009
KDD U2R	DBO	0.9692	0.9850	0.0158
KDD probe	RFW	1.0000	0.9999	−0.0001
KDD probe	DBO	0.9060	0.9431	0.0371
Ann-thyroid 1	RFW	0.9993	0.9982	−0.0011
Ann-thyroid 1	DBO	0.9554	0.9634	0.0080
Ann-thyroid 2	RFW	0.9996	0.9956	−0.0040
Ann-thyroid 2	DBO	0.7751	0.7508	−0.0243
Shuttle 2	RFW	1.0000	0.9997	−0.0003
Shuttle 2	DBO	0.9903	0.9674	−0.0229
Shuttle 3	RFW	0.9946	0.9912	−0.0034
Shuttle 3	DBO	0.9880	0.9740	−0.0140
Shuttle 5	RFW	1.0000	1.0000	0.0000
Shuttle 5	DBO	0.8166	0.9871	0.1705
Shuttle 6	RFW	0.9828	0.9827	−0.0001
Shuttle 6	DBO	0.9995	0.9995	0.0000
Shuttle 7	RFW	0.6341	0.6325	−0.0016
Shuttle 7	DBO	0.9993	0.9994	0.0001

RFW: random forests weighted; DBO: distance-based outlier.

ROC: receiver operator characteristics; AUC: area under curve.

which is higher than 0.9692 with unpartitioned data, as shown in Table 2. The results for the other datasets show that partitioning with ensemble voting results are typically very close to those using unpartitioned methods. RFW typically outperforms DBO.

Table 3 shows the semi-supervised ROC AUC results for one-class support vector machine (SVM) and distance-based outlier (DBO) methods when only examples of the normal class were used for training. The DBO results show that partitioning with ensemble voting results are typically very close to those using unpartitioned methods. Figure 2 shows how the SVM 20 partitions detection rate is higher than the SVM no partitions rate for the initial low false alarm rate of the KDD U2R dataset. The SVM results show that partitioning and ensemble voting is either very close or superior to those without partitioning. DBO used in a one-class or semi-supervised setting typically outperforms supervised DBO for no partitions, while 20 partitions depends on the dataset tested.

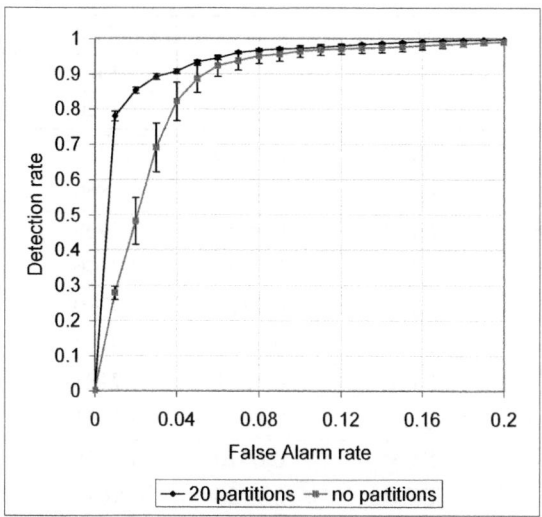

Fig. 1. Modified KDD Cup U2R ROC curves for DBO using 20 partitions and without partitioning on the training group. Curves are vertically averaged over both groups of 30 random splits. Confidence bars are shown for the 99% confidence region of the ROC mean.

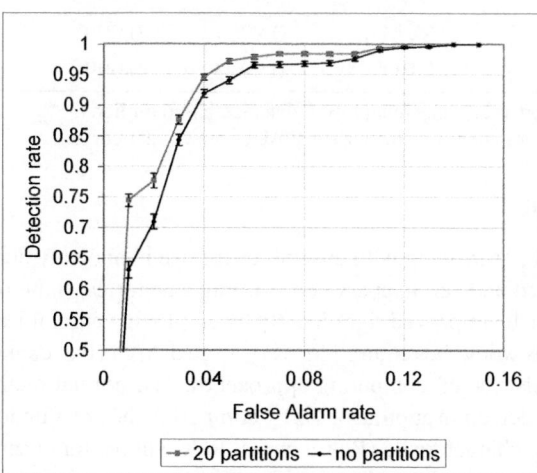

Fig. 2. Modified KDD Cup U2R ROC curves for SVM using 20 partitions and without partitioning on the training group. Curves are vertically averaged over both groups of 30 random splits. Confidence bars are shown for the 99% confidence region of the ROC mean.

Table 3. Average one-class results

Dataset	Method	Mean ROC AUC		
		Partitioning		
		none	20	20 − none
KDD U2R	SVM	0.9850	0.9873	0.0023
KDD U2R	DBO	0.9860	0.9866	0.0006
KDD probe	SVM	0.9971	0.9968	−0.0003
KDD probe	DBO	0.9964	0.9946	−0.0018
Ann-thyroid 1	SVM	0.7420	0.9710	0.2290
Ann-thyroid 1	DBO	0.9659	0.9553	−0.0103
Ann-thyroid 2	SVM	0.4241	0.7188	0.2947
Ann-thyroid 2	DBO	0.8043	0.7201	−0.0842
Shuttle 2	SVM	0.6570	0.9918	0.3348
Shuttle 2	DBO	0.9923	0.9671	−0.0252
Shuttle 3	SVM	0.8534	0.9974	0.1440
Shuttle 3	DBO	0.9924	0.9714	−0.0210
Shuttle 5	SVM	0.9997	0.9989	−0.0008
Shuttle 5	DBO	0.9984	0.9986	0.0002
Shuttle 6	SVM	0.9991	0.9997	0.0006
Shuttle 6	DBO	0.9996	0.9996	0.0000
Shuttle 7	SVM	0.9721	0.9993	0.0171
Shuttle 7	DBO	0.9993	0.9994	0.0001

SVM: support vector machine; DBO: distance-based outlier.
ROC: receiver operator characteristics; AUC: area under curve.

6   Conclusion

An exploration of partitioning with ensembles for use in anomaly detection shows that for both supervised and semi-supervised learning categories, some of the existing approaches can often be improved significantly by employing ensembles. Partitioning and ensemble methods would be required for very large datasets that cannot fit in one memory and require distributed computing approaches. The normal method of comparing anomaly (outlier) detection approaches is by comparing the areas under each approach's receiver operating characteristic (ROC) curve. ROC curves for ensemble methods applied to the datasets here often show significant improvement from increased true positives in the low false positive range.

References

1. Abe, N., Zadrozny, B., Langford, J.: Outlier detection by active learning. In: KDD 2006: Proceedings of the 12th ACM SIGKDD International Conference on Knowledge Discovery and Data Mining, pp. 504–509. ACM, New York (2006)

2. UCI KDD Archive. Kdd cup 1999 data (accessed on, January 1 2010),
    `http://kdd.ics.uci.edu/databases/kddcup99/kddcup99.html`
3. Banfield, R.E., Hall, L.O., Bowyer, K.W., Kegelmeyer, W.P.: A comparison of decision tree
    ensemble creation techniques. IEEE Transactions on Pattern Analysis and Machine Intelli-
    gence, 173–180 (2007)
4. Bay, S.D., Schwabacher, M.: Mining distance-based outliers in near linear time with random-
    ization and a simple pruning rule. In: Proceedings of the Ninth ACM SIGKDD International
    Conference on Knowledge Discovery and Data Mining, pp. 29–38. ACM Press, New York
    (2003)
5. Bradley, A.P.: The use of the area under the roc curve in the evaluation of machine learning
    algorithms. Pattern Recognition 30, 1145–1159 (1997)
6. Breiman, L.: Random forests. Machine Learning 45(1), 5–32 (2001)
7. Chandola, V., Banerjee, A., Kumar, V.: Anomaly detection: A survey. ACM Comput.
    Surv. 41(3), 1–58 (2009)
8. Chang, C.C., Lin, C.J.: Libsvm: a library for support vector machines (accessed on, Novem-
    ber 1 2010), `http://www.csie.ntu.edu.tw/~cjlin/libsvm`
9. Cumming, G., Fidler, F., Vaux, D.L.: Errror bars in experimental biology. The Journal of Cell
    Biology 177(1), 7–11 (2007)
10. Cumming, G., Finch, S.: Inference by eye: Confidence intervals and how to read pictures of
    data. American Psychologist 60(2), 170–180 (2005)
11. Fawcett, T.: An introduction to roc analysis. Pattern Recognition Letters 27(8), 861–874
    (2006), rOC Analysis in Pattern Recognition
12. Giacinto, G., Perdisci, R., Del Rio, M., Roli, F.: Intrusion detection in computer networks by
    a modular ensemble of one-class classifiers. Inf. Fusion 9, 69–82 (2008),
    `http://portal.acm.org/citation.cfm?id=1297420.1297578`
13. Hanley, J.A., McNeil, B.J.: The meaning and use of the area under a receiver operating
    characteristic (roc) curve. Radiology 143, 29–36 (1982)
14. Hempstalk, K., Frank, E., Witten, I.H.: One-class classification by combining density and
    class probability estimation. In: Daelemans, W., Goethals, B., Morik, K. (eds.) ECML
    PKDD 2008, Part I. LNCS (LNAI), vol. 5211, pp. 505–519. Springer, Heidelberg (2008),
    `http://portal.acm.org/citation.cfm?id=1431932&picked=prox&`
    `cfid=19593191&cftoken=93015848,`
    `http://dx.doi.org/10.1007/978-3-540-87479-9_51,`
    ISBN: 978-3-540-87478-2, doi:10.1007/978-3-540-87479-9_51
15. John, B.S., Platt, J.C., Shawe-taylor, J., Smola, A.J., Williamson, R.C.: Estimating the sup-
    port of a high-dimensional distribution. Neural Computation 13, 1443–1471 (2001)
16. Lazarevic, A.: Feature bagging for outlier detection. In: KDD 2005, pp. 157–166 (2005)
17. Niennattrakul, V., Keogh, E., Ratanamahatana, C.A.: Data editing techniques to allow the
    application of distance-based outlier detection to streams. In: IEEE International Conference
    on Data Mining, vol. 0, pp. 947–952 (2010)
18. Perdisci, R., Ariu, D., Fogla, P., Giacinto, G., Lee, W.: Mcpad: A multiple classifier system
    for accurate payload-based anomaly detection. Computer Networks, special issue of Traffic
    Classification and Its Applications to Modern Networks 53(6), 864–881 (2009),
    `http://www.sciencedirect.com/science/article/B6VRG-4V3547G-1/`
    `2/62411af43f5a5f13444f6ab985b9f6ef`
19. Shoemaker, L.: Ensemble Learning With Imbalanced Data. Ph.D. thesis. University of South
    Florida (2010)
20. Tan, P., Steinbach, M., Kumar, V.: Introduction to Data Mining. Addison-Wesley, Reading
    (2006)

Learning to Rank with Nonlinear Monotonic Ensemble

Nikita Spirin[1] and Konstantin Vorontsov[2]

[1] University of Illinois at Urbana-Champaign
[2] Dorodnicyn Computing Center of the Russian Academy of Sciences
spirin2@illinois.edu, vokov@forecsys.ru

Abstract. Over the last decade learning to rank (L2R) has gained a lot of attention and many algorithms have been proposed. One of the most successful approach is to build an algorithm following the ensemble principle. Boosting is the key representative of this approach. However, even boosting isn't effective when used to increase the performance of individually strong algorithms, scenario when we want to blend already successful L2R algorithms in order to gain an additional benefit. To address this problem we propose a novel algorithm, based on a theory of *nonlinear monotonic ensembles*, which is able to blend strong base rankers effectively. Specifically, we provide the concept of defect of a set of algorithms that allows to deduce a popular pairwise approach in strict mathematical terms. Using the concept of defect, we formulate an optimization problem and propose a sound method of its solution. Finally, we conduct experiments with real data which shows the effectiveness of the proposed approach.

1 Introduction

Learning to rank (L2R) has become a hot research topic over the last decades. Numerous amount of methods previously applied to regression and classification have been adapted to L2R. Specifically, one can categorize all L2R algorithms into three big categories: pointwise (reduction of L2R to regression) [4, 8]; pairwise (reduction of L2R to classification) [5, 2, 7, 13, 9]; and listwise (direct optimization) [3, 16, 12].

One of the most popular approach that has been applied in all three categories is boosting [5, 10]. Thus, weak rankers are trained sequentially and then they are blended in a *linear* composition. It is a common knowledge that boosting allows to combine hundreds of base algorithms and isn't inclined to overfitting [10]. However, boosting isn't effective if we want to build an ensemble from a small set of already strong algorithms. Particularly, one cannot build a boosted ensemble over SVM properly. Different methods was developed to cope with this problem and build small ensembles effectively [14, 6]. We adapted methods from [14] to L2R domain and in this paper we propose a novel algorithm solving the L2R problem within a nonlinear monotonic ensemble framework. Monotonic ensembles have expanded the existing variability of ensemble learning methods

C. Sansone, J. Kittler, and F. Roli (Eds.): MCS 2011, LNCS 6713, pp. 16–25, 2011.
© Springer-Verlag Berlin Heidelberg 2011

and allowed to effectively blend a small set of individually strong algorithms. For instance, in our experiments the size of an ensemble varied from 4 to 7 base rankers. Moreover, the algorithm in question is built on a strict mathematical foundation, theoretically consistent with the internal structure of L2R problem and allows to induce pairwise approach merely from theoretical constructions instead of heuristic speculations.

1.1   The Learning to Rank Problem

The L2R problem can be formalized as follows. There is an ordered set of ranks $\mathbb{Y} = \{r_1, \ldots, r_K\}$ and a set of queries $Q = \{q_1, \ldots, q_n\}$. A list of documents $D_q = \{d_{q1}, \ldots, d_{q,n(q)}\}$ is associated with each query $q \in Q$, where $n(q)$ is the number of documents associated with the query q. A factor ranking model is admitted, i.e. each query-document pair (q, d), $d \in D_q$ is represented by the vector of features $\boldsymbol{x}_{qd} = \big(f_1(q, d), \ldots, f_m(q, d)\big) \in \mathbb{R}^m$. Thus, the training set is

$$S = \{\boldsymbol{x}_{qd}, r(q, d)\}, \quad q \in Q, \, d \in D_q,$$

where $r(q, d) \in \mathbb{Y}$ is the corresponding correct relevance score for a (q, d) pair. The objective is to build a ranking function $A \colon \mathbb{R}^m \to \mathbb{Y}$ that maximizes a performance measure on the training set and has a good generalization ability.

2   Our Method: MonoRank

2.1   Nonlinear Monotonic Ensemble: Underlying Theory

Monotonicity constraints often arise in real world machine learning tasks. For example, we can observe such constraints in a credit scoring task where the objective is to build an algorithm that will classify applicants given their responses to questionnaires (e. g. the bigger annual household income and value of property the more probably a customer will pay a loan back). Generally speaking, monotonicity constraints can arise in any task where the factor model is admitted, and the order on targets is in agreement with the order on ordinal features [11].

Another application of this principle is to impose monotonicity constraints not on features but on base algorithms predictions [14, 6]. It is very natural to construct an ensemble of base predictors according to the following principle: if output of a predictor is higher for an object, provided that outputs of other predictors are the same, then the output of the entire ensemble must be also higher for this object. This implies that the aggregating function is to be monotonic. Obviously, linear blending meets the monotonicity restriction if only all the weights are nonnegative. In this paper we use nonlinear monotonic aggregating functions and argue that monotonicity is a more natural and less restrictive principle than the weighted voting, especially for L2R domain.

Let $\Omega = \mathbb{R}^m$ be an object space of query-document feature vectors, according to general factor ranking model; X and Y be partially ordered sets, B be a set of base algorithms $b \colon \Omega \to X$, X is referred to as an estimation space (predictions of base algorithms) and Y as an output space (labels, responses, relevance scores). A training set of object–output pairs $\{(\boldsymbol{x}_k, y_k)\}_{k=1}^{\ell}$ from $\Omega \times Y$ and a set of base algorithms $b_1, \ldots, b_p$ induces a sequence of estimation vectors $\{\boldsymbol{u}_k\}_{k=1}^{\ell}$ from X^p, where $\boldsymbol{u}_k = \left(u_k^1 = b_1(\boldsymbol{x}_k), \ldots, u_k^p = b_p(\boldsymbol{x}_k)\right)$.

Let us define an order on X^p: $(u^1, \ldots, u^p) \leq (v^1, \ldots, v^p)$, if $u^i \leq v^i$ for all $i = 1, \ldots, p$. If vectors $\boldsymbol{u}, \boldsymbol{v} \in X^p$ aren't comparable we will denote it as $\boldsymbol{u} \parallel \boldsymbol{v}$. If $\boldsymbol{u} \neq \boldsymbol{v}$ and $\boldsymbol{u} \leq \boldsymbol{v}$, then $\boldsymbol{u} < \boldsymbol{v}$. A map $F\colon X^p \to Y$ is referred to as monotonic, if $\boldsymbol{u} \leq \boldsymbol{v}$ implies $F(\boldsymbol{u}) \leq F(\boldsymbol{v})$ for all $\boldsymbol{u}, \boldsymbol{v} \in X^p$.

Monotonic ensemble of base algorithms $b_1, \ldots, b_p$ with *monotonic aggregating function* F is a map $a\colon \Omega \to Y$ defined as $a(\boldsymbol{x}) = F\big(b_1(\boldsymbol{x}), \ldots, b_p(\boldsymbol{x})\big), \forall \boldsymbol{x} \in \Omega$.

If base algorithms are fixed, then learning of a monotonic function F from data can be stated as a task of monotonic interpolation. Given a sequence of vectors $\{\boldsymbol{u}_k\}_{k=1}^{\ell}$ from X^p and a sequence of targets $\{y_k\}_{k=1}^{\ell}$ from Y, one should build such a monotonic function F that meets the *correctness condition*

$$F(\boldsymbol{u}_k) = y_k, \quad k = 1, \ldots, \ell. \tag{1}$$

Definition 1. *A pair (i, j) is **defective** for the base algorithm b, if $b(\boldsymbol{x}_i) \geq b(\boldsymbol{x}_j)$ and $y_i < y_j$. A set of all defective pairs of b will be denoted as $\mathbb{D}(b)$. A set $\mathbb{D}(b_1, \ldots, b_p) = \mathbb{D}(b_1) \cap \cdots \cap \mathbb{D}(b_p)$ will be called **cumulative defect** of a set of base algorithms $(b_1, \ldots, b_p)$. Similarly, a pair is **clean** if $b(\boldsymbol{x}_i) < b(\boldsymbol{x}_j)$ and $y_i < y_j$. We will use $\mathbb{C}(b)$ and $\mathbb{C}(b_1, \ldots, b_p)$ for that analogously.*

Directly from this definition it can be derived that for $p = 1$ a monotonic function F exists **iff** $\mathbb{D}(b) = \varnothing$. Thus, the number of defective pairs $|\mathbb{D}(b)|$ can play a role of a quality measure for a base algorithm b. By definition, the cumulative defect $\mathbb{D}(b_1, \ldots, b_p)$ consists of those defective pairs on which all base algorithms fail. So, if we build the next base algorithm b_{p+1} so that it yields the right order $b_{p+1}(\boldsymbol{x}_i) < b_{p+1}(\boldsymbol{x}_j)$ on pairs (i, j) from the cumulative defect, then $\mathbb{D}(b_1, \ldots, b_p, b_{p+1}) = \varnothing$ and a monotonic function F satisfying the condition (1) exists. It is worth mentioning that artificial "emptyfication" of defect (for example, by taking two base algorithms with inverted predictions) won't give practically useful results (generalization ability will be poor). Instead, base algorithms should be trained in succession so that they all be individually strong and latter base algorithms corrected predictions of former ones. From practical point of view we need to analyze a defect of an ensemble $\mathbb{D}(F(b_1, \ldots, b_p))$, but not a defect of a set of base algorithms. The next two statements from [14] show the relationship between the sets $\mathbb{D}(F(b_1, \ldots, b_p))$ and $\mathbb{D}(b_1, \ldots, b_p)$.

Lemma 1. *For each p-ary monotonic aggregating function F, $\mathbb{D}(F(b_1, \ldots, b_p)) \supseteq \mathbb{D}(b_1, \ldots, b_p)$.*

Theorem 2. *The cumulative defect $\mathbb{D}(b_1, \ldots, b_p)$ is empty **iff** there exist a monotonic aggregating function F such that $\mathbb{D}(F(b_1, \ldots, b_p)) = \varnothing$.*

From the statements above it can be derived that if we build a set of base algorithms with zero defect, then we will gain a correct, on a training set, algorithm. The stronger statement on convergence is valid [14], i.e. we need only a finite number of steps (base algorithms) in order to build a correct algorithm.

From this, an iterative strategy of building a monotonic ensemble follows. In order to minimize the size p of a composition the choice of the next algorithm b_{p+1} should be guided by the minimization of the number of defective pairs produced by all preceding base algorithms:

$$b_{p+1}(\boldsymbol{x}_i) < b_{p+1}(\boldsymbol{x}_j) : \; (i,j) \in \mathbb{D}(b_1, \ldots, b_p). \tag{2}$$

However, the correctness condition (1) is too restrictive and may result in overfitting (generalization might be poor). The trick is to stop iterations earlier using a stopping criterion like a degradation of quality on a validation set.

Similar to arching and boosting algorithms, we propose to enrich the optimization task (2) with weights in order to add more flexibility to our model.

$$b_{p+1}(\boldsymbol{x}_i) < b_{p+1}(\boldsymbol{x}_j) \; \text{with} \; w_{ij} : \; (i,j) \in \mathbb{D}(b_1, \ldots, b_p), \tag{3}$$

where w_{ij} is a weight of a defective pair (i,j). So, the task is to find the heaviest consistent sub-system of inequalities. In Section 2.2 we restate this problem in terms of quality functional and analyze its properties, crucial for L2R.

2.2   The Algorithm

Inspired by outstanding performance of monotonic ensembles on classification problems [14,6], we applied the notion of monotonic aggregation to L2R problem and developed an algorithm for it. The algorithm is referred to as MonoRank and the pseudocode is presented in Algorithm 1.

Let us briefly go over all key stages of the algorithm and then we will discuss each stage in detail. First, we train the first base algorithm using the entire training set (line 3). Then we reweigh pairs (line 8) according to the strategies discussed in the Section 2.4. It is worth noting that a monotonic aggregating function isn't needed after the first step, because we have only one base algorithm. Then using updated weights we train the second base algorithm (line 3). Here, all pairs are used but weights already aren't uniform. Having built two base algorithms, we train a monotonic aggregating function in $\mathbb{R}^2$ (lines 4-6), according to the logic we describe in Section 2.5. Then we compute current ensemble performance on a validation set and save it for future reference (line 7). Then the process is repeated: we reweigh pairs based on a current cumulative defect, train the next base algorithm and then fit a monotonic aggregating function. Stopping condition (line 9) is determined by performance of the algorithm on an independent validation set, standard criterion in machine learning research.

The problem (3), restated as a minimization of a quality functional $\mathscr{Q}$ with base algorithms $b_1, \ldots, b_p$ fixed, will look like:

$$\mathscr{Q}(b_1, \ldots, b_p, b_{p+1}) = \sum_{q \in Q} \sum_{(d,d')} w_{qdd'} \big[b_{p+1}(\boldsymbol{x}_{qd}) \geq b_{p+1}(\boldsymbol{x}_{qd'}) \big] \rightarrow \min_{b_{p+1}}, \tag{4}$$

where (d, d') are all documents from D_q such that $(qd, qd') \in \mathbb{D}(b_1, \ldots, b_p)$. Note that in L2R task the indices i, j from (3) become qd, qd' respectively, and only those documents d, d' are comparable that corresponds to the same query q.

Algorithm 1. MonoRank pseudocode

Input: training set $S = \{x_{qd}, r(q,d)\}, q \in Q, d \in D_q$;

    δ — number of unsuccessful iterations before stop;

Output: nonlinear monotonic ensemble of rankers $M_T(x_{qd})$ of size T;

 1: initialize weights $w_{qdd'} = 1$, $q \in Q$, $d, d' \in D_q$;

 2: **for** $t = 1, \ldots, n$ **do**

 3:    train base ranker $b_t(x_{qd})$ using weights $\{w_{qdd'}\}$;

 4:    get predictions $b_t(x_{qd})$ of b_t on a training set S;

 5:    monotonize $\{(b_1(x_{qd}), \ldots, b_t(x_{qd})), r(q,d)\}$;

 6:    build a composition $M_t(x_{qd})$ of size t;

 7:    $T = \operatorname*{argmin}_{p:\, p \leq t} \mathscr{Q}(M_p)$;

 8:    update $\{w_{qdd'}\}$ using M_t;

 9:    **if** $t - T \geq \delta$ **then**

10:       **return** M_T;

In classification and regression tasks special efforts are to be made to reduce the pairwise criterion $\mathscr{Q}$ to usual pointwise empirical risk [14]. In L2R such tricks are needless as long as base rankers can be learned directly from $\mathscr{Q}$ minimization; this is the reason why monotonic ensembles fit so well to L2R.

2.3 Adaptation of Base Rankers for Usage in Monotonic Ensemble

Now we will briefly discuss ways to modify existing L2R algorithms with the objective to use them effectively later in a monotonic ensemble. The general idea is to define weights on pairs of query-document feature vectors and hence guide the learning process accordingly. In RankSVM we only have to add weights $w_{qdd'}$ whenever we come across slack variables in the objective function. Rank-Boost already contains weight distribution over the set of document pairs $q \in Q$, $d, d' \in D_q$, which is by default uniform. Weights $w_{qdd'}$ can also be easily inserted in FRank [13] and RankNet [2].

2.4 Reweighing Strategies

Now let us discuss the initialization of weights $w_{qdd'}$ for algorithms described above in order for them to form a strong and diversified set of base rankers. Particularly, below we will describe various reweighing strategies.

1. The weight for a defective pair equals one. The weight for a nondefective pair equals zero. This is the most natural strategy that is induced from the general theory of monotonic aggregating functions and can also be referred to as *the defect minimization principle*. If we train the next base algorithm using only defective pairs, we will minimize the number of base algorithms and reach the state of empty cumulative defect. Thus, according to theorem 2, we will be able to build a monotonic function on predictions of base rankers. However, this approach has a significant shortcoming. If we train our base algorithms only using defective pairs the generalization ability of the entire algorithm will be poor and hence it won't be practically applicable.

2. The weight for a defective pair is nonzero. The weight for a nondefective pair equals zero. According to our experiments it doesn't allow to gain any rise in quality and only increases the complexity of the model. Our conclusion is in agreement with conclusions made in related research for classification [6].

3. The weight for a defective and clean pair is nonzero. The weight for an incomparable pair is zero. Then the strategy will look like:

$$
w_{qdd'} = \begin{cases} w_{\mathbb{D}}(t), & (qd, qd') \in \mathbb{D}(b_1, \ldots, b_p); \\ 1, & (qd, qd') \in \mathbb{C}(b_1, \ldots, b_p); \\ 0, & \text{otherwise,} \end{cases}
$$

This is the most successful strategy according to our experiments. Moreover, this strategy combines *the defect minimization principle* and *complete cross-validation minimization* that characterizes the generalization ability of the entire algorithm [15]. In this case the weight for a clean pair equals one. And the weight $w_{\mathbb{D}}(t)$ for a defective pair may depend on iteration $t = 1, \ldots, T$. Particularly, it may increase from iteration to iteration in order to lead the training algorithm to turn out the defect on these pairs. We used $w_{\mathbb{D}}(t) = 2^{t-1}$ in our experiments.

2.5 Monotonic Aggregating Function

In this section we present the core part of our approach — how to blend base rankers $b_1, \ldots, b_p$ with a nonlinear monotonic aggregating function $F(b_1, \ldots, b_p)$. We will build our algorithm so as to minimize the quality functional induced by the cumulative defect $|\mathbb{D}(b_1, \ldots, b_p)|$. To begin with, we describe general constructions inherent to regression and classification following [14], and then turn to analysis of structures specific for ranking. It is worth noting that we don't impose any constraints on a set of base rankers while learning them. Therefore, according to the theorem 2 monotonic function F might not exist, because a sequence of base predictions might not be monotonic. To cope with this problem we use monotonization based on isotonic regression [1]. Given a nonmonotonic sequence $\{(\boldsymbol{u}_k, y_k)\}_{k=1}^{\ell}$, where $\boldsymbol{u}_k \in \mathbb{R}^p$ is a vector of base algorithms predictions and $y_k \in \mathbb{R}$ is the corresponding target, monotonization algorithm finds $\{y_k'\}_{k=1}^{\ell}$ minimizing $\sum_{i=1}^{\ell}(y_k' - y_k)^2$ subject to $y_i' \leq y_j'$ for all (i, j) such that $\boldsymbol{u}_i \leq \boldsymbol{u}_j$.

So, let us have a monotonic sequence $\{(\boldsymbol{u}_k, y_k)\}_{k=1}^{\ell}$. The task is to build a monotonic function F that meets the correctness condition (1).

Definition 2. *For any vector $\boldsymbol{u} \in \mathbb{R}^p$ denote its upper and lower set respectively by $M^{\triangle} = \{\boldsymbol{v} \in \mathbb{R}^p \colon \boldsymbol{u} \leq \boldsymbol{v}\}$ and $M^{\triangledown} = \{\boldsymbol{v} \in \mathbb{R}^p \colon \boldsymbol{v} \leq \boldsymbol{u}\}$.*

Consider a continuous function $\mu \colon \mathbb{R}^p \to [0, +\infty)$ nondecreasing by any argument. For example, one can take $\mu(\boldsymbol{x}) = \sum_{i=1}^{p} x^i$ or $\mu(\boldsymbol{x}) = \max\{x^1, \ldots, x^p\}$.

Definition 3. *For any vector $\boldsymbol{u} \in \mathbb{R}^p$ denote the distance from $\boldsymbol{u}$ to an upper and lower set of a vector $\boldsymbol{u}_i$ respectively by*

$$r_i^\triangle = \mu\big((u_i^1 - u^1)_+, \ldots, (u_i^p - u^p)_+\big),$$
$$r_i^\triangledown = \mu\big((u^1 - u_i^1)_+, \ldots, (u^p - u_i^p)_+\big),$$

where $(z)_+ = z$ if $z \geq 0$ and $(z)_+ = 0$ if $z < 0$.

Now let us define functions $h^\triangle(\boldsymbol{u}, \theta)$ and $h^\triangledown(\boldsymbol{u}, \theta)$ that estimate the distance from a vector $\boldsymbol{u} \in \mathbb{R}^p$ to a nearest vector from upper and lower sets:

$$h^\triangle(\boldsymbol{u}, \theta) = \min_{i:\, y_i > \theta} r_i^\triangle(\boldsymbol{u}), \quad h^\triangledown(\boldsymbol{u}, \theta) = \min_{i:\, y_i \leq \theta} r_i^\triangledown(\boldsymbol{u}).$$

Then, define a relative distance from a vector $\boldsymbol{u}$ to the union of all upper sets:

$$\Phi(\boldsymbol{u}, \theta) = \frac{h^\triangledown(\boldsymbol{u}, \theta)}{h^\triangledown(\boldsymbol{u}, \theta) + h^\triangle(\boldsymbol{u}, \theta)}, \text{ where } \boldsymbol{u} \in \mathbb{R}^p, \ \theta \in \mathbb{R}. \tag{5}$$

The first two functions can be used immediately for two-class classification. Specifically, an object is prescribed to the first class if $h^\triangle(\boldsymbol{u}, \theta) > h^\triangledown(\boldsymbol{u}, \theta)$ and to the zero class otherwise, where θ can be any number in $(0, 1)$, e.g. $\theta = \frac{1}{2}$. The third function is a regression stair that equals 0 on a union of lower sets, equals 1 on a union of upper sets, and is a continuous, monotone non-decreasing, piecewise bilinear function in between.

Theorem 3. *Let $\{(\boldsymbol{u}_k, y_k)\}_{k=1}^\ell$ be a monotonic sequence, and a set $\{y_k\}_{k=1}^\ell$ is sorted in ascending order. Then the function $F \colon \mathbb{R}^p \to \mathbb{R}$ defined below is continuous, monotone non-decreasing, and meets the correctness condition (1).*

$$F(\boldsymbol{u}) = y_1 + \sum_{k=1}^{\ell-1} (y_{k+1} - y_k) \Phi(\boldsymbol{u}, y_k)$$

Learning to rank. At first, it is worth noting that due to the structure of the training set comparable documents are only those which are associated with the same query. Another distinction from the above cases is that we don't really need to meet the correctness condition (1) in the case of ranking. The only constraint to meet is to keep the right ordering of documents on the training set.

Provided that documents associated with different queries aren't comparable at all, the quality functional can be rewritten as the sum of functionals, counting defect only for a particular query. We will denote the defect of the entire algorithm with the aggregating function F on a query q as $\mathscr{Q}_q(F)$. Then the optimal aggregating function F must be a solution for a minimization problem

$$\sum_{q \in Q} \mathscr{Q}_q(F) \to \min_F.$$

To give an approximate but computationally efficient solution to this hard problem we propose to use an averaging heuristic. We solve $|Q|$ problems separately:

$$F_q = \arg\min_F \mathscr{Q}_q(F), \quad q \in Q.$$

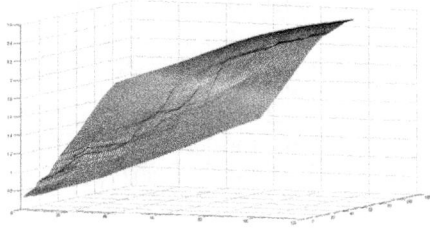
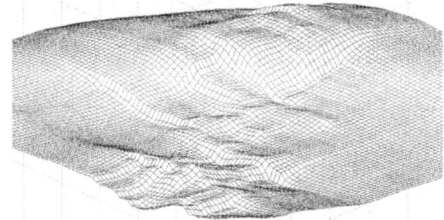

Fig. 1. Monotonic aggregating function for ranking with 2 base algorithms

Fig. 2. Texture of the surface of monotonic function from the left figure

Then we define an aggregating function F by averaging all F_q':

$$F = \sum_{q \in Q} F_q'. \tag{6}$$

where F_q' is a normalized to $[0, 1]$ function $\frac{F_q}{|D_q|}$. Here we use normalization with the value equal to $|D_q|$ to avoid bias towards queries with the large number of associated documents. Obviously, the function F is monotonic being the sum of monotonic functions. We call a set of base rankers, trained following the logic described above, together with the monotonic aggregating function described in this section as a nonlinear monotonic ensemble for learning to rank.

We provide a few pictures of a monotonic aggregating function for ranking, built according to the theory described above. Due to the large number of queries in a training set and hence due to averaging, the monotonic function from the fig. 1 looks like a plane. However, having changed the scale one can observe a complicated texture of the surface, fig. 2. According to our experiments in Section 3.3 the increase in quality takes place directly thanks to this tiny asperities. It is also interesting to notice that asperities appear only above the diagonal. This can be explained as follows. We use strong base rankers, like RankBoost and RankSVM, that's why their predictions are highly correlated (nevertheless, we can blend them effectively) and lie along the diagonal.

3   Experimental Results

3.1   Yahoo! LETOR 2010 Challenge Dataset

This is a dataset provided by Yahoo! company for L2R competition. There are 34815 query-document pairs and 1266 unique queries. Relevance grades are discrete from range $[0, 4]$. Each query-document pair is described by a vector with 575 components. We used 5-fold cross validation to calculate the performance of algorithms. As a base algorithm for MonoRank we used RankBoost. The results are reported in table 1 and in a graphical form in fig. 3.

Table 1. Yahoo! LETOR 2010 results

Metric	MonoRank	RankBoost	RankSVM
NDCG@1	**0.8149**	0.8029	0.8057
NDCG@3	**0.7783**	0.7424	0.7711
NDCG@5	**0.7754**	0.7692	0.7678
NDCG@10	**0.7973**	0.7935	0.7949

Table 2. OSHUMED LETOR 3.0 results

Metric	MonoRank	RankBoost	RankSVM
NDCG@1	**0.5231**	0.4632	0.4958
NDCG@3	**0.4602**	0.4555	0.4207
NDCG@5	**0.4500**	0.4494	0.4164
NDCG@10	**0.4337**	0.4302	0.4140

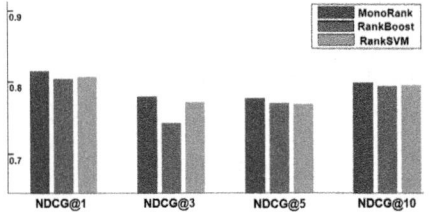

Fig. 3. Performance on Yahoo! LETOR

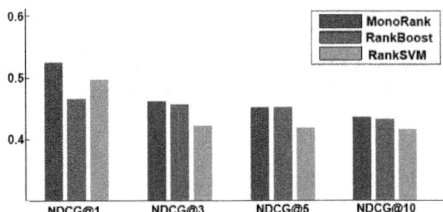

Fig. 4. Performance on OSHUMED

3.2   OSHUMED LETOR 3.0 Dataset

This is a dataset from LETOR repository[1], prepared by Microsoft Research Asia. There are 16140 query-document pairs and 106 unique queries. Relevance grades are discrete from range $[0, 2]$. There are 36 features. In order to guarantee the consistency of algorithms comparison, we used evaluation scripts from LETOR project. The results are reported in table 2 and in fig. 4.

3.3   Experiment Analysis

Now we will briefly discuss the key interesting feature of the monotonic surface we built. Having seen the fig. 1 one might think why we should use so complex construction to blend base rankers. Why couldn't we just use a simple linear combination of base rankers? Of course, this is a reasonable speculation but we have set an experiment to test the hypothesis. We approximated our "wavy" surface with a hyperplane by the least squares method and evaluated the performance on a Yahoo! LETOR 2010 dataset. The results are in table 3.[2]

Table 3. Comparison of "wavy" monotonic aggregating function with its linear approximation

Metric	MonoRank	MonoRank-linearized
NDCG@1	**0.8149**	0.8106
NDCG@3	**0.7783**	0.7735
NDCG@5	**0.7754**	0.7720
NDCG@10	**0.7973**	0.7898

---

[1] http://research.microsoft.com/en-us/um/beijing/projects/letor/
[2] Same results was observed on OSHUMED dataset.

4 Conclusion

In this paper we proposed a new algorithm for L2R problem, following the ensemble principle. The algorithm is referred to as MonoRank and employs the theory of nonlinear monotonic ensembles in ranking model building. The core of the algorithm are nonlinear monotonic aggregating functions that enable to blend strong algorithms effectively. The algorithm is based on sound mathematical constructions that are aligned with the popular pairwise approach for L2R. According to computational results MonoRank shows high accuracy in ranking and outperforms existing algorithms, like RankBoost and RankSVM.

References

1. Barlow, R., Bartholomew, D., Bremner, J., Brunk, H.: Statistical inference under order restrictions; the theory and application of isotonic regression (1972)
2. Burges, C., Shaked, T., Renshaw, E., Lazier, A., Deeds, M., Hamilton, N., Hullender, G.: Learning to rank using gradient descent. In: ICML 2005 (2005)
3. Cao, Z., Qin, T., Liu, T.Y., Tsai, M.F., Li, H.: Learning to rank: From pairwise approach to listwise approach. In: ICML 2007, pp. 129–136 (2007)
4. Cooper, W.S., Gey, F.C., Dabney, D.P.: Probabilistic retrieval based on staged logistic regression. In: SIGIR 1992 (1992)
5. Freund, Y., Iyer, R.D., Schapire, R.E., Singer, Y.: An efficient boosting algorithm for combining preferences. Journal of Machine Learning Research (2003)
6. Guz, I.S.: Nonlinear monotonic compositions of classifiers. In: MMRO 13 (2007)
7. Joachims, T.: Optimizing search engines using click through data. In: SIGIR 2002 (2002)
8. Li, P., Burges, C.J., Wu, Q.: Learning to rank with nonsmooth cost functions. In: Advances in NIPS 19, pp. 193–200 (2006)
9. Qin, T., Liu, T.Y., Lai, W., Zhang, X.D., Wang, D., Li, H.: Ranking with multiple hyperplanes. In: SIGIR 2007, pp. 279–286 (2007)
10. Schapire, R.E.: Theoretical views of boosting and applications. In: Watanabe, O., Yokomori, T. (eds.) ALT 1999. LNCS (LNAI), vol. 1720, p. 13. Springer, Heidelberg (1999)
11. Sill, J., Abu-Mostafa, Y.: Monotonicity hints. In: Advances in NIPS 9 (1997)
12. Taylor, M., Guiver, J., Robertson, S., Minka, T.: Softrank: Optimising non-smooth rank metrics. In: WSDM 2008 (2008)
13. Tsai, M.F., Liu, T.Y., Qin, T., Chen, H.H., Ma, W.Y.: Frank: A ranking method with fidelity loss. In: SIGIR 2007 (2007)
14. Vorontsov, K.: Optimization methods for linear and monotone correction in the algebraic approach to the recognition problem. Comp. Math and Mat. Phys. (2000)
15. Vorontsov, K.: Combinatorial bounds for learning performance. Doklady Mathematics 69(1), 145 (2004)
16. Weimer, M., Karatzoglou, A., Le, Q., Smola, A.: Cofirank — maximum margin matrix factorization for collaborative ranking. In: Advances in NIPS 19 (2007)

A Bayesian Approach for Combining Ensembles of GP Classifiers

C. De Stefano[1], F. Fontanella[1], G. Folino[2], and A. Scotto di Freca[1]

[1] Università di Cassino
Via G. Di Biasio, 43 02043 Cassino (FR) – Italy
{destefano,fontanella,a.scotto}@unicas.it
[2] ICAR-CNR Istituto di Calcolo e Reti ad Alte Prestazioni
Via P. Bucci 87036 Rende (CS) – Italy
folino@icar.cnr.it

Abstract. Recently, ensemble techniques have also attracted the attention of Genetic Programing (GP) researchers. The goal is to further improve GP classification performances. Among the ensemble techniques, also bagging and boosting have been taken into account. These techniques improve classification accuracy by combining the responses of different classifiers by using a majority vote rule. However, it is really hard to ensure that classifiers in the ensemble be appropriately diverse, so as to avoid correlated errors. Our approach tries to cope with this problem, designing a framework for effectively combine GP-based ensemble by means of a Bayesian Network. The proposed system uses two different approaches. The first one applies a boosting technique to a GP–based classification algorithm in order to generate an effective decision trees ensemble. The second module uses a Bayesian network for combining the responses provided by such ensemble and select the most appropriate decision trees. The Bayesian network is learned by means of a specifically devised Evolutionary algorithm. Preliminary experimental results confirmed the effectiveness of the proposed approach.

1 Introduction

In the last years, in order to further improve classification performance, ensemble techniques [11] have been taken into account in the Genetic Programming (GP) field [8]. The GP approach uses the evolutionary computation paradigm to evolve computer programs, according to a user-defined fitness function. When dealing with classification problems, GP–based techniques exhibited very interesting performance [12]. In this context, the decision tree [14] data structure is typically adopted since it allows to effectively arrange in a tree-structured plans the set of attributes chosen for pattern representation. Successful examples of ensemble techniques applied to GP can be found in [1] [5]. In [7], bagging and boosting techniques have been used for evolving ensembles of decision trees. In [4] a novel GP–based classification system, named *Cellular GP for data Classification* (CGPC), has been presented. Such approach, inspired by cellular automata

C. Sansone, J. Kittler, and F. Roli (Eds.): MCS 2011, LNCS 6713, pp. 26–35, 2011.
© Springer-Verlag Berlin Heidelberg 2011

model, enables a fine-grained parallel implementation of GP. In [5], an extension of CGPC based on the use of two ensemble techniques is presented: the first technique is the Breiman's bagging algorithm [11], while the second one is the AdaBoost.M2 boosting algorithm by Freund and Schapire [11]. The experimental results presented in these papers show that CGPC represents an effective classification algorithm, whose performance have been further improved using ensemble techniques. In this framework, it is generally agreed that a key issue is to ensure that classifiers in the ensemble be appropriately diverse, so as to avoid correlated errors [11]. In fact, as the number of classifiers increases, it may happen that a correct classification provided by some classifiers is overturned by the convergence of other classifiers on the same wrong decision. This event is much more likely in case of highly correlated classifiers and may reduce the performance obtainable with any combination strategy.Classifier diversity for bagging and boosting have been experimentally investigated in [10,9]. The results have shown that these techniques do not ensure to obtain sufficiently diverse classifiers. As regards boosting, in [9] it has been observed that while at first steps highly diverse classifiers are obtained, as the boosting process proceeds classifier diversity strongly decreases.

In a previous work [2] an attempt to solve this problem has been made by reformulating the classifier combination problem as a pattern recognition one, in which the pattern is represented by the set of class labels provided by the classifiers when classifying a sample. Following this approach, the role of the combiner is that of estimating the conditional probability of each class, given the set of labels provided by the classifiers for each sample of a training set. In this way, it is possible to automatically derive the combining rule through the estimation of the conditional probability of each class. It it also possible to identify redundant classifiers, i.e. classifiers whose outputs do not influence the output of the combiner: the behavior of such classifiers is very similar to that of other classifiers in the ensemble and they may be eliminated without affecting the overall performance of the combiner, thus overcoming the main drawback of the combining methods discussed above. In [3] a Bayesian Network (BN) [13] has been used to automatically infer the joint probability distributions between the outputs of the classifiers and the classes. The BN learning has been performed by means of an evolutionary algorithm using a direct encoding scheme of the BN structure. Such encoding scheme is based on a specifically devised data structure, called *Multilist*, which allows an easy and effective implementation of the genetic operators.

In this paper we present a new classification system that merges the two aforementioned approaches. We have combined the BoostCGPC algorithm [5], which produces a high performing ensemble of decision tree classifiers, with the BN based approach to classifier combination. Our system tries to exploit the advantages provided by both techniques and allows to identify the minimum number of independent classifiers able to recognize the data at hand.

In order to assess the effectiveness of the proposed system, several experiments have been performed. More specifically, four data sets, having different sizes, number of attributes and classes have been considered. The results have been compared with those obtained by the BoostCGPC approach using a weighted majority vote combining rule.

The remainder of the paper is organized as follows. In Section 2 the BN based combining technique is presented. In Section 3 the architecture of the proposed system is described. In Section 4 the experimental results are illustrated, while discussion and some concluding remarks are eventually left to Section 5.

2   Bayesian Networks for Combining Classifiers

The problem of combining the responses provided by a set of classifiers can be also faced considering the joint probability $p(c, e_1, ..., e_L)$, where e_i represents the response of the i–th classifier and c is the class to be assigned to the sample taken into account. The problem of computing the joint probability $p(c, e_1, ..., e_L)$ may be effectively solved by using a Bayesian Network (BN). In particular, in [2], a BN has been used for for combining the responses of more classifiers in a multi expert system.

A BN is a probabilistic graphical model that allows the representation of a joint probability distribution of a set of random variables through a Direct Acyclic Graph (DAG). The nodes of the graph correspond to variables, while the arcs characterize the statistical dependencies among them. An arrow from a node i to a node j has the meaning that j is conditionally dependent on i, and we can refer to i as a *parent* of j. For each node, a conditional probability quantifies the effect that the parents have on that node. Once the statistical dependencies among variables have been estimated and encoded in the DAG structure, each node e_i is associated with a conditional probability function exhibiting the following property:

$$p(e_i \mid pa_{e_i}, nd_{e_i}) = p(e_i \mid pa_{e_i}) \qquad (1)$$

where pa_{e_i} indicates the set of nodes which are parents of node e_i, and nd_{e_i} indicates all the remaining nodes. This property allows the description of the joint probability of a set of variables $\{c, e_1, \ldots, e_L\}$ as follows:

$$p(c, e_1, \ldots, e_L) = p(c \mid pa_c) \prod_{e_i \in E} p(e_i \mid pa_{e_i}) \qquad (2)$$

It is worth noticing that the node c may be parent of one or more nodes of the DAG. Therefore, it may be useful to divide the e_i nodes of the DAG in two groups: the first one, denoted as E_c, contains the nodes having the node c among their parents, and the second one, denoted as $E_{\bar{c}}$, the remaining ones. With this assumption, Eq. (2) can be rewritten as:

$$p(c, e_1, \ldots, e_L) = p(c \mid pa_c) \prod_{e_i \in E_c} p(e_i \mid pa_{e_i}) \prod_{e_i \in E_{\bar{c}}} p(e_i \mid pa_{e_i}) \qquad (3)$$

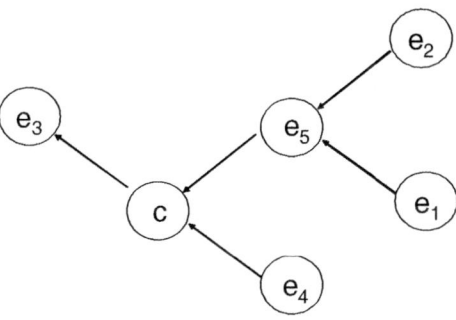

Fig. 1. An example of a BN. The DAG structure induces the factorization of the joint probability $p(c, e_1, e_2, e_3, e_4, e_5) = p(e_3|c)p(c|e_4, e_5)p(e_5|e_1, e_2)p(e_1)p(e_2)p(e_4)$. In this case $E_c = \{e_3\}$, $E_{\bar{c}} = \{e_1, e_2, e_4, e_5\}$ and hence $\hat{c} = \arg\max_{c \in C} p(e_3|c)p(c|e_4, e_5)$.

Since the third term in Eq. (3) does not depend on c, Eq. (6) assumes the form:

$$\hat{c} = \arg\max_{c \in C} p(c, e_1, \ldots, e_L) = \arg\max_{c \in C} p(c \,|\, pa_c) \prod_{e_i \in E_c} p(e_i \,|\, pa_{e_i}) \quad (4)$$

For instance, the BN reported in Fig. 1 considers only the responses of the experts e_3, e_4 and e_5, while the experts e_1 and e_2 are not taken into account. Thus, this approach allows to detect a reduced set of relevant experts, namely the ones connected to node c, whose responses are actually used by the combiner to provide the final output, while the set $E_{\bar{c}}$ of experts, which do not add information to the choice of $\hat{c}$, are discarded.

Using a BN for combining the responses of more classifiers requires that both the network structure, which determines the statistical dependencies among variables, and the parameters of the probability distributions be learned from a training set of examples. The structural learning, is aimed at capturing the relation between the variables, and hence the structure of the DAG. It can be seen as an optimization problem which requires the definition of a search strategy in the space of graph structures, and a scoring function for evaluating the effectiveness of candidate solutions. A typical scoring functions is the posterior probability of the structure given the training data. More formally, if D and S^h denote the training set and the structure of a candidate BN, respectively, the scoring function to be maximized is the likelihood of D given the structure S^h. Once the DAG structure S^h has been determined, the parameters of the conditional probability distributions are computed from training data.

The exhaustive search of the BN structure which maximizes the scoring function is a NP-hard problem in the number of variables. This is the reason why standard algorithms search for suboptimal solutions by maximizing at each step a local scoring function which takes into account only the local topology of the DAG. Moving from these considerations, we have proposed an alternative approach in which the structure of the BN is learned by means of an Evolutionary

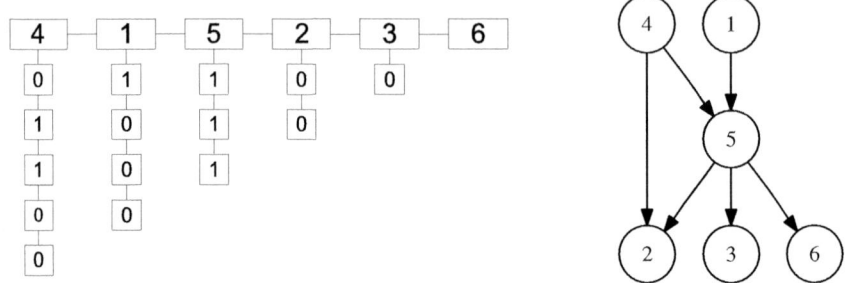

Fig. 2. A multilist (right) and the encoded DAG's(left)

algorithm, using a direct encoding scheme. The algorithm is based on a specifically devised data structure for encoding DAG, called *multilist* (ML), which consists of two basic lists. The first one, called *main list*, contains all the nodes of the DAG arranged in such a way that source nodes occupy the first positions, and sink node, the last ones. Moreover, nodes having both incoming and outgoing arcs are inserted in the *main list* after their parents. To each node of the *main list* is associated a second list called *sublist*, representing the outgoing connections among that node and the other nodes in the DAG. More specifically, if s_i is the *sublist* associated to the $i - th$ element of the *main list*, then it contains information about the outgoing arcs possibly connecting the $i - th$ element and the other elements following it in the *main list*, ordered according to the position of such elements. Since an arc may be present or not, each element of a *sublist* contains a binary information: 1 if the arc exists, 0 otherwise (see figure 2). This ML data structure allows an easy and effective implementation of genetic operators. Moreover, since the above definition ensures that a ML intrinsically represents a DAG structure, the application of such operators always produces valid offspring.

As regards the genetic operators, we have defined two mutation operators which can modify a ML in two different ways. The m mutation changes a ML by swapping two elements of the main list, whereas the s mutation adds and/or deletes one or more arcs in a sub list.

The m–mutation performs a permutation on the elements of the main list, but leaves unchanged the connection topology of the ML. This mutation consists of two steps:

(i) randomly pick two elements in the main list and swap their positions.
(ii) modify sublist elements in such a way to restore the connection topology as it was before the step (i).

It is worth noticing that the m–mutation generates a new ordering of the variables, which modifies the directions of the existing arcs in the DAG, but preserves dependencies between variables. If we consider the DAG in figure 2, for instance, the swap between the second and the fourth node in the main list changes only the directions of the arcs connecting the couples of nodes $(1, 5)$ and $(5, 2)$. This operator is applied according to a predefined probability value p_m.

The s–mutation, instead, modifies the values of the *sublist* elements. For each element of the sublists, p_s represents the probability of changing its value from 0 to 1, or vice versa. Thus the effect of this operator is that of adding or deleting arcs in the DAG. Such an operation is applied with probability p_s. Further details about ML data structure and the genetic operators can be found in [3].

The evolutionary algorithm starts by randomly generating an initial population of P individuals. Afterward, the fitness of each individual is evaluated by computing the scoring function. At each generation, the best e individuals are selected and copied in the new population in order to implement an elitist strategy. Then, the tournament is used to select $(P - e)$ individuals and the m and s mutation operators are applied to each selected individual according to the probabilities p_m and p_s, respectively. Finally these individuals are added to the new population. This process is repeated for n_g generations.

3   System Architecture

The proposed system consists of two main modules: the first one builds an ensemble of decision tree classifiers (experts); the second one implements the combining rule that produces the final classification result of the whole system.

The first module, called *BoostCGPC*, builds decision tree [14] using a Genetic Programming (GP) technique [8], which is an evolutionary computation-based technique able to evolve computer programs according to a user-defined fitness function. The output ensemble is learned by implementing a modified version of the algorithm *AdaBoost.M2* [6]. Such an implementation allows to run the algorithm on distributed memory parallel computer, making the system able to deal with large data sets. Further details about this algorithm can be found in [5].

The second module uses the approach described in the previous section for combining the responses provided by the classifiers making up the ensemble built in the first module. More specifically, let us denote with N the number of classes to be discriminated, with L the number of decision tree classifiers included the ensemble and with $E = \{e_1, \dots, e_L\}$ the set of responses provided by such classifiers for a given input sample. Let us assume that such responses constitute the input to the combiner module. In this module, the combining technique operates as a "higher level" classifier, working on a L-dimensional discrete-valued feature space, which is trained by using a supervised learning procedure. This procedure requires to observe both the "true" class label c, and the set of responses provided by the classifiers for each sample of a training set, in order to estimate the conditional probability $p(c|e_1, \dots, e_L)$. Once this conditional probability has been learned, the combiner evaluates the most probable class $\widehat{c}$ of an unknown input sample, given the expert observations, as follows:

$$\widehat{c} = \arg \max_{c \in C} p\left(c|e_1, \dots, e_L\right) \tag{5}$$

where C is the set of classes. Considering the definition of conditional probability and omitting the terms not depending on the variable c to be maximized, Eq. (5) can be rewritten as:

$$\widehat{c} = \arg \max_{c \in C} p\left(c, e_1, ..., e_L\right). \tag{6}$$

that involves only the joint probabilities $p\left(c, e_1, ..., e_L\right)$. This problem may be effectively solved by using a Bayesian Network, according to the approach outlined in the previous section.

Note that the devised system recognizes unknown samples using a two–step procedure: (i) the feature values describing the unknown sample are provided to each of the ensemble classifiers built by the BoostCGPC module; (ii) the set of responses produced is given in input to the BN module. Such module labels the sample with the most likely class, among those of the problem at hand, given the responses collected by the first module [1]. Note that, for some samples, the BN is not able to assign them a label. This case occurs when two or even more classes are equally likely. In this case, the unknown sample is labeled using the majority vote rule, applied to the first module responses.

4   Experimental Results

The proposed approach has been tested on four data sets: *Census*, *Segment*, *Adult* and *Phoneme*. The size and class distribution of these data sets are described in Table 1. They present different characteristics in the number and type (continuous and nominal) of attributes, two classes versus multiple classes and number of samples. In particular, Census and Adult, are real large data set containing census data collected by the U.S. Census Bureau. The Segment contains image data. Finally, the Phoneme data set contains data distinguishing between nasal and oral vowels. For each data set, two statistically independent sets of equal size, have been built randomly splitting the samples of each class. The first set has been used for training, while the second set for the test.

All the experiments were performed on a Linux cluster with 16 Itanium2 1.4GHz nodes each having 2 GBytes of main memory and connected by a Myrinet

Table 1. The data sets used in the experiments

datasets	attr.	samples	classes
Adult	14	48842	2
Census	4	299285	2
Phoneme	5	5404	2
Segment	36	2310	6

---

[1] Note that the second step does not require any further computation with respect to the Majority Voting rule. In fact, it only needs to read tables storing class probabilities.

high performance network. As regards the boostGCPC algorithm, it has been run on five nodes, using standard GP parameters and a population of 100 individuals for node. The original training set has been partitioned among the nodes and respectively 5 and 10 rounds of boosting, with 100 generations for round, have been used to produce respectively 25 and 50 classifiers on 5 nodes. It is worth to remember the algorithm produce a different classifier for each round on each node.

All results were obtained by averaging over 30 runs. For each run of the BoostCGPC module, a run of the BN module has been carried out. Each BN run has been performed by using the responses, on the whole training set, provided by the classifiers learned in the corresponding BoostCGPC run. The results on the test set has been obtained by first submitting each sample to the learned decision trees ensemble. The ensemble responses have been then provided to the learned BN. Finally, the BN output label has been compared with the true one of that sample.

The results achieved by the our approach (hereafter BN-BoostCGPC) have been compared with those obtained by BoostCGPC approach, which uses the Weighted Majority rule for combining the ensemble responses. The comparison results are shown in Tab. 2. The second column shows the ensembles (25 or 50 classifiers), while the columns 3 and 6 shows the training set errors of the BoostCGPC and BN-BoostCGPC, respectively. Similarly, the columns 4 and 7 show the test set errors of the BoostCGPC and BN-BoostCGPC, respectively. The columns 5 and 8 contain the number of classifiers actually used by both approaches. It is worth noticing that for the BoostCGPC approach such number equals the number of classifier making up the ensemble (25 or 50). The BN-BoostCGPC, instead, uses only the classifiers that are directly connected to the output node in the DAG.

In order to statistically validate the comparison results, we performed the two–tailed t–test($\alpha = 0.05$) over the 30 carried out runs. The values in bold in the test set error columns highlight, for each ensemble, the results which are significantly better according to the two–tailed t–test. The proposed approach achieves better performance on the majority of the considered ensembles while,

Table 2. Comparison results

Datasets	ens.	BoostCGPC			BN-BoostCGPC		
		Train	Test	# sel.	Train	Test	# sel.
Adult	25	15.90	16.94	25	15.85	**16.28**	3.4
	50	16.88	18.23	50	16.55	**16.99**	3.8
Segment	25	11.82	12.69	25	10.82	**11.68**	2.9
	50	10.39	12.06	50	10.34	11.99	2.9
Phoneme	25	16.41	18.87	25	17.70	19.23	3.2
	50	16.90	20.04	50	17.23	19.51	7.8
Census	25	8.81	8.89	25	5.14	**5.27**	4.3
	50	8.88	9.07	50	5.27	**5.37**	3.9

in the remaining cases, the performance are comparable. It is also worth noticing that the most significant improvements have been obtained on Adult and Census data sets, which are the largest ones among those considered. This result is due to the fact that larger data sets allow the BN learning process to better estimate the conditional probabilities to be modeled. Finally, is worth to remark that the results of our system are always achieved by using only a small number of the available classifiers.

5   Conclusions and Future Work

We presented a novel approach for improving the performance of derivation tree ensembles, learned by means of a boosted GP algorithm. The approach consists of two modules, the first one uses a boosted GP algorithm to generate ensembles of decision trees. The second module, instead, employs Bayesian networks to effectively combine the responses provided by the ensemble decision trees.

The experimental results have shown that the proposed system further improves the performance achieved by using the boosted GP algorithm. Moreover, such performances are obtained by using a reduced number of classifiers. Finally, the presented approach seems to be particularly suited to deal with very large data sets. Future work will include testing on several ensembles, having a different number of classifiers. Furthermore, larger data sets will be taken into account, to further investigate the capability of the presented system to deal with very large data sets.

References

1. Cantú-Paz, E., Kamath, C.: Inducing oblique decision trees with evolutionary algorithms. IEEE Transaction on Evolutionary Computation 7(1), 54–68 (2003)
2. De Stefano, C., D'Elia, C., Scotto di Freca, A., Marcelli, A.: Classifier combination by bayesian networks for handwriting recognition. Int. Journal of Pattern Rec. and Artif. Intell. 23(5), 887–905 (2009)
3. De Stefano, C., Fontanella, F., Marrocco, C., Scotto di Freca, A.: A hybrid evolutionary algorithm for bayesian networks learning: An application to classifier combination. In: EvoApplications (1). pp. 221–230 (2010)
4. Folino, G., Pizzuti, C., Spezzano, G.: A cellular genetic programming approach to classification. In: Proc. Of the Genetic and Evolutionary Computation Conference (GECCO 1999), pp. 1015–1020. Morgan Kaufmann, Orlando (1999)
5. Folino, G., Pizzuti, C., Spezzano, G.: Gp ensembles for large-scale data classification. IEEE Transaction on Evolutionary Computation 10(5), 604–616 (2006)
6. Freund, Y., Shapire, R.: Experiments with a new boosting algorithm. In: Proceedings of the 13th Int. Conference on Machine Learning, pp. 148–156 (1996)
7. Iba, H.: Bagging, boosting, and bloating in genetic programming. In: Proc. Of the Genetic and Evolutionary Computation Conference (GECCO 1999), pp. 1053–1060. Morgan Kaufmann, Orlando (1999)
8. Koza, J.R.: Genetic Programming: On the Programming of Computers by means of Natural Selection. MIT Press, Cambridge, MA (1992)

9. Kuncheva, L., Shipp, C.: An investigation into how adaboost affects classifier diversity. In: Proc. of IPMU (2002)
10. Kuncheva, L., Skurichina, M., Duin, R.P.W.: An experimental study on diversity for bagging and boosting with linear classifiers. Information Fusion 3(4), 245–258 (2002)
11. Kuncheva, L.I.: Combining Pattern Classifiers: Methods and Algorithms. Wiley Interscience, Hoboken (2004)
12. Nikolaev, N., Slavov, V.: Inductive genetic programming with decision trees. In: Proceedings of the 9th International Conference on Machine Learning, Prague, Czech Republic (1997)
13. Pearl, J.: Probabilistic Reasoning in Intelligent Systems: Networks of Plausible Inference. Morgan Kaufmann, San Francisco (1988)
14. Quinlan, J.R.: C4.5 Programs for Machine Learning. Morgan Kaufmann, San Mateo (1993)

Multiple Classifiers for Graph of Words Embedding

Jaume Gibert[1], Ernest Valveny[1], Oriol Ramos Terrades[1], and Horst Bunke[2]

[1] Computer Vision Center, Universitat Autònoma de Barcelona
Edifici O Campus UAB, 08193 Bellaterra, Spain
{jgibert,ernest,oriolrt}@cvc.uab.es
[2] Institute for Computer Science and Applied Mathematics, University of Bern,
Neubrückstrasse 10, CH-3012 Bern, Switzerland
bunke@iam.unibe.ch

Abstract. During the last years, there has been an increasing interest in applying the multiple classifier framework to the domain of structural pattern recognition. Constructing base classifiers when the input patterns are graph based representations is not an easy problem. In this work, we make use of the graph embedding methodology in order to construct different feature vector representations for graphs. The graph of words embedding assigns a feature vector to every graph by counting unary and binary relations between node representatives and combining these pieces of information into a single vector. Selecting different node representatives leads to different vectorial representations and therefore to different base classifiers that can be combined. We experimentally show how this methodology significantly improves the classification of graphs with respect to single base classifiers.

1 Introduction

A multiple classifier system tries to combine several base classifiers in such a way that the resulting classification performance improves the accuracy rates of the underlying individual classifiers [1]. A common way to build base classifiers for further combination is by randomly selecting different subsets of features and training classifiers on those subsets [2]. By feature subset selection one is usually able to obtain classifiers with enough diversity, in terms of their discriminative power. Such procedures can be rather easily implemented and have been widely studied for statistical feature vectors. However, when it comes to graph based representations, the construction of single base classifiers has not a straightforward solution.

Based on the idea of feature subset selection, just a few works aiming at constructing multiple classifier systems for graph based representations have been proposed. While for feature vectors the idea is to select a subset of features, for graph representations the way to proceed is to construct base classifiers by selecting subgraphs of the training graphs. For instance, in [3] the authors are able to define different classifiers by randomly removing nodes and their incident

C. Sansone, J. Kittler, and F. Roli (Eds.): MCS 2011, LNCS 6713, pp. 36–45, 2011.

edges from the graphs. This methodology has the problem that the need arises of comparing several instances of labelled graphs, which makes the combination of multiple classifiers a highly complex task. In [4], a more efficient way is proposed by decomposing a set of labelled graphs into several unlabelled subgraphs based on their labelling information. All these subgraphs are compared one by one with respect to different labels which create different base classifiers. Finally, in [5], in order to build several base classifiers, the authors claim for the beneficial use of full graphs instead of just subgraphs. They transform labelled full graphs into different unlabelled graphs by removing information from the nodes and re-weighting the edges based on the linked nodes' information. Although the graphs are altered, the topology is preserved and several base classifiers can be constructed.

The main drawback of all the approaches described above is that they necessarily need to work in the graph domain when comparing graphs and, therefore, the base classifiers to be used are restricted to be of the kNN type. In this work we adopt another approach to construct several base classifiers for graph based representations. We embed every graph into different vector spaces in such a way that several feature vectors are associated to every graph. From these sets of feature vectors, different based classifiers can be trained and then combined under any of the common statistical combination frameworks. The idea is somewhat similar to the one in [6] where the authors transform a graph into a feature vector by computing edit distances to a predefined set of prototypes. As a result, for these vectors, more complex learning machines such as SVMs are used. Then, by using different sets of prototypes, the authors create different populations of vectors leading to several base classifiers that can be combined.

Besides the work of [6], other examples of graph embeddings can be found in the literature. For instance, in [7], the authors approach the problems of graph clustering and graph visualization by extracting different features from an eigen-decomposition of the adjacency matrices of the graphs. In [8], the nodes of the graph are embedded into a metric space and then the edges are interpreted as geodesics between points on a Riemannian manifold. The problem of matching nodes to nodes is viewed as the alignment of the resulting point sets. Finally, in [9], to solve the problem of molecules classification, the authors associate a feature vector to every molecule by counting unary and binary statistics in the molecule; these statistics indicate how many times every atomic element appears in the molecule, and how often there is a bond between two specific atoms.

In this paper, we propose to extend the idea of [9] to the case of graphs with continuous attributes and, as in [6], describe ways of constructing different feature vector representations for every graph. In the next section we define the way of transforming a node labelled graph into a feature vector by means of the Graph of Words Embedding. Then, in Section 3, we formally recall some basic concepts from the field of multiple classifiers systems and describe how we can derive different vector representations of graphs for their further combination. In Section 4, we describe and discuss the experimental results that have been obtained. Finally, Section 5 concludes the article.

2   Graph of Words Embedding

Although the embedding of graphs into vector spaces provides a way to apply statistical pattern analysis techniques to the domain of graphs, the existing methods still suffer from the main drawback that the classical graph matching techniques also did, this is, their computational cost. The Graph of Words Embedding tries to avoid these problems by just visiting nodes and edges instead of, for instance, travelling along paths in the graphs, computing edit distances or performing the eigen-decomposition of the adjacency matrix. In this section we first briefly explain the motivation of this approach and then formally define the procedure.

2.1   Motivation

In image classification, a well-known image representation technique is the so-called bag of visual features, or just bag of words. It first selects a set of feature representatives, called words, from the whole set of training images and then characterizes each image by a histogram of appearing words extracted from the set of salient points in the image [10].

The graph of words embedding proceeds in an analogous way. The salient points in the images correspond to the nodes of the graphs and the visual descriptors are the node attributes. Then, one also selects representatives of the node attributes (words) and counts how many times each representative appears in the graph. This leads to a histogram representation for every graph. To take profit of the edges in the original graphs, one also counts the frequency of the relation between every pair of words. The resulting information is combined with the representatives' histogram in a final vector.

2.2   Embedding Procedure

A graph is defined by the 4-tuple $g = (V, E, \mu, \nu)$, where V is the set of nodes, $E \subseteq V \times V$ is the set of edges, μ is the nodes labelling function, assigning a label to every node, and ν is the edges labelling function, assigning a label to every edge in the graph. In this work we just consider graphs whose node attributes are real vectors, this is, $\mu : V \rightarrow \mathbb{R}^d$ and whose edges remain unattributed, this is, $\nu(e) = \varepsilon$ for all $e \in E$ (where ε is the null label).

Let $\mathcal{P}$ be the set of all node attribute vectors in a given dataset of graphs $\mathcal{G} = \{g_1, \ldots, g_M\}$. From all vectors in $\mathcal{P}$ we derive n representatives, which we shall call *words*, in analogy to the bag of words procedure. Let this set of words be $\mathcal{V} = \{w_1, \ldots, w_n\}$ and be called *vocabulary*. Let furthermore λ be the node-to-word assignment function $\lambda(v) = \arg\min_{w_i \in \mathcal{V}} d(\mu(v), w_i)$, this is, the function that assigns a node to its closest word. Before assigning a vector to each graph, we first construct an intermediate graph that will allow us an easier embedding. This intermediate graph, called *graph of words* $g' = (V', E', \mu', \nu')$ of $g = (V, E, \mu, \nu) \in \mathcal{G}$ with respect to $\mathcal{V}$, is defined as:

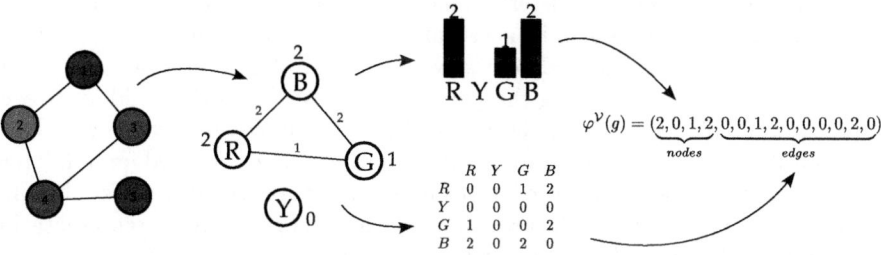

Fig. 1. Example of the graph of words embedding. The graph on the left is assigned to the vector on the right by considering the vocabulary $V = \{R, Y, G, B\}$. Nodes 1 and 5 are assigned to the R word, 2 and 3 to the B word and 4 to the G word. Note that none is assigned to the Y word. The histogram of words is considered as well as the adjacency matrix. The resulting vector is the concatenation of both types of information.

- $V' = V$
- E' is defined by: $(w, w') \in E' \Leftrightarrow$ there exists $(u, v) \in E$ such that $\lambda(u) = w$ and $\lambda(v) = w'$
- $\mu'(w) = |\{v \in V \mid w = \lambda(v)\}|$
- $\nu'(w, w') = |\{(u, v) \in E \mid \lambda(u) = w, \lambda(v) = w'\}|$.

Once the graph of words is constructed, we easily convert the original graph into a vector by combining the node and edge information of the graph of words, by keeping both the information of the appearing words and the relation between these words. We consider the histogram

$$\phi_a^V(g) = (\mu'(w_1), \ldots, \mu'(w_n)). \tag{1}$$

and a flattened version of the adjacency matrix of the graph of words $A = (a_{ij})$, with $a_{ij} = \nu'(w_i, w_j)$:

$$\phi_b^V(g) = (a_{11}, \ldots, a_{ij}, \ldots, a_{nn}), \quad \forall i \leq j \tag{2}$$

The final graph of words embedding is the concatenation of both pieces of information,

$$\varphi^V(g) = (\phi_a^V(g), \phi_b^V(g)). \tag{3}$$

In Figure 1, there is an example of the graph of words procedure for a simple vocabulary of size equal to 4.

2.3   Vocabulary Selection and Dimensionality Reduction

The final configuration of the vectors after the embedding clearly depends on the set of words that have been chosen and this fact is what will actually give us different and diverse vector representations of the graphs. In this paper we decided to use the kMeans algorithm in order to build the vocabulary. The initialization points of the clustering under kMeans are usually taken randomly. In our case, in order to avoid uncertainty in the results, we chose to start by selecting uniformly distributed points over the range of node attributes. The way we proceed is by first selecting the median vector of $\mathcal{P}$ and then adding at each iteration the point which is furthest away from the already selected ones. This is done until a predefined number of words is obtained, and repeated for a different number of words each time. Each of the different sets of representatives obtained by the kMeans algorithm will provide us with a different vector representation of graphs, and thus with a different base classifier.

Another important issue of the graph of words configuration is the quadratic increase of its dimensionality with respect to the vocabulary size. To avoid sparsity problems and to reduce problems that arise from having to deal with such high dimensional vectors, in this article we have applied Kernel PCA [11] in conjunction with the RBF kernel to the original vectors. More details of the dimensionality reduction procedures applied to graph of words embedded vectors can be found in [12].

3   Multiple Classifiers for Graph of Words Embedding

Let us review in this section a few important concepts from the domain of multiple classifier systems and then present their application to the graph of words embedding.

3.1   Multiple Classifier Methods

In [13], the authors originally put together various classifiers under the following taxonomy, based on the output information they are able to supply. The first level of classifiers, called the *abstract level*, outputs a unique label for every pattern to be classified. The combination of such classifiers is usually done by voting strategies, which assign the final decision based on the plurality of votes of all the available classifiers [1]. The second level, the *rank level*, outputs a ranked list of labels for every pattern. Borda Count is the common methodology to combine these rankings. The rankings from all classifiers are combined by ranking functions assigning votes to the classes based on their positions in the classifiers' rankings. The final decision is taken as the minimum of the sum of these rankings. Finally, the *measurement level*, outputs a set of confidence values defining the degree of belongingness to each class. Bayesian combinations are commonly used by which different rules (product, sum, mean, etc.) are applied to the confidences of each class [14]. Finally, non-Bayesian combinations can also be applied such that a weighted linear combination of classifiers is learnt using optimization techniques [15].

3.2   Multiple Classifiers for Graph of Words Embedding

Summarizing what we have explained so far, we represent a set of graphs by vectors using the graph of words embedding. A set of node attribute representatives is chosen and unary and binary statistics of each graph are computed and combined in a single histogram representation. This representation lets us train an SVM, or any other classifier. By selecting sets of representatives of different size we are able to create a population of vectors of different dimensionality. In this article, we have chosen vocabulary sizes from only 2 words up to 100 words, obtained a set of 99 different vector representations for every graph. Thus, a different base classifier can be trained for every different choice of the vocabulary size, giving the possibility of building ensembles of classifiers under the common statistical combination frameworks.

It seems reasonable to follow this direction since every choice of the vocabulary is giving a semantically different representation of the graphs. For instance, a vocabulary including just a few words would express global relations between the nodes of the graphs, while a large vocabulary would describe the input graphs in terms of their local structure.

3.3   Classifier Selection

An important issue that deserves our attention is the question of how to build the final ensemble from the available classifiers. In this article, we have addressed this problem by a simple forward selection strategy [1]. From the set of N classifiers that we have available, we will construct N ensembles. The first ensemble will be constituted by one classifier, the second by two classifiers, etc., until the last ensemble that is going to be the ensemble of all classifiers that we have trained. These ensembles are iteratively build by, first, taking the best single classifier as the first ensemble, and then adding to the k-th ensemble the classifier that best fits the previous ensemble, in terms of the accuracy of the combination on a validation set. As a result, the final ensemble that is applied to the test set is the one with the highest accuracy rate.

4   Experimental Results

The main objective in this work is to evaluate the improvements obtained from the combination of several classifiers with respect to the accuracy of the individual ensemble members. In this section, we describe the graph datasets we have been working with, provide a description of the experimental setup, and present the results that have been obtained.

4.1   Databases

We have chosen three different datasets from the IAM Graph Database Repository [16], posing classification problems on both synthetic and real data. The *Letter Database* represents distorted letter drawings. Starting from a manually constructed prototype of every of the 15 Roman alphabet letters that consist of straight lines only, different degrees of distortion are applied. Each ending point of a line is represented by a node of the graph and attributed with its

(x, y) coordinates. The second graph dataset is the *GREC Database*, which represents architectural and electronic drawings under different levels of noise. In this database, intersections and corners constitute the set of nodes. These nodes are attributed with their position on the 2-dimensional plane. Finally, the *Fingerprint Database* consists of graphs that are obtained from a subset of the NIST-4 fingerprint image database [17] by means of particular image processing operations. Ending point and bifurcations of the skeleton of the processed images constitute the (x, y)-attributed nodes of the graphs. Each of these datasets is split into a training set, a validation set and a test set.

4.2   Experimental Setup

As already said before, for every graph in the datasets described above, we have constructed several vector representations with the graph of words embedding methodology using different vocabulary sets. More precisely, the vocabulary size is chosen to be from only 2 words up to 100, leading to 99 different vectorial representations. Each of these sets of vectors is reduced using kPCA, the parameters of which are properly optimized using the validation set.

Then, a single base classifier for each vocabulary size is trained. Support Vector Machine [18] is an important learning machine that lately has been gaining significant attention due to its good results on real data. The main idea of the method is to separate classes using hyperplanes in the implicit Hilbert space of the kernel function in use. We have used the implementation described in [19], which provides a way to extract outputs on the three different levels that have been described in Section 3.1. Here again, the kernel functions, their associated parameters and the SVM's own parameters have been tuned using the validation set.

In Figure 2(a), we show the accuracy rates of the single SVM classifiers on the validation set for the Fingerprint dataset. Every classifier is based on a different vocabulary of different size (represented on the horizontal axis), and we can see how the different SVM classifiers lead to different results, supporting the assumption of diversity of the representation of the different graph of words configurations. The goal of this paper, and of the MCS methodology in general, is to outperform the best of these individual base classifiers.

With regards to the combination rules that have been used, we have worked with the voting combination strategy for the first level of classifiers, with the Borda Count for the second, with the product and the mean rules in the Bayesian combination framework, and with the *IN* and the *DN* approaches in the case of the Non-Bayesian methodology (the former has a closed optimal solution, while the latter is optimized by numerical minimization).

In Figure 2(b), we show the effect of the classifier selection strategy on the validation set of the Fingerprint database. The different combination strategies of the three different classifier levels are shown in the figure. As expected, the tendency is an increasing performance with just a few classifiers, and then, once more and more classifiers are added to the combination, the accuracy rates go down, leading to worse results than with the best single classifier alone.

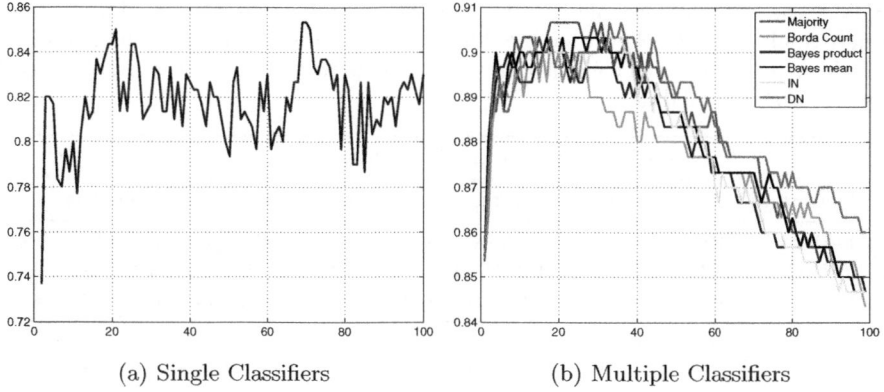

(a) Single Classifiers   (b) Multiple Classifiers

Fig. 2. Results on the validation set for the Fingerprint dataset. (a) Classification rates of the single classifiers. On the horizontal axis the number of words in the vocabulary set is shown. (b) Combination of all single classifiers after classifier selection. On the horizontal axis we show the number of classifiers in the combination. The vertical axis shows the classification rate of the combination. Note that the maximum on (a) is the starting point on (b).

4.3   Results on the Test Set

Once the system and all the associated parameters are properly selected using the validation set, we take the best performing ensemble and test it on the test set of the three databases. In Table 1, we show the results. The table is split into single and multiple classifiers. On the single classifier part we show the results of the best of the single SVMs that we have trained on our resulting graph of words embedded vectors. For the multiple classifiers, we show the results for voting, Borda Count, the product rule in the case of Bayesian combination and the *IN* solution for the Non-Bayesian one. The mean rule and the *DN* strategies lead to lower recognition rates (yet statistically non-significant) and this is the reason why they are not shown.

In the case of the Letter dataset the results of the different combinations do not show a significant improvement over the reference system. The combination of classifiers for the Borda Count methodology and the Bayes combination is obtaining better results than the single classifier, but not statistically significant ones. It even happens that the combination of classifiers gets lower results than the single base classifier (non-Bayes methodologies). This is obviously due to the fact that the final configuration of the system is chosen as the one that gives the best results on the validation set and this configuration does not necessarily lead to the best result on the test set.

On the other hand, we can see that the improvements obtained on the Fingerprint and GREC databases are all significant, no matter which combination strategy is chosen. Obviously, the different configurations of the graph of words vectors are sufficiently diverse to learn different characteristics of the graphs

Table 1. Results on the test set for different classifiers (accuracy rates in %). kPCA-SVM is the best SVM classifier on the reduced graph of words vectors. Multiple Classifiers results are also shown for the different combination strategies. The number of classifiers for the optimal combination is shown in parenthesis. The best result on each dataset is shown bold face.

	Single Classifier	Multiple Classifiers			
Database	kPCA-SVM	Voting	Borda C.	Bayes	Non-Bayes
Letter	98.8	98.8 (4)	**99.2** (5)	**99.2** (4)	98.4 (2)
GREC	93.7	**99.0** (33) ⋆	98.6 (3) ⋆	98.6 (7) ⋆	**99.0** (5) ⋆
Fingerprint	80.8	83.1 (18) ⋆	82.9 (14) ⋆	**83.9** (7) ⋆	83.6 (31) ⋆

⋆ Statistically significant improvement over the single classifier (Z-test using $\alpha = 0.05$).
○ Statistically significant deterioration over the single classifier (Z-test using $\alpha = 0.05$).

and to obtain important improvements over the reference system. In particular, for the GREC database and the Non-Bayes combination, the members of the best ensemble are the graph of words configurations of vocabulary size $\{19, 39, 5, 2, 49\}$ (in order of decreasing importance), while for the case of the Fingerprint dataset and the Bayes combination they are $\{69, 24, 71, 67, 73, 19, 77\}$. The variety of these sizes supports the idea that using different configurations of the proposed methodology is a way to properly discover semantically different information content among the graph structures.

5   Conclusions

In this article we have presented a novel way to construct combinations of several classifiers for graph-based representations of patterns. Every graph is embedded into different vectorial spaces by means of the graph of words embedding. Every different selection of the vocabulary set in the graph of words leads to a different vectorial representation and thus to a different base classifier. The individual SVM base classifiers have been combined using several standard methods for classifier combination and a forward selection strategy. Experiments have shown significant improvements on two out of three datasets.

Since the methodology seems suitable for improving the classification rates on graph classification problems, future work will consider other vocabulary selection methods, which could lead to other base classifier to further enrich the ensembles of classifiers.

References

1. Kuncheva, L.I.: Combining Pattern Classifiers: Methods and Algorithms. John Wiley, Chichester (2004)
2. Ho, T.K.: The Random Subspace Method for Constructing Decision Forests. IEEE Trans. on Pattern Analysis and Machine Intelligence 20(8), 832–844 (1998)

3. Schenker, A., Bunke, H., Last, M., Kandel, A.: Building Graph-Based Classifier Ensembles by Random Node Selection. In: Roli, F., Kittler, J., Windeatt, T. (eds.) MCS 2004. LNCS, vol. 3077, pp. 214–222. Springer, Heidelberg (2004)
4. Lee, W.J., Duin, R.P.W.: A Labelled Graph Based Multiple Classifier System. In: Benediktsson, J.A., Kittler, J., Roli, F. (eds.) MCS 2009. LNCS, vol. 5519, pp. 201–210. Springer, Heidelberg (2009)
5. Lee, W.J., Duin, R.P.W., Bunke, H.: Selecting Structural Base Classifiers for Graph-based Multiple Classifier Systems. In: El Gayar, N., Kittler, J., Roli, F. (eds.) MCS 2010. LNCS, vol. 5997, pp. 155–164. Springer, Heidelberg (2010)
6. Riesen, K., Bunke, H.: Classifier Ensembles for Vector Space Embedding of Graphs. In: Haindl, M., Kittler, J., Roli, F. (eds.) MCS 2007. LNCS, vol. 4472, pp. 220–230. Springer, Heidelberg (2007)
7. Luo, B., Wilson, R.C., Hancock, E.R.: Spectral embedding of graphs. Pattern Recognition 36(10), 2213–2230 (2003)
8. Robles-Kelly, A., Hancock, E.R.: A Riemannian approach to graph embedding. Pattern Recognition 40(3), 1042–1056 (2007)
9. Gibert, J., Valveny, E., Bunke, H.: Graph of Words Embedding for Molecular Structure-Activity Relationship Analysis. In: Bloch, I., Cesar Jr., R.M. (eds.) CIARP 2010. LNCS, vol. 6419, pp. 30–37. Springer, Heidelberg (2010)
10. Dance, C., Willamowski, J., Fan, L., Bray, C., Csurka, G.: Visual categorization with bags of keypoints. In: ECCV International Workshop on Statistical Learning in Computer Vision, pp. 1–22 (2004)
11. Schölkopf, B., Smola, A., Müller, K.R.: Nonlinear Component Analysis as a Kernel Eigenvalue Problem. Neural Computation 10, 1299–1319 (1998)
12. Gibert, J., Valveny, E., Bunke, H.: Dimensionality Reduction for Graph of Words Embedding. In: Jiang, X. (ed.) GbRPR 2011. LNCS, vol. 6658, pp. 22–31. Springer, Heidelberg (2011)
13. Xu, L., Krzyzak, A., Suen, C.Y.: Methods of Combining Multiple Classifiers and Their Applications to Handwriting Recognition. IEEE Trans. on Systems, Man and Cybernetics 22(3), 418–425 (1992)
14. Kittler, J., Hatef, M., Duin, R.P.W., Matas, J.: On Combining Classifiers. IEEE Trans. on Pattern Analysis and Machine Intelligence 20(3), 226–239 (1998)
15. Ramos Terrades, O., Valveny, E., Tabbone, S.: Optimal Classifier Fusion in a Non-Bayesian Probabilistic Framework. IEEE Trans. on Pattern Analysis and Machine Intelligence 31, 1630–1644 (2009)
16. Riesen, K., Bunke, H.: IAM Graph Database Repository for Graph Based Pattern Recognition and Machine Learning. In: da Vitoria Lobo, N., Kasparis, T., Roli, F., Kwok, J.T., Georgiopoulos, M., Anagnostopoulos, G.C., Loog, M. (eds.) S+SSPR 2008. LNCS, vol. 5342, pp. 287–297. Springer, Heidelberg (2008)
17. Watson, C., Wilson, C.: NIST Special Database 4, Fingerprint Database. National Institute of Standards and Technology (1992)
18. Schölkopf, B., Smola, A.J.: Learning with Kernels: Support Vector Machines, Regularization, Optimization, and Beyond. MIT Press, Cambridge (2002)
19. Chang, C.C., Lin, C.J.: LIBSVM: A library for Support Vector Machines. Software (2001), http://www.csie.ntu.edu.tw/~cjlin/libsvm

A Dynamic Logistic Multiple Classifier System for Online Classification

Amber Tomas

Department of Statistics, The University of Oxford, UK
`tomas@stats.ox.ac.uk`

Abstract. We consider the problem of online classification in nonstationary environments. Specifically, we take a Bayesian approach to sequential parameter estimation of a logistic MCS, and compare this method with other algorithms for nonstationary classification. We comment on several design considerations.

1   Introduction

There have been many studies which show that combining the outputs of a number of classifiers can result in better classification performance than using a single classifier. Popular examples of such methods include Bagging [1], Boosting [13] and Random Forests [2], and there is much current research in the area. However, many of the algorithms assume that the population of interest is not changing over time, and that a set of observations from this population is available on which the classification algorithm can be trained. In practice, many populations do change over time and so the assumption of stationarity is not always reasonable. Therefore, it is natural to wonder if the success of MCSs for stationary, batch learning problems can be extended to dynamic online learning problems.

This problem has already been approached in several different ways, and [7] gives a good overview of current methods. In this paper we propose a model-based method for combining the outputs of a number of component classifiers, and for sequentially updating the classification rule as new observations become available. In the language of [7], the algorithm we propose is a "dynamic combiner" - changes in the population are modelled by updating the parameters of the combining rule, and the component classifiers remain fixed. The only restriction we place on the component classifiers is that they must output an estimate of the conditional class distribution, not only a class label.

We present our model in Section 2, then in Section 3 discuss how this model can be used to implement classification. In Section 4 we compare the performance of the suggested algorithm to other algorithms for dynamic classification on two simulated examples.

2   A Dynamic Logistic Model for Combining Classifiers

We assume that the population of interest consists of K classes, labelled $1, \ldots, K$. At some time t an observation $\boldsymbol{x}_t$ and label y_t are generated according to the

C. Sansone, J. Kittler, and F. Roli (Eds.): MCS 2011, LNCS 6713, pp. 46–55, 2011.

joint probability distribution $P_t(\boldsymbol{X}_t, Y_t)$. Given an observation $\boldsymbol{x}_t$, we denote the estimate output by the ith component classifier of $\text{Prob}\{Y_t = k|\boldsymbol{x}_t\}$ by $\hat{p}_i(k|\boldsymbol{x}_t)$, for $k = 1, \ldots, K$ and $i = 1, \ldots, M$, where M denotes the number of component classifiers.

Our final estimate of $\text{Prob}\{Y_t = k|\boldsymbol{x}_t\}$ is obtained by combining the component classifier outputs according to the multiple logistic model

$$\hat{p}_t(k|\boldsymbol{x}_t) = \frac{\exp(\boldsymbol{\beta}_t^T \boldsymbol{\eta}_k(\boldsymbol{x}_t))}{\sum_{j=1}^K \exp(\boldsymbol{\beta}_t^T \boldsymbol{\eta}_j(\boldsymbol{x}_t))}, \quad k = 1, \ldots, K, \tag{1}$$

where $\boldsymbol{\beta}_t = (\beta_{t1}, \beta_{t2}, \ldots, \beta_{tM})$ is a vector of parameters, the ith component of $\boldsymbol{\eta}_k(\boldsymbol{x}_t), \eta_{ki}(\boldsymbol{x}_t)$, is a function of the output of the ith component classifier, and $\boldsymbol{\eta}_1(\boldsymbol{x}_t) = 0$ for all $\boldsymbol{x}_t$. Changes in the population are modelled via changes in the parameters $\boldsymbol{\beta}_t$. It can be seen that this model is a dynamic generalised linear model, as defined by [16].

Similar models have been used for binary classification by [12] and [10], however rather than combine the outputs of a number of classifiers they used the raw inputs $\boldsymbol{x}_t$ in place of $\boldsymbol{\eta}(\boldsymbol{x}_t)$.

Before the classifier can be implemented, there are three steps to be taken:

1. a set of component classifiers must be determined;
2. the form of the functions $\boldsymbol{\eta}_k(\cdot)$ must be specified; and
3. an algorithm for updating the weights must be chosen.

In this section we comment on the first two points, and make suggestions for point 3 in Section 3.

2.1 Choosing the Component Classifiers

As has been documented for other MCSs, the choice of component classifiers is absolutely vital to the performance of an implementation of this model. If the component classifiers are all very similar, then the MCS is unlikely to be sufficiently flexible to discriminate well in all future scenarios. This suggests that the component classifiers should be relatively diverse compared to a stationary scenario, to increase the range of decision boundaries that can be represented at any future time. If there is prior knowledge about the extent of change expected, then this can be used when training the component classifiers. In the absence of prior knowledge, a reasonable strategy may be to train component classifiers on small, random subsets of available data, or else to use artificial data to generate a wider range of classifiers.

2.2 Specification of the $\eta_k(\cdot)$

The form of the functions $\boldsymbol{\eta}_k(\cdot)$ was deliberately unspecified in the general formulation of the model. Using different forms is likely to result in models with different properties, and *apriori* there is no reason to suggest that one form will dominate all the others in terms of classification performance. However, some choices seem more sensible than others, and in this paper we suggest two options:

1. $\eta_{ki}(\boldsymbol{x}_t) = \hat{p}_i(k|\boldsymbol{x}_t) - \hat{p}_i(1|\boldsymbol{x}_t)$, $i = 1, \ldots, M$ and $\qquad\qquad(2)$

2. $\eta_{ki}(\boldsymbol{x}_t) = \log\left(\dfrac{\hat{p}_i(k|\boldsymbol{x}_t)}{\hat{p}_i(1|\boldsymbol{x}_t)}\right)$, $i = 1, \ldots, M$. $\qquad\qquad(3)$

Both options allow $\eta_{ki}(\boldsymbol{x}_t)$ to take either positive or negative values, which means that the probabilities $\hat{p}(k|\boldsymbol{x}_t)$ in equation (1) can take all values in $(0, 1)$ even if the parameters are constrained to be non-negative. This allows us to consider non-negativity constraints on the parameters, which would not be appropriate if we defined $\eta_{ki}(\boldsymbol{x}_t) = \hat{p}_i(k|\boldsymbol{x}_t)$, for example. It is also interesting to note that at any time t, if (2) is used then the decision boundary is equivalent to that for the weighted averaging rule with weights given by $\boldsymbol{\beta}_t$ (or equivalently weights $c\boldsymbol{\beta}_t, c > 0$). If all the component classifiers are trained by linear discriminant analysis, then using (3) constrains the decision boundary of the final classifier also to be linear [14].

2.3   General Applicability

As with other dynamic combiners (an MCS in which the classifiers are fixed but the combining rule is updated over time [7]), we do not expect this model to perform universally well. If the population changes too much from what is expected when training the component classifiers then the model (1) will not necessarily be very accurate for any value of the parameters. However, given that the changes are not too severe, this model should be equally suited to slow and rapid changes providing that the method used for updating the parameters is appropriate.

3   Algorithms for Parameter Updating and Classification

In this section we propose methods for implementing a classifier based on the model introduced in Section 2, assuming that the functions $\boldsymbol{\eta}_k(\cdot)$ and the component classifiers are already specified. Firstly we discuss the classification rule, and then suggest how this can be implemented.

3.1   The Predictive Approach to Classification

Many model-based methods for classification estimate the parameter values directly, and then plug these estimates in to the model to produce a final classification. However, for classification problems we are typically more interested in estimating the conditional class distribution than the parameters themselves. Therefore, we follow the predictive approach to classification which produces estimates of the conditional class distribution and deals with uncertainty about the parameters by averaging over all possible values. Specifically, suppose that at time $t - 1$ our knowledge about the parameters $\boldsymbol{\beta}_{t-1}$ is summarised by the posterior distribution $p(\boldsymbol{\beta}_{t-1}|z_{1:t-1})$, where

$$z_{1:t-1} \triangleq \{(x_1, y_1), (x_2, y_2), \ldots, (x_{t-1}, y_{t-1})\}.$$

After having received an observation x_t we are interested in the probabilities

$$\tilde{p}(Y_t = k|x_t, z_{1:t-1}) = \int_{\boldsymbol{\beta}_t} p(Y_t = k, \boldsymbol{\beta}_t|x_t, z_{1:t-1})d\boldsymbol{\beta}_t, \; k = 1, \ldots, K$$

$$= \int_{\boldsymbol{\beta}_t} p(Y_t = k|x_t, \boldsymbol{\beta}_t)p(\boldsymbol{\beta}_t|z_{1:t-1})d\boldsymbol{\beta}_t, \; k = 1, \ldots, K, \; (4)$$

where $p(\boldsymbol{\beta}_t|z_{1:t-1})$ can be expressed as

$$p(\boldsymbol{\beta}_t|z_{1:t-1}) = \int_{\boldsymbol{\beta}_{t-1}} p(\boldsymbol{\beta}_t|\boldsymbol{\beta}_{t-1})p(\boldsymbol{\beta}_{t-1}|z_{1:t-1})d\boldsymbol{\beta}_{t-1}. \qquad (5)$$

We will use $\tilde{p}(k|x_t)$ to denote $\tilde{p}(Y_t = k|x_t, z_{1:t-1})$. To evaluate (5) the posterior distribution $p(\boldsymbol{\beta}_{t-1}|z_{1:t-1})$ is assumed known from time $t - 1$, and we also need information about $p(\boldsymbol{\beta}_t|\boldsymbol{\beta}_{t-1})$. Assuming for now that we have this information, (5) can then be substituted into (4), which can be evaluated by using the model (1) to specify the probabilities $p(Y_t = k|x_t, \boldsymbol{\beta}_t)$.

In order to calculate the expression $p(\boldsymbol{\beta}_t|\boldsymbol{\beta}_{t-1})$ in (5), we must specify a model for how the parameters change (or "evolve") over time. Following [16], we model the parameter evolution according to a Gaussian random walk:

$$\boldsymbol{\beta}_t = \boldsymbol{\beta}_{t-1} + \boldsymbol{\omega}_t, \qquad (6)$$

where $\boldsymbol{\omega}_t \sim N(\mathbf{0}, V_t)$. With this assumption we then have all the information needed to compute (4) and hence the probabilities $\tilde{p}(k|x_t), k = 1, 2, \ldots, K$. Having calculated these probabilities, we classify x_t to the class with the largest value of $\tilde{p}(k|x_t)$, so

$$\hat{c}(x_t) = \text{argmax}_k \; \tilde{p}(k|x_t). \qquad (7)$$

Suppose that we then observe the true label y_t. We can update the distribution of $\boldsymbol{\beta}_t$ according to the relationship

$$p(\boldsymbol{\beta}_t|z_{1:t}) \propto p(y_t|x_t, \boldsymbol{\beta}_t)p(\boldsymbol{\beta}_t|z_{1:t-1}), \qquad (8)$$

where $p(\boldsymbol{\beta}_t|z_{1:t-1})$ is given by (5), and $p(y_t|x_t, \boldsymbol{\beta}_t)$ by (1). This classification and updating process can then be repeated for each new observation.

3.2   Implementation

In order to simplify implementation of the prediction and updating procedure above, it is assumed that the parameter evolution variance matrix $V_t = v_t I$, where I is the identity matrix. This is a common assumption in dynamic modelling (see for example [5], [12] and [6]). Under this assumption, the rate of change of the parameter corresponding to each component classifier is assumed to be equal. v_t can be thought of as a "forgetting" parameter - a larger value of v_t implies that the parameter values are more likely to change by a large amount from one time point to the next, thereby reducing the influence of previous observations. If v_t is too large, then the extra noise in the parameter evolution will lead to added classification error. If v_t is too small, then the updating process may not be able to track the true changes in the parameter values. The performance of the classifier is quite sensitive to the specification of this parameter.

In general, the integrals in (4) and (5) can not be evaluated exactly. In [14] we considered the use of a Gaussian updating procedure to approximate the integrals in (4), and also a method for implementing dynamic generalised models given in [15]. However, we found that using Sequential Monte Carlo (SMC) techniques allows more flexibility and is easily applicable to both binary and multi-class classification. Further, using SMC it is easy to model the evolution of v_t over time. We again use a Gaussian random-walk model

$$v_t = v_{t-1} + \nu_t, \ \nu_t \sim N(0, q_t), \tag{9}$$

and assume that $q_t = q$, a constant which must be specified in advance. This allows the algorithm to adaptively alter the variance of the parameter evolution over time, and so dynamically adjust to periods of slow and rapid change. Although q must be specified in advance, the performance of the algorithm tends to be more robust to misspecification of q than to v.

SMC is computationally intensive, and as a result can only cope with relatively low-dimensional parameter spaces, i.e. with relatively few component classifiers. For the simulations presented in section 4 we used three component classifiers and 1000 particles to represent the corresponding 4D parameter space $\{\beta_1, \beta_2, \beta_3, v_t\}$. To maintain the same density of the particle approximation with 20 component classifiers would require $1000^{21/4} \approx 5.6 \times 10^{15}$ particles! It may be reasonable to assume that only a few component classifiers will be required at any time, in which case the dimensionality problem could be addressed by using appropriate methods of regularisation. However, for this paper we used a simple SIS/R filter following the algorithms contained in [4].

The mean and variance of β_0 can be initialised based on knowledge of how the component classifiers were trained, and any prior information. There is no need for a training set on which to initialise the algorithm. However, there is an identification problem with specifying the mean of β_0, because the decision boundary produced by the model (1) with parameters β is equivalent to that produced with parameters $c\beta$, for $c > 0$. However, this feature means that the algorithm can adapt to misspecified parameter evolution variance - if q is too large compared to β_0, for example, then the parameter values will "inflate" over time to bring this into balance.

4   Experimental Comparison

In this section we show the results of applying the SMC implementation of our method (which we label as DLMCS, for Dynamic Logistic Multiple Classifier

System) to two artificial examples. We compare the performance to several other methods for dynamic classification, namely Winnow [9], oLDC with adaptive learning rate [8] and a naive online adaptation of the Learn^{++}.NSE algorithm [11].

The online adaptation of Learn^{++}.NSE differs from the original algorithm in the way that new batches of training data are obtained. Every w observations a new classifier is trained on a batch consisting of the previous w observations, rather than on a set of i.i.d. observations. To distinguish between this implementation and that for which it was originally intended, we refer to this algorithm as oLearn^{++}.NSE.

4.1 Methods

100 observations from the population at time $t = 0$ are used as a training set. Bootstrap samples of this data are used to train the component classifiers for the DLMCS algorithm and for Winnow. This training data is also used to initialise the oLDC algorithm, and to train the initial classifier for oLearn^{++}.NSE.

For each of 5000 independent runs the training data is generated, algorithms initialised and the errors of the classifiers recorded at each time point. These errors are then averaged to produce an estimate of the error rate of each classification algorithm at every point in time.

To set the tuning parameters of the algorithms we took a naive approach, and assumed no prior knowledge of the type of population change which might occur. The tuning parameters were therefore set to values reported as resulting in a generally good performance. No attempt was made to dynamically alter the value of these parameters over time, although such a scheme is likely to improve performance for some scenarios. For Winnow, the parameter α was set to 2. The window-width parameter w for oLearn^{++}.NSE was set to $w = 10$, and the parameters of the sigmoid function were set to $a = 0.5$ and $b = 10$.

For all examples we specified $\eta_k(\cdot)$ as in equation (3). The SMC algorithm was implemented using 1000 particles with a re-sampling threshold of 500, the prior distribution of v was $N(0.01, 0.002)$ and q was set to 0.001. 20 component classifiers were used for Winnow and only three for the SMC algorithm, due to computational constraints. The size of the bootstrap samples used to train the component classifiers of the Winnow and DLMCS algorithms was equal to 8 - chosen to be reasonably small to encourage "diversity" amongst the component classifiers, in the hope that some of the component classifiers would be relevant to later scenarios. All component classifiers were linear discriminant classifiers, trained using the lda function in the R library MASS. To measure the importance of selecting appropriate component classifiers, we also implemented the DLMCS model using three component classifiers which were near-optimal at times $t = 0$, $t = 100$ and $t = 200$. This classifier is labelled as DLMCS*. Because the component classifiers were trained on bootstrap samples, in all cases the prior mean of β_0 was set to $M^{-1}\mathbf{1}$, where M is the number of component classifiers.

4.2 Datasets and Results

We compare the performance of the algorithms on two synthetic examples: slow and sudden change. For both scenarios we have two Gaussian populations in $\mathbb{R}^2$. The mean of class 1 is stationary and centered on $(0,0)$. For the slow change example, the mean of class 2 is initially at $(2,2)$ and drifts to $(2,-2)$ at a constant rate over 200 time steps. For the sudden change example the mean of class 2 is initially at $(2,2)$, then switches to $(2,-2)$ after 70 time-steps, then switches back to $(2,2)$ after 140 time-steps. Both classes have covariance matrix equal to the identity. The optimal decision boundary is therefore linear.

The average error rates of the classification algorithms for the scenarios of slow and sudden change are shown in Figures 1 and 2 respectively. For the slow change scenario shown in Figure 1, there seem to be two groups of algorithms. Those that do the best are oLDC, DLMCS* and oLearn^{++}.NSE, whilst the performance of Winnow and DLMCS are substantially worse[1]. This can be explained by the fact that the poorly performing algorithms were those for which the component classifiers were trained on data only from time $t = 0$. The performance of oLearn^{++}.NSE is significantly better because by adding new, relevant classifiers there is always at least one classifier reasonably well suited to the current population. The good performance of DLMCS* demonstrates that the DLMCS algorithm is hampered not by the parameter updating algorithm,

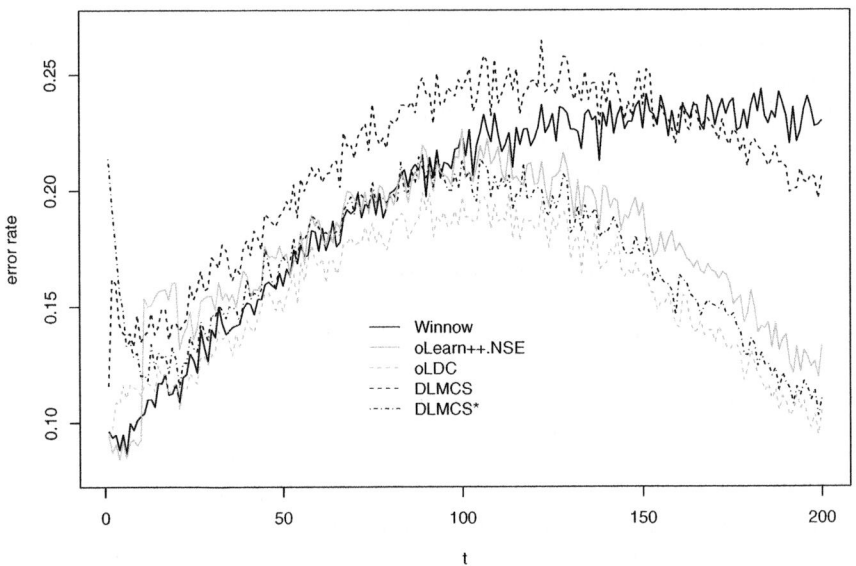

Fig. 1. Average error rates for the different algorithms under slow change

---

[1] All pairwise differences in the total cumulative error are significantly different at the 5% level, as shown in Table 1.

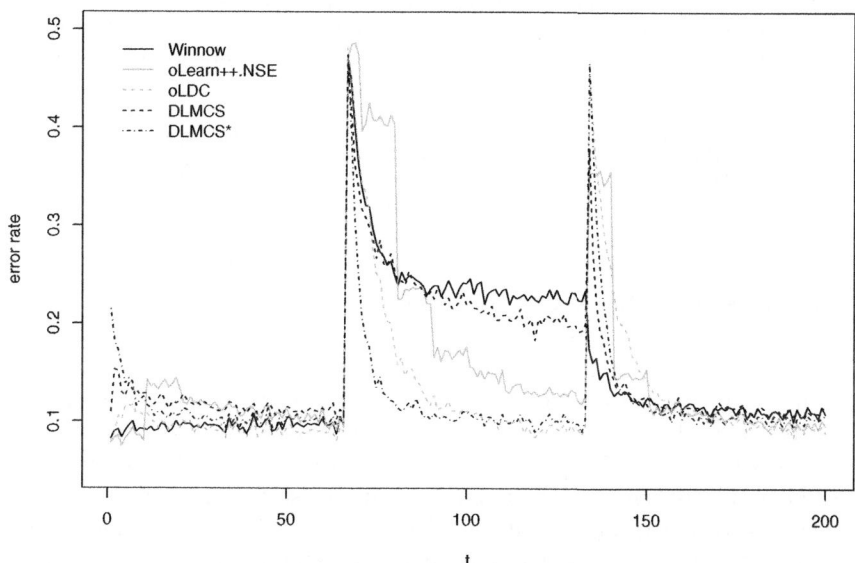

Fig. 2. Average error rates for the different algorithms under rapid change

but rather by the component classifiers. It should also be noted that in these examples oLDC and the DLMCS algorithms have the advantage that they produce linear decision boundaries, whereas Winnow and Learn^{++}.NSE will not. However, the results are similar to those for non-linear scenarios (not shown). It is also interesting that for the first 150 time-steps the DLMCS algorithm performs worse than the Winnow algorithm. This may be because Winnow uses 20 component classifiers, whereas DLMCS has only three. Both DLMCS and DLMCS* took about 10 time-steps to 'burn-in', indicating that the prior specifications were not optimal.

The results for the scenario of sudden change, shown in Figure 2, display a similar pattern. Most algorithms perform similarly for the first 70 time-steps. When the population suddenly changes, those algorithms capable of adapting to the new scenario do so - firstly DLMCS* followed by oLDC and then oLearn^{++}.NSE. This again demonstrates that, conditional on an appropriate choice of component classifiers, the DLMCS algorithm can adapt relatively quickly to sudden changes in the population. For the final 70 times-steps the algorithms again adjust to the change, but interestingly all perform worse than on the original 70 time-steps, perhaps because the information provided by the training set at $t = 0$ has been forgotten. Results from a comparison of the total cumulative error[2] show that oLearn^{++}.NSE, Winnow and DLMCS perform comparably, but not as well as DLMCS* or oLDC.

---

[2] In general, if adjustment to either rapid change or stationarity is of primary importance then it may be more appropriate to compare a weighted cumulative error.

Table 1. Average rank and critical difference (CD) for pairwise testing of a difference between the total cumulative error of any two algorithms at the 5% level of significance and using the Nemenyi test (per [3]). Differences which are not significantly different to zero are shown in bold.

	oLDC	oLearn^{++}.NSE	Winnow	DLMCS	(DLMCS*)	CD
slow	1.61	2.81	3.82	4.28	2.48	0.086
sudden	2.15	**3.70**	**3.70**	**3.75**	1.70	0.086

5　Discussion and Conclusions

We have proposed a method of combining classifier outputs and dynamically updating the way the outputs are combined to produce a classification. Experiments have shown that this can be used to produce an effective algorithm for non-stationary classification. Although not the focus of this paper, this algorithm could also be used for online learning in stationary environments (by setting $v = 0$). For example, with an appropriate choice for the function $\boldsymbol{\eta}_k(\cdot)$ (such as in equation (2)), the method we have presented can be seen as a method for online weighted averaging.

However, the method we propose is not an out-of-the-box classifier. The performance of the algorithm, as with other dynamic combiners, is heavily dependent on the choice of component classifiers. If the population change does not move the decision boundary outside the region of competence of the component classifiers, then the algorithm is capable of tracking change. When this condition is met, we have shown that this algorithm is competitive with other dynamic MCS algorithms. However, if the population changes considerably from that which was expected at the time the component classifiers were chosen, then algorithms such as oLearn^{++}.NSE which continually update the set of component classifiers are likely to perform better. Hence the algorithm presented in this paper is likely to be best suited to seasonal or small changes in the population.

Future work will look more closely at the choice of component classifiers, and incorporating shrinkage into the Sequential Monte Carlo algorithm to remove the influence of irrelevant classifiers and reduce the effective dimensionality of the parameter space. This may include the possibility of combining this model with a heuristic for updating the component classifiers, or using previous data to tune the value of the variance parameters v or q.

Bibliography

[1] Breiman, L.: Bagging predictors. Machine Learning 26, 123–140 (1996)
[2] Breiman, L.: Random forests. Machine Learning 45, 5–32 (2001)
[3] Demšar, J.: Statistical comparisons of classifiers over multiple data sets. Journal of Machine Learning Research 7, 1–30 (2006)
[4] Doucet, A., Godshill, S., Andrieu, C.: On sequential Monte Carlo sampling methods for Bayesian filtering. Statistics and Computing 10, 197–208 (2000)

[5] de Freitas, J.F.G., Niranjan, M., Gee, A.H.: Hierarchical Bayesian models for regularization in sequential learning. Neural Computation 12, 933–953 (2000)
[6] Højen-Sørensen, P., de Freitas, N., Fog, T.: On-line probabilistic classification with particle filters. Neural Networks for Signal Processing X, 2000. In: Proceedings of the 2000 IEEE Signal Processing Society Workshop, vol. 1, pp. 386–395 (2000), citeseer.ist.psu.edu/322567.html
[7] Kuncheva, L.I.: Classifier ensembles for changing environments. In: Roli, F., Kittler, J., Windeatt, T. (eds.) MCS 2004. LNCS, vol. 3077, pp. 1–15. Springer, Heidelberg (2004)
[8] Kuncheva, L.I., Plumpton, C.O.: Adaptive learning rate for online linear discriminant classifiers. In: da Vitoria Lobo, N., Kasparis, T., Roli, F., Kwok, J.T., Georgiopoulos, M., Anagnostopoulos, G.C., Loog, M. (eds.) S+SSPR 2008. LNCS, vol. 5342, pp. 510–519. Springer, Heidelberg (2008)
[9] Littlestone, N.: Learning quickly when irrelevant attributes abound: A new linear-threshold algorithm. Machine Learning 2, 285–318 (1998)
[10] McCormick, T.H., Raftery, A.E., Madigan, D., Burd, R.S.: Dynamic logistic regression and dynamic model averaging for binary classification (submitted)
[11] Muhlbaier, M.D., Polikar, R.: An ensemble approach for incremental learning in nonstationary environments. In: Haindl, M., Kittler, J., Roli, F. (eds.) MCS 2007. LNCS, vol. 4472, pp. 490–500. Springer, Heidelberg (2007)
[12] Penny, W.D., Roberts, S.J.: Dynamic logistic regression. In: International Joint Conference on Neural Networks, IJCNN 1999, vol. 3, pp. 1562–1567 (1999)
[13] Schapire, R.: The strength of weak learnability. Machine Learning 5, 197–227 (1990)
[14] Tomas, A.: A Dynamic Logistic Model for Combining Classifier Outputs. Ph.D. thesis, The University of Oxford (2009)
[15] West, M., Harrison, J.: Bayesian Forecasting and Dynamic Models, 2nd edn. Springer Series in Statistics. Springer, Heidelberg (1997)
[16] West, M., Harrison, P.J., Migon, H.S.: Dynamic generalized linear models and Bayesian forecasting. Journal of the American Statistical Association 80(389), 73–83 (1985)

Ensemble Methods for Reinforcement Learning with Function Approximation

Stefan Faußer and Friedhelm Schwenker

Institute of Neural Information Processing, University of Ulm, 89069 Ulm, Germany
{stefan.fausser,friedhelm.Schwenker}@uni-ulm.de

Abstract. Ensemble methods allow to combine multiple models to increase the predictive performances but mostly utilize labelled data. In this paper we propose several ensemble methods to learn a combined parameterized state-value function of multiple agents. For this purpose the *Temporal-Difference* (TD) and *Residual-Gradient* (RG) update methods as well as a policy function is adapted to learn from joint decisions. Such joint decisions include *Majority Voting* and *Averaging* of the state-values. We apply these ensemble methods to the simple pencil-and-paper game Tic-Tac-Toe and show that an ensemble of three agents outperforms a single agent in terms of the Mean-Squared Error (MSE) to the true values as well as in terms of the resulting policy. Further we apply the same methods to learn the shortest path in a 20×20 maze and empirically show that the learning speed is faster and the resulting policy, i.e. the number of correctly choosen actions is better in an ensemble of multiple agents than that of a single agent.

1 Introduction

In a single-agent problem multiple agents can be combined to act as a committee agent. The aim here is to rise the performance of the single acting agent. In contrast to a multi-agent problem multiple agents are needed to act in the same environment with the same (cooperative) or opposed (competitive) goals. Such multi-agent problems are formulated in a Collaborative Multiagent MDP (CMMDP) model. The *Sparse Cooperative Q-Learning* algorithm has been successfully applied to the distributed sensor network (DSN) problem where the agents cooperatively focus the sensors to capture a target (Kok et al. 2006 [6]). In the predator-prey problem multiple agents are predators (one agent as one predator) and are hunting the prey. For this problem the *Q-learning* algorithm also has been used where each agent maintains its own and independent Q-table (Partalas et al. 2007 [7]). Further an addition to add and to remove agents during learning, i.e. to perform self-organization in a network of agents has been proposed (Abdallah et al. 2007 [8]).

While the *Multi-Agent Reinforcement Learning* (MARL) as described above is well-grounded with research work only little is known for the case where multiple agents are combined to a single agent (committee agent) for single-agent

C. Sansone, J. Kittler, and F. Roli (Eds.): MCS 2011, LNCS 6713, pp. 56–65, 2011.

problems. One reason may be that RL algorithms with state-tables theoretically converge to a global minimum independent of the initialized state-values and therefore multiple runs with distinct state-value initializations result to the same solution with always the same bias and no variance. However in *Ensemble Learning* methods like *Bagging* (Breiman 1996 [3]) the idea is to reduce variance and to improve the ensemble's overall performance. Sun et al. 2009 [5] has studied the partitioning of the input and output space and has developed some techniques using Genetic Algorithms (GA) to partition the spaces. Multiple agents applied the *Q-Learning* algorithm to learn the action-values in subspaces and have been combined through a weighting scheme to a single agent. However extensive use of heuristics and the need of much computation time for the GA algorithm makes this approach unusable for MDPs with large state spaces. In a more recent work (Wiering et al. 2008 [9]) action-values are combined resulting by multiple independently learnt RL algorithms (*Q-Learning, SARSA, Actor-Critic*, etc.) to decide about the best action to take. As *Q-Learning* tend to converge to another fixed point than *SARSA* and *Actor-Critic* the action-value functions therefore have a different bias and variance.

A *Markov Decision Process* (MDP) with a large state space imposes several problems on *Reinforcement Learning* (RL) algorithm. Depending on the number of states it may be possible to use RL algorithm that save the state-values in tables. However for huge state-spaces another technique is to learn a parameterized state-value function by linear or nonlinear function approximation. While the state-values in tables are independent on each other, the function approximated state-values are highly dependent based on the selection of the feature space and may therefore converge faster. In an application to English Draughts (Fausser et al. 2010 [10]) which has about 10^{31} states the training of the parameterized state-value function needed about $5,000,000$ episodes to reach an amateur player level. Although a parameterized state-value function with simple features can be learnt, it may not converge to a global fixed point and multiple runs with distinct initial weights tends to result in functions with different solutions (different bias and large variance) of the state-values.

Our contribution in this paper is to describe several ensemble methods that aim to increase the learning speed and the final performance opposed to that of a single agent. We show the new derived TD update method as well as the new policy to learn from joint decisions. Although we use parameterized state-value functions in order to deal with large state MDPs we have applied the methods to more simple problems to be able to compare performances. Our work differs from others that we are combining multiple agents for a single agent problem and by our general way of combining multiple state-values that enables to target problems with large-state spaces. It can be empirically shown that these ensemble methods improves the overall performance of multiple agents for the pencil-and-paper game Tic-Tac-Toe as well as for several mazes.

2   Reinforcement Learning with Parameterized State-Value Functions

Assume we want to estimate a smooth (differentiable) state-value function $V_{\theta}^{\pi}(s)$ with its parameter vector $\boldsymbol{\theta}$ where $V_{\theta}^{\pi}(s) \approx V^*(s), \forall s$ using the *TD Prediction* method (Sutton & Barto 1998 [1]). Starting in a certain state s_t we take an action a_t defined by a given policy π, observe reward (signal) r_{t+1} and move from state s_t to state s_{t+1}. For this single state transition we can model the Squared TD Prediction error ($TDPE$) as follows:

$$TDPE(\boldsymbol{\theta}) = \left[r_{t+1} + \gamma V_{\theta}(s_{t+1}) - V_{\theta}(s_t) \mid s_{t+1} = \pi(s_t)\right]^2 \qquad (1)$$

The aim is to minimize the above error by updating parameter vector $\boldsymbol{\theta}$. Applying the gradient-descent technique to the $TDPE$ this results in two possible update functions. The first one being *Temporal-Difference* learning with $\gamma V(s_{t+1})$ kept fixed as training signal:

$$\Delta\boldsymbol{\theta}^{TD} := \alpha\left[r_{t+1} + \gamma V_{\theta}(s_{t+1}) - V_{\theta}(s_t) \mid s_{t+1} = \pi(s_t)\right] \cdot \frac{\partial V_{\theta}(s_t)}{\partial\boldsymbol{\theta}} \qquad (2)$$

and the second one being *Residual-Gradient* learning (Baird 1995 [2]) with variable $\gamma V_{\theta}(s_{t+1})$ in terms of θ:

$$\Delta\boldsymbol{\theta}^{RG} := -\alpha\left[r_{t+1} + \gamma V_{\theta}(s_{t+1}) - V_{\theta}(s_t) \mid s_{t+1} = \pi(s_t)\right] \cdot \left[\gamma\frac{\partial V_{\theta}(s_{t+1})}{\partial\boldsymbol{\theta}} - \frac{\partial V_{\theta}(s_t)}{\partial\boldsymbol{\theta}}\right] \qquad (3)$$

In both equations $\alpha > 0$ is the learning rate and $\gamma \in (0, 1]$ discounts future state-values. Now suppose that policy π is a function that chooses one successor state s_{t+1} out of the set of all possible states $S_{successor}(s_t)$ based on its state-value:

$$\pi(s_t) := \underset{s_{t+1}}{\operatorname{argmax}}\left[V_{\theta}(s_{t+1}) \mid s_{t+1} \in S_{successor}(s_t)\right] \qquad (4)$$

It is quite clear that this simple policy can only be as good as the estimations of $V_{\theta}(s_{t+1})$. Thus an improvement of the estimations of $V_{\theta}(s_{t+1})$ results in a more accurate policy π and therefore in a better choice of a successor state s_{t+1}. An agent using this policy tries to maximize its summed high-rated rewards and avoids getting low-rated rewards as much as possible. The optimal state-value function $V^*(s)$ is:

$$V^*(s) = E\left\{\sum_{t=0}^{\infty} \gamma^t r_{t+1} \mid s_o = s\right\} \qquad (5)$$

While a parameterized state-value function can only approximate the optimal state-values to a certain degree it is expected that a function approximation of these state-values result in faster learning, i.e. needs less learning iterates than learning with independent state-values. Furthermore different initializations of the weights $\boldsymbol{\theta}$ may result in different state-values after each learning step.

3   Ensemble Methods

Suppose a single-agent problem is given, e.g. finding the shortest path through a maze. Given a set of M agents $\{A_1, A_2, \ldots, A_M\}$ each with its own nonlinear function approximator, for instance a *Multi-Layer Perceptron* (MLP), and either TD updates (2) or RG updates (3) to adapt the weights then it is possible to independently train the M agents or to dependently train the M agents in terms of their state-value updates and their decisions. Irrespective of the training method the decision of all M agents can be combined as a *Joint Majority Decision*:

$$\pi^{VO}(s_t) := \underset{s_{t+1}}{\operatorname{argmax}} \left[\sum_{i=1}^{M} N_i(s_t, s_{t+1}) \right] \tag{6}$$

where $N_i(s_t, s_{t+1})$ models the willingness of agent i to move from state s_t to state s_{t+1}:

$$N_i(s_t, s_{t+1}) = \begin{cases} 1, & \text{if} \quad \pi_i(s_t) = s_{t+1}, \\ 0, & \text{else} \end{cases} \tag{7}$$

Policy $\pi_i(s_t)$ is equivalent to equation (4) but with a subfix to note which agent, i.e. which function approximator to use. The state-values of all agents can be further combined to an *Average Decision* based on averaging the state-values:

$$\pi^{AV}(s_t) := \underset{s_{t+1}}{\operatorname{argmax}} \left[\frac{1}{M} \sum_{i=1}^{M} V_{\theta_i}(s_{t+1}) \mid s_{t+1} \in S_{successor}(s_t) \right] \tag{8}$$

Here $V_{\theta_i}(s_{t+1})$ is the state-value of agent i using the weights θ_i of this agent. Summed up three policies, namely $\pi_s(s_t)$ (4), $\pi^{VO}(s_t)$ (6) and $\pi^{AV}(s_t)$ (8) are available to decide about the best state-value where only the last two ones include the state-values of the other agents in an ensemble, i.e. perform a joint decision. One way of constructing a RL ensemble is to independently train the M agents and to combine their state-values for a joint decision using one of the above described policies after the training. Another way is to use the joint decision during the learning process. For this case it may be necessary to add some noise to the policies to keep agents (state-value functions) diverge. Another suggestion is to have different starting state positions for each agent in the MDP resulting in a better exploration of the MDP.

3.1   Combining the State-Values

While joint decisions during the learning process implicitly updates the state-values of one agent dependent on the state-values of all other agents $M - 1$ it can be another improvement to explicitly combine the state-values. Assume agent i is currently in state s_t and based on one of the policies described in the last section moves to state s_{t+1} and gets reward r_{t+1}. Independent on the choosen policy the state-values of the successor state s_{t+1} of all agents M can be combined to an *Average Predicted Value*:

$$V^{AV}(s_{t+1}) = \frac{1}{M} \sum_{i=1}^{M} V_{\boldsymbol{\theta}_i}(s_{t+1}) \tag{9}$$

As the weights of the function approximators of all agents differ because of diverse initialization of the weights, exploration, different starting positions of the agents, decision noise and instabilities in the weight updates it is expected that the combination of the state-values results in a more stable and better predicted state-value. Now this single state-transition can be modelled in a Squared Average Predicted TD Error *(ATDPE)* function including the *Average Predicted Value* instead of one state-value of this single agent i:

$$ATDPE(\boldsymbol{\theta}_i) = \left[r_{t+1} + \gamma \frac{1}{M} \sum_{j=1}^{M} V_{\boldsymbol{\theta}_j}(s_{t+1}) - V_{\boldsymbol{\theta}_i}(s_t) \mid s_{t+1} = \pi(s_t) \right]^2 \tag{10}$$

By gradient-descent of the *ATDPE* function like we have done in section 2 with the *TDPE* function this formulates a new combined TD update function:

$$\Delta\boldsymbol{\theta}_i^{CTD} := \alpha \left[r_{t+1} + \gamma V^{AV}(s_{t+1}) - V_{\boldsymbol{\theta}_i}(s_t) \mid s_{t+1} = \pi(s_t) \right] \cdot \frac{\partial V_{\boldsymbol{\theta}_i}(s_t)}{\partial \boldsymbol{\theta}_i} \tag{11}$$

as well as a new combined RG update function:

$$\Delta\boldsymbol{\theta}_i^{CRG} := -\alpha \left[r_{t+1} + \gamma V^{AV}(s_{t+1}) - V_{\boldsymbol{\theta}_i}(s_t) \mid s_{t+1} = \pi(s_t) \right]$$
$$\cdot \left[\frac{1}{M} \gamma \frac{\partial V_{\boldsymbol{\theta}_i}(s_{t+1})}{\partial \boldsymbol{\theta}_i} - \frac{\partial V_{\boldsymbol{\theta}_i}(s_t)}{\partial \boldsymbol{\theta}_i} \right] \tag{12}$$

With one of the above update functions the agents learn from the average predicted state-values. Theoretical this can be further combined with one of the prior described joint decision policies. Using the simple single-decision policy (4) this results in an interesting ensemble where each agent decides based on their own state-values but learns from the average predicted state-values. For this case less noise for the decision functions are required as the agents mainly keep their bias. With one of the joint policies, i.e. *Joint Majority Decision* (6) or *Average Decision* (8) all agents perform joint decisions and learn from the average predicted state-values.

As the combined update functions and the policies for joint decisions only need some additional memory space to save the state-values of all agents and this memory space is far lower than the memory space of the function approximator weights they can be ignored in memory space considerations. Therefore training an ensemble of M agents takes M times memory space and M times computation time of a single agent.

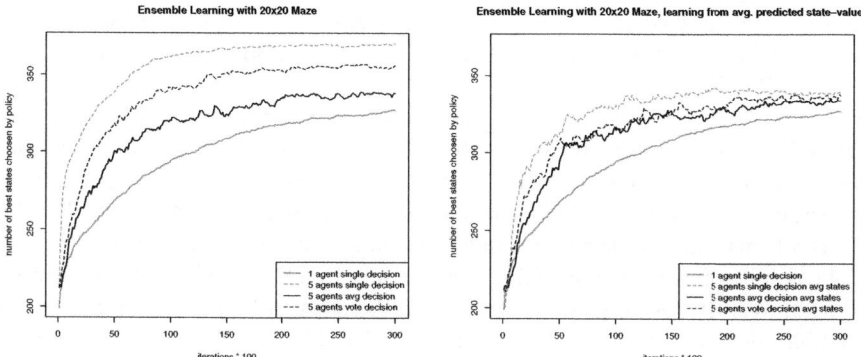

Fig. 1. Empirical results of different ensemble methods applied to five 20×20 mazes. Measured are the number of states that are correctly choosen by the resulting policy where *left* the agents have learnt from joint decisions and *right* the agents have learnt from joint decisions and averaged state-values.

4   Experiments and Results

To evaluate the behaviour of the ensemble methods described in the last sections we have performed experiments with the pencil-and-paper game Tic-Tac-Toe and several mazes. For fair evaluations we have performed multiple runs where we have given the same seed for the pseudo random number generator for all ensemble methods to ensure that the weights of the parameterized state-value function have been identically initialized. For example if we have performed 2 testruns then we have given *seed*1 for all evaluated methods in the first testrun and *seed*2 $\neq$ *seed*1 in the second testrun. The given values are the averaged values of the multiple runs.

4.1   Maze

In the maze-problem an agent tries to find the shortest path to the goal. For our experiments we have created five 20×20 mazes each with randomly positioned 100 barriers. A barrier can be horizontally or vertically set between two states and does not fill out a whole state. Each maze has one goal where the goal position is about upper-left, upper-right, lower-left, lower-right or in the middle of the maze. An agent receives a reward of 1 if he moves to a goal and a reward of 0 otherwise. From each state there are up to 4 possible successor states. The agent cannot move over a barrier or outside the maze. We have applied the Breadth-first search algorithm (Russel & Norvig 2002 [12]) to calculate the true state values and the optimal policy. For the experiments we have designed $M = 5$ agents where each of the agent has an own 2-layer MLP with 8 input neurons, 3 hidden neurons and one output neuron. The input neurons are coded as follows:

1. x position, 2. y position, 3. $20 - x$, 4. $20 - y$, 5. 1 if $x \geq 11$ otherwise 0, 6. 1 if $x \leq 10$ otherwise 0, 7. 1 if $y \geq 11$ otherwise 0, 8. 1 if $y \leq 10$ otherwise 0. For all evaluations the agents had the following common training parameters: (combined) TD update, $\alpha = 0.01$, $\gamma = 0.9$, epsilon-greedy exploration strategy with $\epsilon = 0.3$, tangens hyperbolicus (tanh) transfer functions for hidden layer and output layer and uniformly distributed random noise between ± 0.05 for joint decisions. Each agent has an own randomly initialized start-state and maintains its current state. If one agent reaches a goal or exceeds to reach the goal within 100 iterations then he starts at a new randomly initialized start-state.

The results of an ensemble of five agents compared to a single agent with values averaged over 10 testruns and 5 mazes can be seen in figure 1. Comparing the ensemble methods in terms of the number of states that are correctly choosen by the resulting policy the methods with joint decisions are better than the methods learning from joint decisions and average predicted state-values. Even more a simple combination of five independently trained agents (5 agents single decision curve) seem to be the best followed by a combination of five dependently trained agents with *Joint Majority Voting* decisions. All ensemble methods learn faster and have a better final performance than a single agent within $30,000$ iterations.

4.2   Tic-Tac-Toe

The pencil-and-paper game Tic-Tac-Toe is a competitive 2-player game where each player marks one of maximum nine available spaces turnwise until one player either has three of his own marks horizontal, vertical or diagonal resulting in a win or all spaces are marked resulting in a draw. Tic-Tac-Toe is has 5477 valid states excluding the empty position and starting from the empty position the game always results in a draw if both player perform the best moves. For our experiments we have designed $M = 3$ agents where each of the agent has an own 2-layer MLP with 9 input neurons, 5 hidden neurons and one output neuron. One input neuron binary codes one game space and is -1 if the space is occupied by the opponent, 1 if the space is occupied by the agent or 0 if the space is empty. The weights of the MLP are updated by the (combined) RG update function. A reward of 1 is received if the agent moves to a terminal state where he has won and receives a reward of 0 otherwise, i.e. for a transition to a non-terminal state and to a terminal state where he has lost or reached a draw. For all evaluations the agents had the following common training parameters: $\alpha = 0.0025$, $\gamma = 0.9$, epsilon-greedy exploration strategy with $\epsilon = 0.3$, tangens hyperbolicus (tanh) transfer functions for hidden layer and output layer and uniformly distributed random noise between ± 0.05 for joint decisions.

Each agent learns by Self-Play, i.e. uses the same decision policy and state-values for an inverted position to decide which action the opponent should take. Irrespective of the ensemble methods all agents learn episode-wise and start from the same initial state (the emtpy position). To calculate the true state-values and the optimal policy we have slightly modified the Minimax algorithm (Russel & Norvig 2002 [12]) to include the rewards and the discounting rate γ.

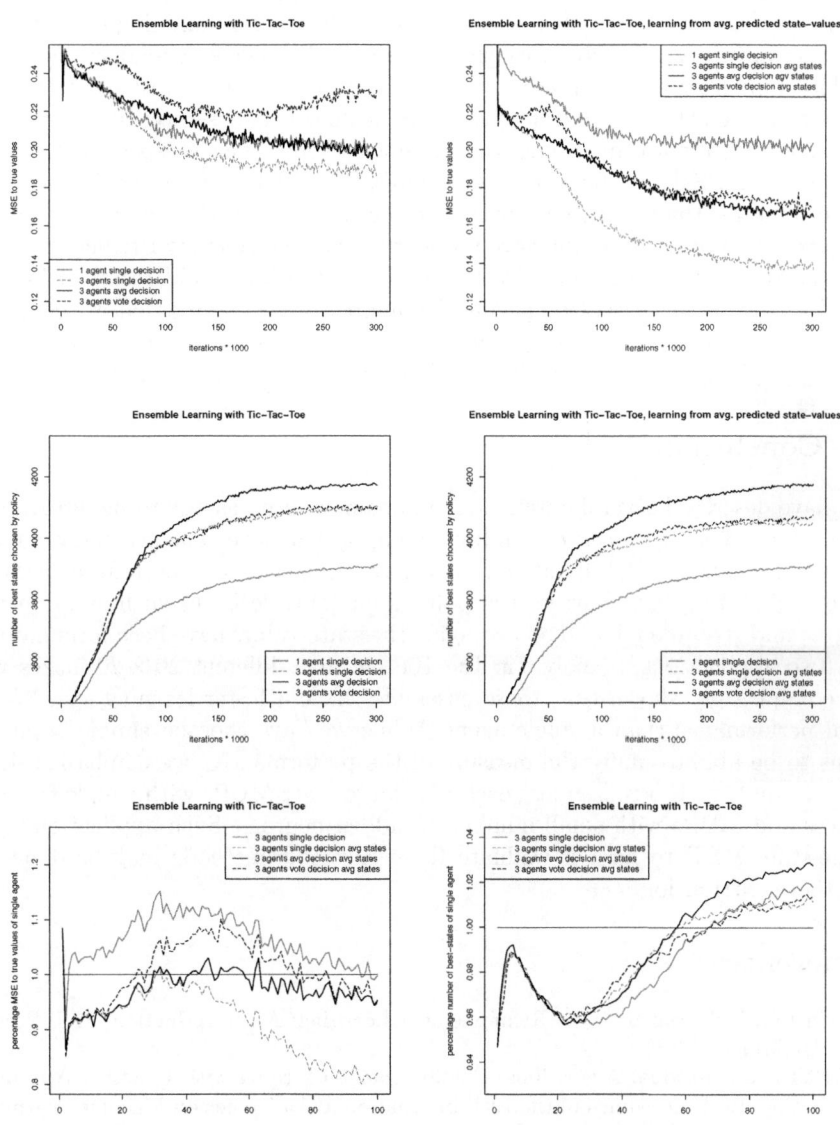

Fig. 2. Empirical results of different ensemble methods applied to Tic-Tac-Toe. The *first two* figures show the Mean-Squared Error (MSE) to the true state-values. The *next two* figures show the number of best states that are choosen by the resulting policy, higher values are better. The *last two* figures compare the MSE to the true state-values (*left*) and the number of best states (*right*) of an ensemble of three agents to a single agent.

The results of an ensemble of three agents compared to a single agent with values averaged over 10 testruns can be seen in figure 2. Examining the MSE to the true values, an ensemble with three agents with single independent decisions and learning from average predicted state-values reaches the lowest error. During the first $100,000$ training episodes the MSE is almost always lower than the MSE of a single agent with three times the training episodes. This is especially true for the first $\approx 20,000$ and the last $\approx 40,000$ iterations. All other ensemble methods perform better than a single agent except the three agents that learnt from joint *Majority Voting* decision but have not learnt from the Average Predicted state-values. Maybe lowering the noise for the joint decision would result in better MSE values for this case. Comparing the number of best states that are choosen by the resulting policy, all ensembles without exception are performing better than a single agent. Consider that Tic-Tac-Toe has 4520 non-terminal states.

5   Conclusion

We have described several ensemble methods new in its aspect to be integrated into *Reinforcement Learning* with function approximation. The necessary extensions to the TD and RG update formulas have been shown to learn from average predicted state-values. Further the policies for joint decisions such as *Majority Voting* and *Averaging* based on averaging the state-values have been formulated. For two applications, namely Tic-Tac-Toe and five different 20×20 mazes we have empirically shown that these ensembles have a faster learning speed and final performance than a single agent. While we have choosen simple applications to be able to unifiy the measure of the performances, we emphasize that our ensemble methods are most useful for large-state MDPs with simple feature spaces and MDPs with small number of hidden neurons. Such application to a large-state MDP to further evaluate these ensemble methods may be done in another contribution.

References

1. Sutton, R.S., Barto, A.G.: Reinforcement Learning: An Introduction. MIT Press, Cambridge (1998)
2. Baird, L.: Residual Algorithms: Reinforcement Learning with Function Approximation. In: Proceedings of the 12th International Conference on Machine Learning pp. 30–37 (1995)
3. Breiman, L.: Bagging Predictors. Machine Learning 24, 123–140 (1996)
4. Schapire, R.E.: The Strength of Learnability. Machine Learning 5(2), 197–227 (1990)
5. Sun, R., Peterson, T.: Multi-Agent Reinforcement Learning: Weighting and Partitioning. Journal on Neural Networks 12(4-5), 727–753 (1999)
6. Kok, J.R., Vlassis, N.: Collaborative Multiagent Reinforcement Learning by Payoff Propagation. Journal of Machine Learning Research 7, 1789–1828 (2006)
7. Partalas, I., Feneris, I., Vlahavas, I.: Multi-Agent Reinforcement Learning using Strategies and Voting. In: 19th IEEE International Conference on Tools with Artificial Intelligence (ICTAI 2007), vol. 2, pp. 318–324 (2007)

8. Abdallah, S., Lesser, V.: Multiagent Reinforcement Learning and Self-Organization in a Network of Agents. In: Proceedings of the Sixth International Joint Conference on Autonomous Agents and Multi-Agent Systems, AAMAS 2007 pp.172–179 (2007)
9. Wiering, M.A., van Hasselt, H.: Ensemble Algorithms in Reinforcement Learning. IEEE Transactions on Systems, Man and Cybernetics, Part B: Cybernetics 38, 930–936 (2008), ISSN 1083-4419
10. Faußer, S., Schwenker, F.: Learning a Strategy with Neural Approximated Temporal-Difference Methods in English Draughts. In: ICPR 2010, pp. 2925–2928 (2010)
11. Bishop, C.M.: Neural Networks for Pattern Recognition. Oxford University Press, Oxford (1995)
12. Stuart, J.R., Norvig, P.: Artificial Intelligence: A Modern Approach, 2nd edn. Prentice-Hall, Englewood Cliffs (2002)

GRASP Forest: A New Ensemble Method for Trees[*]

José F. Diez-Pastor , César García-Osorio , Juan J. Rodríguez ,
and Andrés Bustillo

University of Burgos
jfdiez@beca.ubu.es,{cgosorio,jjrodriguez,abustillo}@ubu.es

Abstract. This paper proposes a method for constructing ensembles of decision trees: GRASP Forest. This method uses the metaheuristic GRASP, usually used in optimization problems, to increase the diversity of the ensemble. While Random Forest increases the diversity by randomly choosing a subset of attributes in each tree node, GRASP Forest takes into account all the attributes, the source of randomness in the method is given by the GRASP metaheuristic. Instead of choosing the best attribute from a randomly selected subset of attributes, as Random Forest does, the attribute is randomly chosen from a subset of selected good attributes candidates. Besides the selection of attributes, GRASP is used to select the split value for each numeric attribute. The method is compared to Bagging, Random Forest, Random Subspaces, AdaBoost and MutliBoost, being the results very competitive for the proposed method.

Keywords: Classifier ensembles, Bagging, Random Subspaces, Boosting, Random Forest, decision trees, GRASP.

1 Introduction

Classifier ensembles [1] are combinations of several classifiers which are called *base classifiers* or *member classifiers*. Ensembles often give better results than individual classifiers. The kind of ensemble most often used is the homogeneous ensemble, in which all the base classifiers are built using the same method. In these ensembles, the diversity is commonly forced by training each base classifier with a variant of the training data set: Bagging [2] uses different random samples of the training set, Random Subspace [3] uses different subsets of attributes, AdaBoost [4] and Multiboost [5] adaptively change the distribution of the training set based on the performance of the previous classifiers, this way, the instances more difficult for the previous classifiers have a higher probability of being in the next training sample. Other methods, like Random Forest [6], increase the randomness by combining the sampling of the training set with the

---

[*] This work was supported by the Project TIN2008-03151 of the Spanish Ministry of Education and Science.

C. Sansone, J. Kittler, and F. Roli (Eds.): MCS 2011, LNCS 6713, pp. 66–75, 2011.

random selection of subsets of attributes in each node of the tree. This way, in each node, the splits only consider the selected subset of attributes. In [7], they use a very simple technique to randomize the election of the split among the twenty best split in each node. Recently, the method called Random Feature Weights [8] proposes to use all the attributes, but with different probabilities of being considered in the splits that depend on a weight associated to the attribute. To assure the diversity, the weights are randomly generated for each tree in the ensemble.

Decision trees are frequently used as base classifiers because they are efficient and unstable, that is, small changes in the training set or in the construction method will produce very different classifiers.

The algorithms for building decision trees are top-down methods. In the root of the tree they use all the instances to find which attribute is the best to split the instances in two subsets assigned to two new nodes[1], children of the root node. This process is recursively repeated in each new node till a stop criteria is verified. The best attribute is determined in each node by evaluating a merit function. Some common split criteria are: Information Gain and Gain Ratio [9], or Gini Index [10]. In this paper the merit function used is Gain Ratio.

The meta heuristic GRASP (Greedy Randomize Adaptive Search Procedure) [11,12], a widely used strategy in optimization problems, has been recently used in [13] to modify the way the attribute is selected in the process of building a binary decision tree. The controlled randomness introduced by GRASP is able to build less complex trees without affecting the accuracy.

This work takes as starting point the idea used in [13], extends it using GRASP also in the selection of the split value for each attribute, and combining these trees in the construction of an ensemble, the GRASP Forest. Using GRASP in the selection of the split values gives an extra level of randomness that helps in increasing the diversity in the ensemble. This increased diversity compensates the loss in accuracy of the individual trees, improving in overall the ensemble accuracy.

The rest of the paper is organised as follows. Next section describes the proposed method. Section 3 describes the experiments and the results. Finally, Section 4 gives the conclusions and presents some future lines of research.

2   Method

Usually, to built a decision tree we have a training dataset D, several attributes $a_1, a_2, \ldots, a_n$ and a merit function $f(a_i, D)$ that gives a value to the i-th attribute. One of the most used merit function is *Information Gain* defined as

$$Gain(D, a) = Entropy(D) - \sum_v \frac{|D_v|}{|D|} Entropy(D_v) \qquad (1)$$

where D is the data set, a is the candidate attribute, v indicates the values of the attribute and D_v is the subset of the data set D formed by the examples,

---

[1] If the attribute a_i is nominal, they create a new branch for each possible value of a_i.

Table 1. Backpack problem (weight limit of the backpack: 10 weight units)

			Ratio
Element	Weight	Value Ratio	Value/Weight
1	10	11	1.10
2	6	9	1.50
3	4	1	0.25

where $a = v$. The entropy is defined as

$$Entropy(D) = \sum_{i=1}^{c} -p_i \log_2(p_i) \qquad (2)$$

where c is the number of classes and p_i the probability of class i.

The GRASP method [11,12] is a iterative process, each iteration has two steps:

1. Build an initial solution using a method that is greedy, random and adaptive.
2. Local search from the built solution trying to improve it.

With each iteration the best found solution is updated and the process ends when a stop condition is reached.

As the building method is greedy, random and adaptive, every time a new element is added to the solution, instead of choosing the best possible element, one is randomly chosen from a short list of good candidates called the *Restricted Candidate List (RCL)*. This list is created with those items whose values are close enough to the value of the best item. This closeness is defined by a percentage α of the best value. The idea behind GRASP is that the best solution in each step does not always lead the process to the global optimal solution of the problem. A good example is the backpack problem, for example, given 3 objects with weights, values and ratios (value/weight) shown in table 1 and a backpack capacity 10, is necessary to select a subset of them that fits in the backpack and that maximizes the value. A greedy method would take in each step the element with best possible value-weight ratio: first element 2, then element 3, with total value of 10; however, the best solution would have been to choose the element 1 that improves by one the previous solution.

The content of the RCL is defined as:

$$RCL = \{i : Value_i/Weight_i \geq \alpha Ratio_{max} + (1 - \alpha)Ratio_{min}\}$$

If $\alpha = 1$ the list would have only one element, the element chosen by the greedy procedure; if $\alpha = 0$ the list would have all possible elements and the selection would be totally random.

In the work described in this paper, from GRASP we only use the construction of the solution by a greedy, random and adaptive procedure. Greedy, as the construction of trees is greedy by nature, random due to the random selection of attributes and split points, and adaptive because depending on the maximum and minimum gain ratio of the attributes the number of these considered for selection is different for each node. The aim is to increase the randomness in the process of building the base classifiers.

The method GRASP Forest (see Algorithm 1) works by using Algorithm 2 to create L decision trees that are added to the ensemble.

Algorithm 1: GRASP Forest

Input: Dataset D_T, set of attributes Attributes, size of ensemble L, value α
        between 0 and 1 to control the level of randomness
Output: Ensemble of decision trees
for $l = 1$ *to* L **do**
 | GTree $\leftarrow$ TrainDecisionTree (D_T,Attributes, α)
 | Add GTree to the ensemble;
end

The way the trees are built is similar to the traditional algorithms. In each node, the merit function is evaluated for the m attributes. With these values a list of candidates is created from which the attribute to be used in the node is chosen. Note that with $\alpha = 1$ this algorithm would choose the same attribute as a traditional method, with $\alpha = 0$ the selection of the attribute is totally random.

Algorithm 2: TrainDecisionTree (for numeric attributes)

Input: Dataset D_T, set of attributes Attributes, value between 0 and 1 to
        control the level of randomness α
Output: Tree
if Attributes *is empty or number of examples $<$ minimum allowed per branch*
then
 | Node.label = most common value label in examples
 | **return** Node;
else
 | **for** $j = 1$ *to* m **do**
 | | model [j]$\leftarrow$ GraspSplit (D_T,Attributes,j,α)
 | **end**
 | maxGain $\leftarrow$ Max(model.gain); minGain $\leftarrow$ Min(model.gain)
 | List $\leftarrow \{j = 1, 2, \ldots, m|$ model [j].gain $\geq \alpha$maxGain $+ (1 - \alpha)$minGain$\}$
 | Randomly choose $j_g \in$ List; Att $= j_g$, splitPoint = model [j_g].splitPoint
 | $D_l \leftarrow \{x \in D_T | x_{ij_g} \leq splitPoint\}$; $D_r \leftarrow \{x \in D_T | x_{ij_g} > splitPoint\}$;
 | Node.son[0] = TrainDecisionTree (D_l,Attributes, α);
 | Node.son[1] = TrainDecisionTree (D_r,Attributes, α);
end

Algorithm 3 shows how the idea of GRASP is also used in the process of choosing the splitting point for numeric attributes. The normal way of selecting the splitting point would be to find the point that maximizes the merit function value (in this work Gain Ratio). However, in *GraspSplit*, we again create a list of good candidates, with all points with value higher than a minimal value determined by α, and one of them is randomly chosen and returned together with its merit function value.

Thus, the randomness in each node is both in selecting the attribute and in the choice of the splitting point within that attribute. With $\alpha = 1$, the generated tree is the same as with a traditional algorithm[2]. With $\alpha = 0$ the generated tree will be completely random as it happens in *Extremely Randomized Trees* [14].

Algorithm 3: GraspSplit

Input: Dataset D_T, set of attributes Attributes, attribute index j, value α
         between 0 and 1 to control the level of randomness
Output: model
/* Compute values for all possible split point and their index   */
infoGain $\leftarrow$ List of all possible split info gains
infoGainIndex $\leftarrow$ List of all possible split indexes
maxGain $\leftarrow$ Max(infoGain); minGain $\leftarrow$ Min(infoGain)
List $\leftarrow \{j = 1, 2, \ldots, m | \text{infoGain}[j] \geq \alpha \text{maxGain} + (1 - \alpha)\text{minGain}\}$
Randomly choose $j_g \in$ List
model.gain = infoGain $[j_g]$, model.splitPoint = infoGainIndex $[j_g]$
return model

---

3   Results

The proposed method was implemented in the Weka library [15] by modifying J48, the Weka implementation of C4.5 [9] in conjunction with Random Committee[3]. The rest of decision trees and other ensembles are from this library. The size of the ensembles was set to 50. Since, when using trees in ensembles, we are interested in increasing diversity, we validate our method with low values of α, from 0.1 to 0.5. We compare our method with the following ensembles, whose settings are the default parameters of Weka unless otherwise indicated:

1. Bagging [2].
2. Boosting: AdaBoost.M1 [4] and Multiboost [5]. In both version the variants with resampling and reweighting were used (in the tables represented with S and W). For Multiboost the approximate number of subcommittees was 10.
3. Random Subspaces [3]: with two different configurations, with 50% and 75% of the original set of attributes.
4. Random Forest [6]: three different configurations, random subsets of attributes of size 1, 2 and base 2 logarithm of the number of attributes in the original set.

---

[2] Except in the case of multiple attributes with the same value of the merit function, the traditional algorithm will always choose the same, according to the method of calculating the maximum, the version using GRASP will choose randomly between them, and similarly in the case of the splitting points.

[3] A Random Committee is an ensemble of randomizable base classifiers. Each base classifier is built using the same data but a different seed for the generation of randomness. The final predicted probabilities are simply the average of the probabilities generated by the individual base classifiers.

Table 2. Summary of the data sets used in the experiments. #N: Numeric features, #D: Discrete features,#E: Examples, #C: Classes.

Dataset	#N	#D	#E	#C	Dataset	#N	#D	#E	#C
abalone	7	1	4177	28	lymphography	3	15	148	4
anneal	6	32	898	6	mushroom	0	22	8124	2
audiology	0	69	226	24	nursery	0	8	12960	5
autos	15	10	205	6	optdigits	64	0	5620	10
balance-scale	4	0	625	3	page	10	0	5473	5
breast-w	9	0	699	2	pendigits	16	0	10992	10
breast-y	0	9	286	2	phoneme	5	0	5404	2
bupa	6	0	345	2	pima	8	0	768	2
car	0	6	1728	4	primary	0	17	339	22
credit-a	6	9	690	2	promoters	0	57	106	2
credit-g	7	13	1000	2	ringnorm	20	0	300	2
crx	6	9	690	2	sat	36	0	6435	6
dna	0	180	3186	3	segment	19	0	2310	7
ecoli	7	0	336	8	shuttle	9	0	58000	7
glass	9	0	214	6	sick	7	22	3772	2
heart-c	6	7	303	2	sonar	60	0	208	2
heart-h	6	7	294	2	soybean	0	35	683	19
heart-s	5	8	123	2	soybean-small	0	35	47	4
heart-statlog	13	0	270	2	splice	0	60	3190	3
heart-v	5	8	200	2	threenorm	20	0	300	2
hepatitis	6	13	155	2	tic-tac-toe	0	9	958	2
horse-colic	7	15	368	2	twonorm	20	0	300	2
hypo	7	18	3163	2	vehicle	18	0	846	4
ionosphere	34	0	351	2	vote1	0	15	435	2
iris	4	0	150	3	voting	0	16	435	2
krk	6	0	28056	18	vowel-context	10	2	990	11
kr-vs-kp	0	36	3196	2	vowel-nocontext	10	0	990	11
labor	8	8	57	2	waveform	40	0	5000	3
led-24	0	24	5000	10	yeast	8	0	1484	10
letter	16	0	20000	26	zip	256	0	9298	10
lrd	93	0	531	10	zoo	1	15	101	7

For all ensembles, both pruned and not pruned trees were used as base classifiers, except for Random Forest, because pruning is not recommended in this case [6]. Binary trees were used for two reasons, first, this was the kind of trees used in [13], second, in the case of nominal attributes, the use of GRASP metaheuristic for selecting the splitting point is only possible for binary trees (in a non-binary tree, a node that uses a nominal attribute has as many children as nominal values, and it is not necessary to create the list of splitting points since only one splitting point is possible). For Random Forest, that does not work with binary nodes, we used the preprocessing described in [10], where nominal attributes with k values are transformed into k binary attributes. It could have been possible to optimize the value of α of our method by using a validation data subset or internal cross-validation,

Table 3. Wins, ties and losses; the comparison of the ensembles with binary and with non-binary trees is shown (U: unpruned trees, P: pruned trees)

Method	All			Significative (0.05)		
	V	T	L	V	T	L
Bagging (P)	17	35	10	3	59	0
Bagging (U)	15	39	8	3	59	0
AdaBoost-W (P)	16	34	12	2	60	0
AdaBoost-W (U)	15	36	11	2	60	0
AdaBoost-S (P)	18	35	9	2	60	0
AdaBoost-S (U)	15	34	13	2	60	0
MultiBoost-W (P)	13	36	13	3	59	0
MultiBoost-W (U)	14	37	11	2	60	0
MultiBoost-S (P)	12	37	13	2	60	0
MultiBoost-S (U)	16	34	12	2	60	0
Random Subspaces 50% (P)	18	36	8	0	62	0
Random Subspaces 50% (U)	17	35	10	0	62	0
Random Subspaces 75% (P)	20	35	7	3	59	0
Random Subspaces 75% (U)	17	36	9	2	60	0
Random Forest $K = \log numAtt$	42	2	18	3	59	0
Random Forest $K = 1$	24	4	34	0	62	0
Random Forest $K = 2$	34	1	27	1	60	1

but this would not have been fair for the other methods, besides we were interested in analysing the global effect of α.

Before comparing the new method, to check that the good results of GRASP Forest over the other ensembles are not due to the use of binary trees degrading their performance, we compared the other ensembles using binary and non binary trees. In Table 3 the wins, ties and losses are shown both total and with a 0.05 significance level for the corrected t paired Student test [16] using 5×2 cross validation [17] and the same 62 datasets used in the next experiments. We can see that the performance is slightly better when the base classifiers are binary trees. Anyway, the results with binary trees are clearly not worse, so the use of binary trees in the comparison of these ensembles with GRASP forest is fair.

Finally, we did the comparison with our method. The experiments were performed using 5×2 cross validation, over 62 data sets from the UCI repository [18] (see table 2). Table 4 shows the results as an average ranking [19][4]. Various configurations for the parameter α obtained favourable results compared to most traditional ensembles.

Figure 1 shows average rankings for different configurations of GRASP forest: using GRASP for both the attribute and splitting point selection, using GRASP only for attribute selection, using pruned trees as base classifiers, using not

---

[4] For each dataset, the methods are sorted according to their performance. The best method has rank 1, the second rank 2, and so on. If several methods have the same result, they are assigned an average value. For each method, its average rank is calculated as the mean across all the datasets.

Table 4. Ensemble methods sorted by average rank (U: unpruned trees, P: pruned trees)

Method	Ranking
1 GRASP Forest $\alpha = 0.2$ (U)	10.0645161290323
2 GRASP Forest $\alpha = 0.3$ (U)	10.4838709677419
3 RandomForest $K = \log numAtt$	11.1129032258065
4 GRASP Forest $\alpha = 0.4$ (P)	11.1935483870968
5 GRASP Forest $\alpha = 0.3$ (P)	11.3306451612903
6 GRASP Forest $\alpha = 0.4$ (U)	11.4274193548387
7 GRASP Forest $\alpha = 0.1$ (U)	11.5887096774194
8 MultiBoost-S (P)	11.9435483870968
9 MultiBoost-S (U)	12.3467741935484
10 GRASP Forest $\alpha = 0.5$ (U)	12.7661290322581
11 GRASP Forest $\alpha = 0.2$ (P)	12.8548387096774
12 GRASP Forest $\alpha = 0.5$ (P)	13.2177419354839
13 AdaBoost-S (P)	13.3709677419355
14 MultiBoostAB-W (U)	13.6290322580645
15 RandomForest $S = 2$	13.6370967741935
16 MultiBoostAB-W (P)	13.6451612903226
17 GRASP Forest $\alpha = 0.1$ (P)	13.6532258064516
18 AdaBoost-S (P)	13.6774193548387
19 AdaBoost-W (P)	14.3145161290323
20 AdaBoost-W (U)	15.1209677419355
21 Random-Subspaces-50% (U)	15.4435483870968
22 RandomForest $K = 1$	16.5806451612903
23 Random-Subspaces-50% (P)	17.1854838709677
24 Bagging (P)	17.7903225806452
25 Bagging (U)	18.0483870967742
26 Random-Subspaces-50% (U)	20.7096774193548
27 Random-Subspaces-75% (P)	20.8629032258064

pruned trees as base classifiers, and using 11 different values of α between 0 and 1. The ensemble size was 50. On the left, the average rankings are calculated from all ensemble configurations, 44 in total (2 variants of GRASP $\times$ 2 different base tree classifiers $\times$ 11 values of α). On the right, the average rankings are calculated for each size of the ensemble and for six different methods.

It is possible to appreciate how, in general, the ensembles that use GRASP in the two steps of the tree construction, both attribute and split point selection, get better results both with prune and not pruned trees. The global optimum for α is around 0.2, the point from which the increase in diversity does not compensate the lost in accuracy of the individual trees in the ensemble. As well, for ensembles with few base classifiers, the best results are obtained with trees with low randomness ($\alpha = 0.5$), but as the size of the ensemble increases, the trees with the best rankings are those with higher level of randomness ($\alpha = 0.2$ and $\alpha = 0.3$).

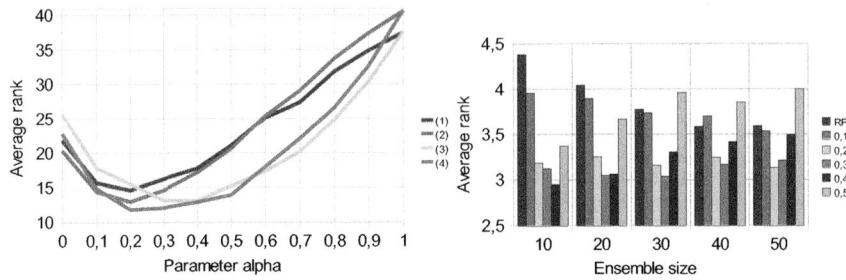

Fig. 1. *Left*: Average rankings in function of α and four different GRASP Forest configurations: (1) and (2) using GRASP only to choose the attribute, (3) and (4) using GRASP both for attribute and splitting point selection; (1) and (3) using pruned trees as base classifiers, (2) and (4) using not pruned trees as base classifiers (average rankings calculated from all 44 configurations). *Right*: Average rankings for different ensemble sizes and six different methods, Random Forest with $K = \log numAtts$ and GRASP Forest with five different α values and unprunned trees (GRASP used both for attribute and splitting point selection).

4   Conclusion and Future Lines

GRASP is a metaheuristic strategy widely used for optimization problems, recently it has been used for the construction of trees less complex than the ones built with deterministic algorithms. In this paper, we propose an evolution of the use of this metaheuristic in the construction of trees that is used to increase the diversity of ensembles. The results are favourable compared with traditional ensembles.

There are several future research lines. This paper presents a method that injects randomness into two steps in the construction of the tree: in the choice of the splitting attribute and in the selection of the splitting point within that attribute. In this work, the parameter that determines the randomness is the same for both steps, a line of future work will be to study how the performance of the method could be affected if independent values are used in these steps. Another aspect to consider is the performance of this new method in combination with traditional ensembles. Rather than considering these methods as competitors against which to measure the GRASP Forest, consider them as allies that the GRASP Forest could improve.

We have shown that when the average ranking is calculated as a function of α, there is a kind of global optimum. However, this does not mean that a value of α could optimally work for all data sets. In [13] is stated that intermediate values of α improve the accuracy of the trees in simple databases and high values improve the accuracy in complex databases. Another line of future work could be to carry out an exhaustive study of what is the effect of the value of α in terms of different meta-features [20] or complexity measures [21] for several datasets. This way, this knowledge could be used in the algorithm to adaptively choose the value of α taking into account the characteristics of the dataset.

Last but not least, we would like to work on adapting GRASP Forest for regression problems.

References

1. Kuncheva, L.I.: Combining Pattern Classifiers: Methods and Algorithms. Wiley Interscience, Hoboken (2004)
2. Breiman, L.: Bagging Predictors. Machine Learning 24, 123–140 (1996)
3. Ho, T.K.: The random subspace method for constructing decision forests. IEEE Transactions on Pattern Analysis and Machine Intelligence 20, 832–844 (1998)
4. Freund, Y., Schapire, R.: Experiments with a new boosting algorithm. In: Proceedings of the Thirteenth International Conference on Machine Learning (ICML), pp. 148–156 (1996)
5. Webb, G.I.: MultiBoosting: A Technique for Combining Boosting and Wagging. Machine Learning 40, 159–196 (2000)
6. Breiman, L.: Random forests. Machine learning 45, 5–32 (2001)
7. Dietterich, T.G.: An experimental comparison of three methods for constructing ensembles of decision trees: Bagging, boosting, and randomization. Machine Learning 40, 139–157 (2000), doi:10.1023/A:1007607513941
8. Maudes, J., Rodríguez, J.J., García-Osorio, C., García-Pedrajas, N.: Random feature weights for decision tree ensemble construction. Information Fusion (2010)
9. Quinlan, R.J.: C4.5: Programs for Machine Learning. Morgan Kaufmann, San Francisco (1993)
10. Breiman, L., Friedman, J., Stone, C.J., Olshen, R.A.: Classification and Regression Trees. Chapman and Hall, Boca Raton (1984)
11. Feo, T., Resende, M.: A probabilistic heuristic for a computationally difficult set covering problem. Operations Research Letters 8, 67–71 (1989)
12. Feo, T., Resende, M.: Greedy randomized adaptive search procedures. Journal of Global Optimization 6, 109–133 (1995)
13. Pacheco, J., Alfaro, E., Casado, S., Gámez, M., García, N.: Uso del metaheurístico GRASP en la construcción de árboles de clasificación. Rect@ 11, 139–154 (2010)
14. Geurts, P., Ernst, D., Wehenkel, L.: Extremely randomized trees. Machine Learning 63, 3–42 (2006)
15. Hall, M., Frank, E., Holmes, G., Pfahringer, B., Reutemann, P., Witten, I.H.: The weka data mining software: an update. SIGKDD Explor. Newsl. 11, 10–18 (2009)
16. Nadeau, C., Bengio, Y.: Inference for the generalization error. Machine Learning 52, 239–281 (2003), doi:10.1023/A:1024068626366
17. Dietterich, T.: Approximate statistical tests for comparing supervised classification learning algorithms. Neural computation 10, 1895–1923 (1998)
18. Frank, A., Asuncion, A.: UCI machine learning repository (2010)
19. Demšar, J.: Statistical comparisons of classifiers over multiple data sets. The Journal of Machine Learning Research 7, 1–30 (2006)
20. Castiello, C., Castellano, G., Fanelli, A.: Meta-data: Characterization of input features for meta-learning. In: Modeling Decisions for Artificial Intelligence, pp. 457–468 (2005)
21. Ho, T., Basu, M.: Complexity measures of supervised classification problems. IEEE Trans. Pattern Anal. Mach. Intell. 24, 289–300 (2002), info:doi:10.1109/34.990132

Ensembles of Decision Trees for Imbalanced Data[*]

Juan J. Rodríguez, José F. Díez-Pastor, and César García-Osorio

University of Burgos, Spain
jjrodriguez@ubu.es, jfdiez@beca.ubu.es, cgosorio@ubu.es

Abstract. Ensembles of decision trees are considered for imbalanced datasets. Conventional decision trees (C4.5) and trees for imbalanced data (CCPDT: Class Confidence Proportion Decision Tree) are used as base classifiers. Ensemble methods, based on undersampling and over-sampling, for imbalanced data are considered. Conventional ensemble methods, not specific for imbalanced data, are also studied: Bagging, Random Subspaces, AdaBoost, Real AdaBoost, MultiBoost and Rotation Forest. The results show that the ensemble method is much more important that the type of decision trees used as base classifier. Rotation Forest is the ensemble method with the best results. For the decision tree methods, CCPDT shows no advantage.

Keywords: Imbalanced data, Decision Trees, Bagging, Random Subspaces, Boosting, Rotation Forest.

1 Introduction

A dataset is imbalanced if the proportion of the classes is rather different. Conventional classification methods are usually biased towards the classes with more examples. Hence, several specific methods have been proposed for this type of data [1].

One of the approaches for dealing with imbalanced datasets is to modify the construction method of a classifier, in a way that this imbalance is taken into account. There are some variants of decision trees for imbalanced data [2,3].

Decision trees are among the most used methods as base classifiers for ensemble methods. They are unstable [4], this is useful for obtaining diverse classifiers. Moreover, they are fast [5], this is convenient because in ensemble methods many classifiers are constructed and used for classification.

Given that decision trees are good candidates for base classifiers and that there are decision tree methods for imbalanced data, ensembles of these trees can be used to tackle imbalance problems. This work explores the performance of ensembles of these decision trees. In principle, any ensemble method could be used, even if it is not specifically designed for imbalanced data, because the base classifier already deals with this issue.

[*] This work was supported by the Project TIN2008-03151 of the Spanish Ministry of Education and Science.

C. Sansone, J. Kittler, and F. Roli (Eds.): MCS 2011, LNCS 6713, pp. 76–85, 2011.

Some ensemble methods, like Boosting, can work well with imbalanced data, although they are not designed specifically for this problem [6]. If the base classifiers are biased in favour of the majority class, there will be more errors in the minority class, so its examples will have more weight than the examples of the majority class in the next iterations.

Other approach for dealing with imbalanced datasets is modifying the dataset. There are two main approaches: undersampling the majority class and oversampling the minority class [7,8,1]. These methods lead immediately to ensembles: each base classifier is trained with a different random sample. Undersampling and oversampling have been combined with other ensemble methods, such as Boosting and Random Subspaces [7,6,9].

The rest of the paper is organised as follows. Next section describes the used methods: base classifiers and ensembles. Section 3 describes the experiments and the results. Finally, Section 4 presents the conclusions.

2   Methods

2.1   Decision Trees

Let tp, fp, tn, fn be, respectively, the number of true positives, false positives, true negatives and false negatives. The *Information Gain*, as used in C4.5, for a two-class problem and a binary split can be defined using the *relative impurity* (Imp) as in [3,10]:

$$\text{InfoGain}_{C4.5} = \text{Imp}(tp + fn, fp + tn)$$
$$-(tp + fp) * \text{Imp}\left(\tfrac{tp}{tp+fp}, \tfrac{fp}{tp+fp}\right)$$
$$-(fn + tn) * \text{Imp}\left(\tfrac{fn}{fn+tn}, \tfrac{tn}{fn+tn}\right)$$

where $\text{Imp}(p, n)$ is defined as $-p \log p - n \log n$.

In *CCP* (*Class Confidence Proportion*) decision trees (*CCPDT*), the main idea is to use true positive and false positive rates (tpr and fpr) instead of the number of examples of each type (tp and fp). In these trees, the used measure is [3]:

$$\text{InfoGain}_{CCP} = \text{Imp}(tp + fn, fp + tn)$$
$$-(tpr + fpr) * \text{Imp}\left(\tfrac{tpr}{tpr+fpr}, \tfrac{fpr}{tpr+fpr}\right)$$
$$-(2 - tpr - fpr) * \text{Imp}\left(\tfrac{1-tpr}{2-tpr-fpr}, \tfrac{1-fpr}{2-tpr-fpr}\right)$$

If several attributes give a gain similar to the maximum gain, one of them is selected according to their *Hellinger distance* [3,2]. This distance is based on the square root difference of tpr and fpr ($|\sqrt{tpr} - \sqrt{fpr}|$) instead of their proportion, as used in CCP.

CCPDT can also be pruned. The used method is based on Fisher's exact test [3].

In [3] CCPDT obtains favourable results when compared against C4.5, CART, HDDT (*Helliger Distance Decision Tree*) [2] and sampling methods [11].

2.2   Ensemble Methods

In Bagging [4], each base classifier is trained with a random sample, *with replacement*, of the original training dataset. The size of the sample, by default, is the same as the original dataset. Some examples could appear several times in the sample, while other could be excluded.

In Random Subspaces [12], the base classifiers are trained using all the training examples but in a random subspace. That is, using only a random subset of the features.

AdaBoost [13] is an iterative method. The training examples are assigned a weight, initially the same. The base classifiers are constructed taking into account these weights. The weights are adjusted during training: misclassified examples are given greater weights. So, in the next iteration, they will be more important and the next classifier will put more effort in their classification. The base classifiers also have weights, better classifiers have greater weights. The predictions are obtained using a weighted vote.

Real AdaBoost [14] is a variant of AdaBoost. The difference is that it takes into account the probabilities assigned by the base classifier to the different training examples, instead of only considering whether the prediction is correct or not.

MultiBoost [15] is a variant of AdaBoost, the modification is based on Bagging. The ensemble is formed by several sub-committees. These sub-committees are constructed using AdaBoost. When a sub-committee is constructed, example weights are assigned a random value using the continuous Poisson distribution. These are the initial weights for the next sub-committee.

Random Undersampling [8,6] is a technique for dealing with imbalanced data. The examples of the majority class are sampled, while all the examples of the minority class are used.

SMOTE [16] is an oversampling method. Synthetic examples of the minority class are generated by randomly selecting a minority example and one of its nearest neighbours, and choosing a point in the segment that connects both.

In Rotation Forest [17,18], each base classifier is trained on a rotated dataset. For transforming the dataset, the attributes are randomly split in groups. In each group, a random non empty subset of the classes is selected, the examples of the classes that are not in the subset are removed. From the remaining examples, a sample is taken that is used to calculate PCA. All the components from all the groups are the features of the transformed dataset. All the examples in the original training data are used for constructing the base classifier, the selection of classes and examples is used to calculate the PCA projection matrices (one per group), but then *all the data* is transformed according with these matrices. The base classifier receives all the information available in the dataset, because all the examples and all the components obtained from PCA are used.

Rotation Forest is not designed for imbalance data. Although the base classifier could deal with imbalance, this imbalance is not taken into account when rotating the dataset. In this method, for each attribute group, a non-empty subset of the classes is selected. For a two-class imbalanced problem, this subset can be formed by the minority class, the majority class, or both. The imbalance is only present in the subset with the two classes. We propose a balanced version of Rotation Forest. When a subset of classes is formed by two classes, the majority class is undersampled. The size of this sample is the number of examples in the minority class. Note that this undesampling is only done for calculating the principal components, the base classifier is trained with all the training examples.

3 Experiments

3.1 Datasets

Table 1 shows the datasets used in the study. All of them are two-classes datasets. Most of them are from the UCI repository [19]. The source for the rest is shown in the table.

Table 1. Characteristics of the datasets

Dataset	Examples	Attributes		Minority percentage	References
		Numeric	Nominal		
adult	48842	6	8	23.93	[19]
breast-w	699	9	0	34.48	[19,9]
breast-y	286	0	9	29.72	[19,9]
credit-g	1000	7	13	30.00	[19,9]
ecg[1]	200	304	0	33.50	[20]
fourclass[2]	862	2	0	35.61	[3,9]
haberman	306	3	0	26.47	[19]
heart-s	123	5	8	6.50	[19]
heart-v	200	5	8	25.50	[19,9]
hypo	3163	7	18	4.77	[19,9]
laryngeal2[3]	692	16	0	7.66	[21]
musk-2	6598	166	0	15.41	[19]
phoneme[4]	5404	5	0	29.35	[3,9]
pima	768	8	0	34.90	[19,3,9]
sick	3772	7	22	6.12	[19]
svmguide1[2]	3089	4	0	35.25	[22,3,9]
tic-tac-toe	958	0	9	34.66	[19,9]
wafer[1]	1194	1188	0	10.64	[20]

1:   http://www.cs.cmu.edu/~bobski/pubs/tr01108.html
2:   http://www.csie.ntu.edu.tw/~cjlin/libsvmtools/datasets/
3:   http://www.bangor.ac.uk/~mas00a/activities/real_data.htm
4:   http://www.dice.ucl.ac.be/neural-nets/Research/Projects/ELENA/databases/REAL/phoneme/

3.2   Settings

Weka [23] was used for the experiments. The results were obtained using 5×2 folds stratified cross validation. Ensembles were formed by 100 decision trees. The number of considered methods is 56: 4 decision tree configurations that are considered as individual methods and as base classifiers for 13 ensemble methods. Default options from Weka were used, unless otherwise specified.

The decision trees were constructed using J48, Weka's implementation of C4.5 [24], and CCPDT [3]. For both methods, trees with and without pruning were considered; they are denoted as *(P)* and *(U)*. For the 4 configurations Laplace smoothing [3] was used on the leaves.

For Random Subspaces, two subspace sizes were considered: 50% and 75% of the original space size. Reweighting and resampling [13] were considered for AdaBoost, Real AdaBoost and MultiBoost. In the former, denoted with the suffix -W, the base classifiers are constructed with weighted examples. In the latter, denoted with the suffix -S, the base classifiers are constructed with a sample from the training data. This sample is obtained according to the weight distribution of the examples. For MultiBoost, the number of sub-committees is 10 and each sub-committee is formed by 10 decision trees.

For undersampling, the sample size was the number of examples of the minority class. For SMOTE, the number of artificial examples of the minority class was the original number of examples in the minority class, that is, the number of examples is doubled. The resulting dataset can also still be imbalanced, but less than the original. Artificial data is less reliable than the original data, it could be detrimental to have more artificial examples than real examples for the minority class. The amount of artificial data could be optimized for each dataset, but this would increase substantially the computation time. Other considered methods also have parameters that could be optimized, optimizing the parameters for one of the method would require optimizing the parameters from all the methods.

3.3   Results

Average ranks [25] were used for summarizing the results. For each dataset, the methods are sorted according to their performance. The best method has rank 1, the second rank 2, and so on. If several methods have the same result, they are assigned an average value. For each method, its average rank is calculated as the mean across all the datasets.

Table 2 shows the average ranks obtained from the accuracies of the different methods for the different datasets. Table 3 shows the average ranks, obtained from the areas under the ROC curves (AUC) [26]. The values of these average ranks are organized in matrices, where the row indicates the ensemble method and the column indicates the base classifier. The average value for each ensemble (row) and base classifier (column) is also shown in the tables. The ensemble methods are sorted according to these averages.

For both tables, the differences among the averages of the base classifiers are much smaller than the differences among the averages of the ensemble methods.

Table 2. Average ranks, using the accuracy as the performance measure

Ensemble	J48		CCPDT		Average
	(U)	(P)	(U)	(P)	
Balanced Rotation Forest	12.33	13.11	12.42	15.75	13.40
Rotation Forest	12.31	13.72	14.53	15.61	14.04
MultiBoost-W	20.36	25.61	21.36	24.69	23.01
MultiBoost-S	19.53	27.22	20.42	26.36	23.38
AdaBoost-W	22.92	25.25	23.86	24.81	24.21
Bagging	24.42	27.03	23.33	28.69	25.87
AdaBoost-S	22.56	28.75	23.58	28.64	25.88
RealAdaBoost-S	25.28	28.17	23.17	30.75	26.84
RealAdaBoost-W	27.19	29.47	26.53	26.39	27.40
Random Subspaces-75%	29.31	32.64	29.97	35.22	31.78
Random Subspaces-50%	30.97	31.06	33.44	36.89	33.09
SMOTE	36.78	42.81	33.64	41.17	38.60
Single	42.53	46.14	36.14	43.92	42.18
Undersampling	48.39	49.39	48.11	51.39	49.32
Average	26.78	30.03	26.46	30.73	

Table 3. Average ranks, using the AUC as the performance measure

Ensemble	J48		CCPDT		Average
	(U)	(P)	(U)	(P)	
Balanced Rotation Forest	7.28	9.50	9.39	10.72	9.22
Rotation Forest	7.94	9.44	9.22	11.06	9.42
Bagging	17.11	22.78	17.72	24.33	20.49
Undersampling	18.72	24.22	18.72	27.75	22.35
Random Subspaces-50%	21.19	18.81	29.39	25.81	23.80
SMOTE	24.50	28.83	24.39	31.33	27.26
Random Subspaces-75%	25.44	24.75	31.64	29.42	27.81
AdaBoost-W	33.67	30.89	30.61	29.83	31.25
AdaBoost-S	32.94	35.11	34.39	35.72	34.54
RealAdaBoost-W	33.44	38.78	34.00	36.61	35.71
RealAdaBoost-S	34.72	38.17	34.06	38.17	36.28
MultiBoost-W	38.64	38.64	38.67	35.14	37.77
MultiBoost-S	38.83	38.58	35.58	38.14	37.78
Single	41.25	42.94	46.94	50.11	45.31
Average	26.84	28.67	28.19	30.30	

Hence, an interesting conclusions is that the selection of the ensemble method is much more important than the selection of the decision tree method. Trees without pruning have a better average than trees with pruning.

The comparison of CCPDT with J48 gives mixed results. For accuracy, the best average (across all the ensemble methods) is for CCPDT without pruning (although the difference is very small), but for AUC it is J48 without pruning. If we consider single trees, CCPDT is clearly better than J48 for accuracy but clearly worst for AUC.

Tables 2 and 3 have in common the two ensemble methods in the top rows: Balanced Rotation Forest and Rotation Forest. The difference between the balanced version and the normal version is very small. There is a wide gap between the average ranks of the Rotation Forest methods and the other methods. Moreover, the average ranks of the different variants of Rotation Forest (balanced or not, pruned or not, J48 or CCPDT) are always better than the average ranks of all the variants of all the other methods.

As expected, the average ranks for single trees are among the worst, although for accuracy it is better than undersampling. The remaining methods have different positions in the ranks for the accuracy and the AUC.

According to the Nemenyi test [25], *"the performance of two classifiers is significantly different if the corresponding average ranks differ by at least the critical difference"*. For 18 datasets, 56 configurations and a confidence level of $\alpha = 0.05$, the value of this critical difference is 21.99. Hence, the best configuration (balanced Rotation Forest, unpruned J48) is significantly different from all the configurations based on Boosting and single trees, but not from the remaining configurations. It must be noted that this test is conservative, can have little power and adjust the critical difference for making comparisons among all the pairs of classifiers [25]. Hence, it can be adequate for comparing few classifiers using many datasets, but in this experiment the situation is the opposite. On the other hand, *"average ranks by themselves provide a fair comparison of the algorithms"* [25].

Figure 1 shows a scatter plot with these average ranks. Smaller values are better, methods that appear at the bottom-left have handled the imbalance well. Several groups can be identified:

- Boosting variants: they have better average ranks for accuracy than for AUC. Their positions in the ranks are very uniform. For AUC, the 6 variants are in consecutive positions, they are the worst ensemble methods, only single trees have worst results. For accuracy, they are not in consecutive positions but only Bagging is among them. Excluding Rotation Forests and single trees, MultiBoost is the best method for accuracy but the worst for AUC.
- Bagging and Random Subspaces: they have better average ranks for AUC than for accuracy. After Rotation Forest, Bagging is the best method for AUC.
- SMOTE and Undersampling: these are ensemble methods specific for imbalanced problems. They are among the worst for accuracy, but they are much better for AUC. Nevertheless, Bagging has better average ranks than these methods.

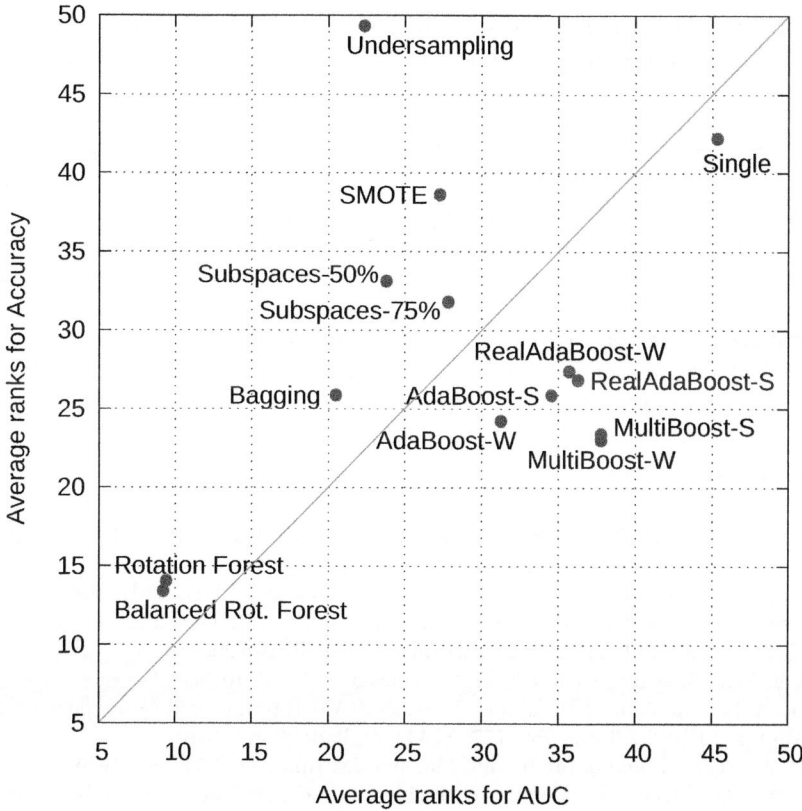

Fig. 1. Average ranks for the ensemble methods, according to the AUC and Accuracy

4   Conclusions and Future Work

The performance of several ensemble methods using decision trees as base classifiers have been studied over 18 imbalanced datasets. Accuracy and AUC have been used as performance measures.

Two types of decision trees have been considered: J48 (Weka's implementation of C4.5) and CCPDT (a specific method for imbalanced datasets). Some of the considered ensemble methods are not specific for imbalanced data: Bagging, Random Subspaces, AdaBoost, MultiBoost and Rotation Forest. The specific methods are based on undersampling and SMOTE. Moreover, a balanced version of Rotation Forest was also considered.

One of the conclusions of this study is that the ensemble method is much more relevant than the type of tree used as base classifier. The decision tree method for imbalanced data does not give better results than the standard decision tree method. Generally, it is better to use trees without pruning.

The ensemble method with the best results is Rotation Forest. According to the average ranks, the balanced version has better results than the original version, but the difference is very small. The original version handles class imbalance better than the other methods, including specific methods for imbalanced data.

This work has not considered combinations of different ensemble methods. Some of these combinations have been previously considered: Boosting and SMOTE [7], Boosting and undersampling [6], Random Subspaces and SMOTE or undersampling [9] . . . Comparing these combinations with Rotation Forest and combining other ensemble methods with Rotation Forest will be considered in future work.

Acknowledgements. We wish to thank the developers of Weka [23] and CCPDT [3]. We also express our gratitude to the donors of the different datasets and the maintainers of the UCI Repository [19].

References

1. He, H., Garcia, E.A.: Learning from Imbalanced Data. IEEE Transactions on Knowledge and Data Engineering 21, 1263–1284 (2009)
2. Cieslak, D., Chawla, N.: Learning decision trees for unbalanced data. In: Daelemans, W., Goethals, B., Morik, K. (eds.) ECML PKDD 2008, Part I. LNCS (LNAI), vol. 5211, pp. 241–256. Springer, Heidelberg (2008)
3. Liu, W., Chawla, S., Cieslak, D.A., Chawla, N.V.: A Robust Decision Tree Algorithm for Imbalanced Data Sets. In: 10th SIAM International Conference on Data Mining, SDM 2010, pp. 766–777. SIAM, Philadelphia (2010)
4. Breiman, L.: Bagging predictors. Machine Learning 24, 123–140 (1996)
5. Opitz, D., Maclin, R.: Popular ensemble methods: An empirical study. Journal of Artificial Intelligence Research 11, 169–198 (1999)
6. Seiffert, C., Khoshgoftaar, T.M., Van Hulse, J., Napolitano, A.: RUSBoost: A hybrid approach to alleviating class imbalance. IEEE Transactions on Systems, Man, and Cybernetics - Part A: Systems and Humans 40, 185–197 (2010)
7. Chawla, N.V., Lazarevic, A., Hall, L.O., Bowyer, K.W.: SMOTEBoost: Improving prediction of the minority class in boosting, pp. 107–119 (2003)
8. Liu, X.Y., Wu, J., Zhou, Z.H.: Exploratory Undersampling for Class-Imbalance Learning. IEEE Transactions on Systems, Man, and Cybernetics, Part B (Cybernetics) 39, 539–550 (2009)
9. Hoens, T., Chawla, N.: Generating Diverse Ensembles to Counter the Problem of Class Imbalance. In: Zaki, M.J., Yu, J.X., Ravindran, B., Pudi, V. (eds.) PAKDD 2010. LNCS, vol. 6119, pp. 488–499. Springer, Heidelberg (2010)
10. Flach, P.: The geometry of ROC space: understanding machine learning metrics through ROC isometrics. In: Proc. 20th International Conference on Machine Learning (ICML 2003), pp. 194–201. AAAI Press, Menlo Park (2003)
11. Chawla, N., Cieslak, D., Hall, L., Joshi, A.: Automatically countering imbalance and its empirical relationship to cost. Data Mining and Knowledge Discovery 17, 225–252 (2008)
12. Ho, T.K.: The random subspace method for constructing decision forests. IEEE Transactions on Pattern Analysis and Machine Intelligence 20, 832–844 (1998)
13. Freund, Y., Schapire, R.E.: A decision-theoretic generalization of on-line learning and an application to boosting. Journal of Computer and System Sciences 55, 119–139 (1997)

14. Friedman, J., Hastie, T., Tibshirani, R.: Additive logistic regression: a statistical view of boosting. Annals of Statistics 95, 337–407 (2000)
15. Webb, G.I.: Multiboosting: A technique for combining boosting and wagging. Machine Learning 40, 159–196 (2000)
16. Chawla, N.V., Bowyer, K.W., Hall, L.O., Kegelmeyer, W.P.: SMOTE: Synthetic minority over-sampling technique. Journal of Artificial Intelligence Research 16, 321–357 (2002)
17. Rodríguez, J.J., Kuncheva, L.I., Alonso, C.J.: Rotation forest: A new classifier ensemble method. IEEE Transactions on Pattern Analysis and Machine Intelligence 28, 1619–1630 (2006)
18. Kuncheva, L.I., Rodríguez, J.J.: An experimental study on rotation forest ensembles. In: Haindl, M., Kittler, J., Roli, F. (eds.) MCS 2007. LNCS, vol. 4472, pp. 459–468. Springer, Heidelberg (2007)
19. Frank, A., Asuncion, A.: UCI machine learning repository (2010), http://archive.ics.uci.edu/ml
20. Olszewski, R.T.: Generalized Feature Extraction for Structural Pattern Recognition in Time-Series Data. PhD thesis, Computer Science Department, Carnegie Mellon University (2001)
21. Kuncheva, L.I., Hadjitodorov, S.T., Todorova, L.P.: Experimental comparison of cluster ensemble methods. In: FUSION 2006, Florence, Italy (2006)
22. Hsu, C.W., Chang, C.C., Lin, C.J.: A practical guide to support vector classification. Technical report, Department of Computer Science, National Taiwan University (2003)
23. Hall, M., Frank, E., Holmes, G., Pfahringer, B., Reutemann, P., Witten, I.H.: The WEKA data mining software: An update. SIGKDD Explorations 11 (2009)
24. Quinlan, J.R.: C4.5: Programs for Machine Learning. Machine Learning. Morgan Kaufmann, San Mateo (1993)
25. Demšar, J.: Statistical comparisons of classifiers over multiple data sets. Journal of Machine Learning Research 7, 1–30 (2006)
26. Fawcett, T.: An introduction to ROC analysis. Pattern Recognition Letters 27, 861–874 (2006)

Compact Ensemble Trees for Imbalanced Data

Yubin Park and Joydeep Ghosh

Department of Electrical and Computer Engineering
The University of Texas at Austin, Austin, TX-78712, USA

Abstract. This paper introduces a novel splitting criterion parametrized by a scalar 'α' to build a class-imbalance resistant ensemble of decision trees. The proposed splitting criterion generalizes information gain in C4.5, and its extended form encompasses Gini(CART) and DKM splitting criteria as well. Each decision tree in the ensemble is based on a different splitting criterion enforced by a distinct α. The resultant ensemble, when compared with other ensemble methods, exhibits improved performance over a variety of imbalanced datasets even with small numbers of trees.

1 Introduction

Imbalanced datasets are pervasive in real-world applications, including fraud detection, risk management, text classification, medical diagnosis etc. Despite their frequent occurrence and huge impact in day to day applications, many standard machine learning algorithms fail to address this problem properly since they assume either balanced class distributions or equal misclassification costs [10]. There have been various approaches proposed to deal with imbalanced classes, including: over/undersampling [13], [17], SMOTE (synthetic minority oversampling technique), cost-sensitive [15], modified kernel-based, and active learning methods [1], [8].

Several authors have tried to theoretically address the nature of the class imbalance problem [3], [11], [18]. Their results suggest that the degree of imbalance is not the only factor hindering the learning process [10]. Rather, the difficulties reside with various other factors such as overlapping classes, lack of representative data, small disjuncts etc, that get amplified when the distribution of classes is imbalanced. In this paper, we approach the imbalanced learning problem by combining multiple decision trees. If these different "base" classifiers can focus on different features of the data and handle complex objectives collaboratively, then an ensemble of such trees can perform better for datasets with class imbalance.

Breiman had observed that the most challenging classification problem is how to increase *simplicity* and *understanding* without losing *accuracy* [16]. Also, it has been shown that a small variety of strong learning algorithms are typically more effective than using a large number of *dumbed-down* models [9]. So a second goal is to build robust, imbalance-resistant ensembles using only a few classifiers.

While many ensemble trees induce diversity using random selection of features or data points, this paper proposes a novel splitting criterion parametrized by

C. Sansone, J. Kittler, and F. Roli (Eds.): MCS 2011, LNCS 6713, pp. 86–95, 2011.

a scalar α. By varying α, we get dissimilar decision trees in the ensemble. This new approach results in ensembles that are reasonably simple yet accurate over a range of class imbalances.

We briefly summarize the main contributions of this paper here:

1. We introduce a new decision tree algorithm using α-divergence. A generalized tree induction formula is proposed, which includes Gini, DKM, and C4.5 splitting criteria as special cases.
2. We propose a systematic ensemble algorithm using a set of α-Trees covering a range of α. The ensemble shows consistent performance across a range of imbalance degrees. The number of classifiers needed in the method is far less than Random Forest or other ensembles for a comparable performance level.

Related Work. Several approaches try to tackle the imbalanced learning problem by oversampling or generating synthetic data points in order to balance the class distributions [14]. An alternative is to employ cost-sensitive methods that impose different misclassification costs. Even though these methods have shown good results, their performance depends on heuristics that need to be tuned to the degree of imbalance.

Ensemble methods generally outperform single classifiers [5], and decision trees are popular choices for the base classifiers in an ensemble [2]. In recent years, the Random Forest has been modified to incorporate sampling techniques and cost matrices to handle class-imbalance [6]. Though this modified Random Forest shows superior performance over other imbalance-resistant classifiers, its complexity increases too.

Some of the earlier works such as [4] by L. Breiman investigate various splitting criteria - Gini impurity, Shannon entropy and twoing in detail. Dietterich *et al* [7] showed that the performance of a tree can be influenced by its splitting criteria and proposed a criterion called DKM which results in lower error bounds based on the Weak Hypothesis Assumption. Karakos *et al* proposed Jensen-Rényi divergence parametrized by a scalar α as a splitting criterion [12], but the determination of the "best" α was based on heuristics.

This paper applies a novel splitting criterion(α-divergence) to ensemble methods to solve the class imbalance problem with a small number of base trees. Decision trees based on distinct α values possess different properties, which in turn increases diversity in the ensemble.

2   Preliminaries

α-Divergence. Decision tree algorithms try to determine the best split based on a certain criterion. However, the "best" split usually depends on the characteristics of the problem. For example, for some datasets we might want 'low precision'-'high recall' results and for some the other way around. For this to be resolved it's better to have a criterion that can be adapted by easy manipulation.

Our metric, α-divergence, which generalizes KL-divergence [19], easily achieves this feat.

$$D_\alpha(p\|q) = \frac{\int_x \alpha p(x) + (1-\alpha)q(x) - p(x)^\alpha q(x)^{1-\alpha} dx}{\alpha(1-\alpha)} \qquad (1)$$

where p, q are any two probability distributions and α is a real number. Some special cases are:

$$D_{\frac{1}{2}}(p\|q) = 2\int_x (\sqrt{p(x)} - \sqrt{q(x)})^2 dx \qquad (2)$$

$$\lim_{\alpha \to 1} D_\alpha(p\|q) = KL(p\|q) \qquad (3)$$

$$D_2(p\|q) = \frac{1}{2}\int_x \frac{(p(x) - q(x))^2}{q(x)} dx \qquad (4)$$

Equation (2) is Hellinger distance, and equation (3) is KL-divergence. α-Divergence is always positive and is 0 if and only if $p = q$. This enables α-divergence to be used as a (dis)similarity measure between two distributions.

Splitting Criterion using α-Divergence. The splitting criterion function of C4.5 can be written using α-divergence as :

$$I(X;Y) = \lim_{\alpha \to 1} D_\alpha(p(x,y)\|p(x)p(y)) = KL(p(x,y)\|p(x)p(y)) \qquad (5)$$

where $p(x, y)$ is a joint distribution of a feature X and the class label Y, and $p(x)$ and $p(y)$ are marginal distributions. To maintain consistency with the C4.5 algorithm, a new splitting criterion function is proposed as follows:

Definition 1. *α-Divergence splitting criterion is $D_\alpha(p(x,y)\|p(x)p(y))$, where $0 < \alpha < 2$.*

Note that $\alpha = 1$ gives the information gain in C4.5.

Using this splitting criterion, a splitting feature is selected, which gives the maximum α-divergence splitting criterion.

Constructing an α-Tree. Using the proposed decision criterion the decision tree induction follows in **algorithm 1**. Let us call this new tree as α-**Tree**. In algorithm 1 'Classify' can be either 'majority voting' or 'probability approximation' depending on the purpose of the problem. This paper uses 'probability approximation' as 'majority voting' might cause overfitting for the imbalanced data. The effect of varying α will be discussed in Section 4.2.

Algorithm 1. Grow Single α-Tree

Input: Training Data (features $X_1, X_2, ..., X_n$ and class Y), $\alpha \in (0, 2)$
Output: α-Tree
Select the best feature X^*, which gives the maximum α-**divergence criterion**
if (no such X^*) or (number of data points < cut-off size) **then**
    return Classify(Training Data)
else
    partition the training data into m subsets, based on the value of X^*
    for for $i = 1$ to m **do**
        ith child = Grow Single α-Tree (ith partitioned data, α)
    end for
end if

---

3   Properties of α-Divergence Criterion

Properties of α-Divergence. If both $p(x, y)$ and $p(x)p(y)$ are properly defined probability distributions, then the above α-divergence becomes:

$$D_\alpha(p(x,y)||p(x)p(y)) = E_X[D_\alpha(p(y|x)||p(y))]. \tag{6}$$

Consider two Bernoulli distributions, $p(x)$ and $q(x)$ having the probability of success θ_p, θ_q respectively, where $0 < \theta_p, \theta_q < 1/2$. Then the α-divergence from $p(x)$ to $q(x)$ and its 3rd order Taylor expansion w.r.t. θ_p is:

$$D_\alpha(p||q) = \frac{1 - \theta_p{}^\alpha \theta_q{}^{1-\alpha} - (1-\theta_p)^\alpha(1-\theta_q)^{1-\alpha}}{\alpha(1-\alpha)} \tag{7}$$

$$\approx A(\theta_p - \theta_q)^2 + B(\alpha - 2)(\theta_p - \theta_q)^3 \tag{8}$$

where $A = \frac{1}{2}(\frac{1}{\theta_q} + \frac{1}{1-\theta_q})$, $B = \frac{1}{6}(\frac{1}{\theta_q{}^2} - \frac{1}{(1-\theta_q)^2})$ and $A, B > 0$. Then, given $0 < \alpha < 2$ and $\theta_p > \theta_q$, the 3rd order term in equation (8) is negative. So by increasing α the divergence from p to q increases. On the other hand if $\theta_p < \theta_q$ the 3rd order term in equation (8) is positive and increasing α decreases the divergence. This observation motivates proposition 1 below. Later we describe proposition 2 and its corollary 1.

Proposition 1. *Assume that we are given Bernoulli distributions $p(x)$, $q(x)$ as above and $\alpha \in (0,2)$. Given $\theta_q < 1/2$, $\exists \epsilon > 0$ s.t. $D_\alpha(p||q)$ is a monotonic 'increasing' function of α where $\theta_p \in (\theta_q, \theta_q + \epsilon)$, and $\exists \epsilon' > 0$ s.t. $D_\alpha(p||q)$ is a monotonic 'decreasing' function of α where $\theta_p \in (\theta_q - \epsilon', \theta_q)$. (Proof. This follows from equation (8).)*

Proposition 2. *$D_\alpha(p||q)$ is convex w.r.t. θ_p. (Proof. 2nd derivative of equation (7) w.r.t θ_p is positive.)*

Corollary 1. *Given binary distributions, $p(x)$, $q(x)$, $r(x)$, where $0 < \theta_p < \theta_q < \theta_r < 1$, $D_\alpha(q||p) < D_\alpha(r||p)$ and $D_\alpha(q||r) < D_\alpha(p||r)$. (Proof. Since $D_\alpha(s(x)||t(x)) \geq 0$ and is equal if and only if $s(x) = t(x)$, using proposition 2, corollary 1 directly follows.)*

Effect of varying α. Coming back to our original problem, let us assume that we have a binary classification problem whose positive class ratio is θ_c where $0 < \theta_c \ll 1/2$ (imbalanced class). After a split, the training data is divided into two subsets: one with higher ($> \theta_c$) and the other with lower ($< \theta_c$) positive class ratio. Let us call the subset with higher positive class ratio as *positive*, and the other as *negative subset*. Without loss of generality, suppose we have binary features $X_1, X_2, ..., X_n$ and $p(y = 1|x_i = 0) < p(y = 1) < p(y = 1|x_i = 1)$ and $p(x_i) \approx p(x_j)$ for any i, j. From equation (6) the α-divergence criterion becomes:

$$p(x_i = 1)D_\alpha(p(y|x_i = 1)||p(y)) + p(x_i = 0)D_\alpha(p(y|x_i = 0)||p(y)) \qquad (9)$$

where $1 \leq i \leq n$. From 'proposition 1' we observe that increase in α increases $D_\alpha(p(y|x_i = 1)||p(y))$ and decreases $D_\alpha(p(y|x_i = 0)||p(y))$ (lower-bounded by 0).

$$(9) \approx p(x_i = 1)D_\alpha(p(y|x_i = 1)||p(y)) + const. \qquad (10)$$

From 'corollary 1', increasing α shifts our focus to high $p(y = 1|x_i = 1)$. In other words, increasing α results in the splitting feature having higher $p(y = 1|x_i = 1)$, *positive predictive value* (PPV) or *precision*. On the other hand reducing α results in lower $D_\alpha(p(y|x_i = 1)||p(y))$ and higher $D_\alpha(p(y|x_i = 0)||p(y))$. As a result, reducing α gives higher $p(y = 0|x_i = 0)$, *negative predictive value* (NPV) for the splitting features.

The effect of varying α appears clearly with an experiment using real datasets. For each α value in the range of $(0, 2)$, α-Tree was built based on 'sick' dataset from UCI thyroid dataset. α-Trees were grown until '3rd level depth', as fully-grown trees deviate from the above property. Note that this analysis is based on a single split, not on a fully grown tree. As the tree grows, a subset of data on each node might not follow the imbalanced data assumption. Moreover, the performance of a fully grown tree is affected by not only 'α', but also other heuristics like 'cut-off size'. 5-fold cross validation is used to measure each performance. Averaged PPV and NPV over 5-cv are plotted in Figure 1.

By varying the value of α we can control the selection of splitting features. This is a crucial factor in increasing 'diversity' among decision trees. The greedy nature of decision trees means that even a small change in α may result in a substantially different tree.

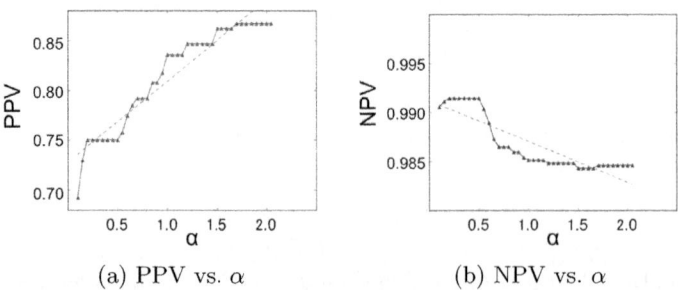

(a) PPV vs. α (b) NPV vs. α

Fig. 1. Effect of varying α. Dotted lines are linearly regressed lines.

Note that the above analysis is based on Taylor expansion of α-divergence that holds true when $p(y|x_i) \approx p(y)$, which is the case when datasets are imbalanced. This property may not hold if $p(y|x_i)$ differs a lot from $p(y)$.

Connection to DKM and CART. The family of α-divergence naturally includes C4.5's splitting criterion. But the connection to DKM and CART is not that obvious. To see the relation between α-divergence and the splitting functions of DKM and CART, we extend the definition of α-divergence (equation (6)) as follows:

Definition 2. *Extended α-divergence is defined as $E_X[D_\alpha(p(y|x)||q(y))]$ where $q(y)$ is any arbitrary probability distribution.*

Definition 2 is defined by replacing $p(y)$ with any arbitrary distribution $q(y)$, which serves as a reference distribution. The connection to DKM and CART is summarized in the following two propositions:

Proposition 3. *Given a binary classification problem, if $\alpha = 2$ and $q(y) = (\frac{1}{2}, \frac{1}{2})$ then the extended α-divergence splitting criterion gives the same splitting feature as the Gini impurity criterion in CART. (Proof. See Appendix A.)*

Proposition 4. *Given a binary classification problem, if $\alpha = \frac{1}{2}$ and $q(y) = p(\bar{y}|x)$ then the extended α-divergence splitting criterion gives the same splitting feature as the DKM criterion. (Proof. See Appendix A.)*

CART implicitly assumes a balanced reference distribution while DKM adaptively changes its reference distribution for each feature. This explains why CART generally performs poorly on imbalanced datasets and DKM provides a more skew-insensitive decision tree.

4   Bootstrap Ensemble of α-Trees

In this section, we propose the algorithm for creating an ensemble of α-Trees. The **BEAT** (**B**ootstrap **E**nsemble of **A**lpha **T**rees) algorithm for an ensemble of k trees is illustrated in Algorithm 2. Observe that the parameters (a, b) for Beta distribution and the number of trees are design choices. The parameters (a, b) can be chosen using a validation set.

Algorithm 2. Bootstrap Ensemble of α-Trees (**BEAT**)

Input: Training Data D (features $X_1, X_2, ..., X_n$ and class Y) and parameters (a, b).

for for $i = 1$ to k **do**
    Sample $\alpha/2 \sim Beta(a, b)$.
    Sample D_i from D with replacement (Bootstrapping).
    Build an α-Tree C_i from D_i using **algorithm 1**.
end for
for for each test record $t \in T$ **do**
    $C^*(t) = Avg(C_1(t), C_2(t), ..., C_k(t))$
end for

BEAT uses Bootstrapping when making its base classifiers. Like other Bagging methods, BEAT exhibits better performance as the number of trees in BEAT increases. The test errors of BEAT and Bagged-C4.5 ($C4.5^B$) are shown in Figure 2 (a). The experiment is performed based on 'glass' dataset from UCI repository. The 'headlamps' class in the dataset is set as positive class, and the other classes are set as negative class (13.5% positive class ratio). 5×2 cross validation is used. The performance of BEAT is comparable with $C4.5^B$.

As the value of α affect the performance of α-Tree, the parameters (a, b), which determine the distribution of α, change the performance of BEAT. Misclassification rate generally doesn't capture the performance on imbalanced datasets. Although the misclassification rate of BEAT doesn't vary much from $C4.5^B$, the improvement can be seen apparently in 'precision' and 'recall', which are crucial when dealing with imbalanced datasets. This property is based on the observation in Section 3, but the exact relationship between the parameters and the performance is usually data-dependent. Figure 2 (b) shows the averaged 'f-score' result based on the same 'glass' dataset. Unlike the error rate result, the f-scores of BEAT and $C4.5^B$ show clear distinction. Moreover the resultant average ROC curves of BEAT (Figure 2 (c), (d)) changed as the parameters (a, b) change. The ability to capture different ROC curves allows great flexibility on choosing different decision thresholds.

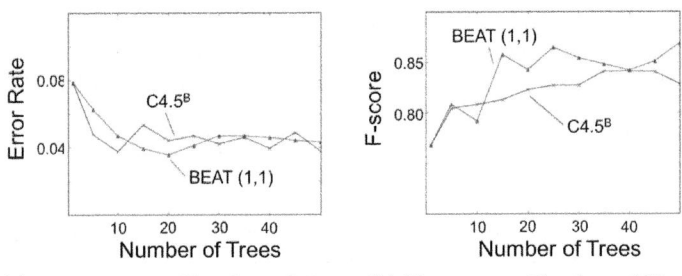

(a) Error rate vs. Number of Trees   (b) F-score vs. Number of Trees

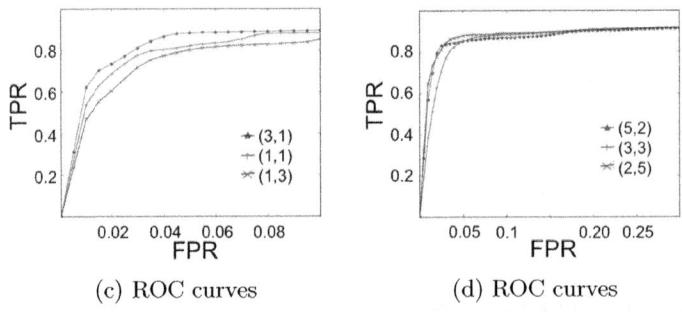

(c) ROC curves   (d) ROC curves

Fig. 2. Properties of BEAT. ROC curves are measured using 30 α-Trees.

Experimental Evaluation. All the datasets used in this paper are from the UCI Repository. Datasets with multiple classes are converted into 2-class problems. 5×2 cross validation instead of 10-fold cross validation is used due to highly imbalanced data. Aside from the stated modifications each dataset is used "as is".

A comparative evaluation of BEAT with C4.5, $C4.5^B$, and Balanced Random Forest (BRF) [6], was performed. All trees are binary/fully grown and 30 base trees are used. No prunning is applied, and for features having more than 2 categories, dummy coding scheme is used to build a binary tree. Table 1 reports the average f-score over 5×2 cross validation. For BRF, the number of random attributes are fixed to 2 ($m = 2$). BRF generally needs more number of base classifiers to perform stably.

Table 1. F-score results on real datasets. Parameters (a, b) of BEAT are indicated in the parenthesis. The test errors are shown in the parenthesis below f-scores. Although BRF has the highest f-score in 'wpbc' dataset, BRF records the highest error rate among ensemble trees. $C4.5^B$ and BEAT show comparable test errors, but BEAT outperforms in f-scores.

Dataset	C4.5	$C4.5^B$	BRF	BEAT(1,1)	BEAT(1,3)	BEAT(3,1)
glass	0.756	0.846	nan	**0.868**	0.84	0.865
	(6.0%)	(3.7%)	(4.3%)	(3.7%)	(4.2%)	(3.6%)
allhypo	0.901	0.956	0.644	0.958	**0.96**	0.956
(Thyroid)	(1.5%)	(0.69%)	(7.7%)	(0.67%)	(0.63%)	(0.69%)
allhyper	0.57	0.711	0.434	0.715	0.689	**0.728**
(Thyroid)	(2.2%)	(1.5%)	(6.6%)	(1.4%)	(1.5%)	(1.4%)
sick	0.826	0.871	0.580	0.866	**0.876**	0.866
(Thyroid)	(2.1%)	(1.5%)	(8.5%)	(1.6%)	(1.5%)	(1.6%)
allrep	0.74	0.869	0.413	0.876	0.870	**0.884**
(Thyroid)	(1.6%)	(0.83%)	(8.8%)	(0.84%)	(0.82%)	(0.76%)
wpbc	0.293	0.387	*0.434*	**0.425**	0.355	0.315
(Breast Cancer)	(33.7%)	(22.3%)	*(31.7%)*	(22.3%)	(22%)	(23.6%)
page blocks	0.791	0.860	0.746	0.863	**0.865**	0.858
	(4.3%)	(2.7%)	(6.8%)	(2.7%)	(2.7%)	(2.9%)

5   Concluding Remarks

In this paper, we presented the BEAT approach incorporating a novel decision criterion parametrized by α. Experimental results show that BEAT is stronger and more robust for imbalanced data, compared to other tree based ensemble methods.

Even though our algorithm gives consistent results for various cases, a joint optimization hasn't been achieved yet with respect to the parameters (a, b) and the number of trees. Moreover, the extended α-divergence criterion needs to be further investigated as well.

Acknowledgements. This research was supported by SK Telecom, IC^2 Institute, NHARP and NSF IIS-1016614. We are also grateful to Abhimanu Kumar for some useful discussions.

References

1. Akbani, R., Kwek, S., Japkowicz, N.: Applying support vector machines to imbalanced datasets. In: Proceedings of the 15th European Conference on Machine Learning (2004)
2. Banfield, R.E., Hall, L.O., Bowyer, K.W., Kegelmeyer, W.P.: A comparison of decision tree ensemble creation techniques. IEEE Transactions on Pattern Analysis and Machine Intelligence (2006)
3. Batista, G.E., Prati, R.C., Monard, M.C.: A study of the behavior of several methods for balancing machine learning training data. ACM SIGKDD Explorations Newsletter 6, 20–29 (2004)
4. Breiman, L.: Technical note: Some properties of splitting criteria. Machine Learning 24, 41–47 (1996)
5. Chawla, N.V.: Many are better than one: Improving probabilistic estimates from decision trees. In: Machine Learning Challenges, pp. 41–55 (2006)
6. Chen, C., Liaw, A., Breiman, L.: Using random forest to learn imbalanced data. Tech. rep. Dept. of Statistics, U.C. Berkeley (2004)
7. Dietterich, T., Kearns, M., Mansour, Y.: Applying the weak learning framework to understand and improve c4.5. In: Proceedings of the Thirteenth International Conference on Machine Learning, pp. 96–104 (1996)
8. Ertekin, S., Huang, J., Giles, C.L.: Learning on the border: Active learning in imbalanced data classification. In: Proceedings of the 30th Annual International ACM SIGIR conference, pp. 823–824 (2007)
9. Gashler, M., Giraud-Carrier, C., Martinez, T.: Decision tree ensemble: Small heterogeneous is better than large homogeneous. In: The 7th International Conference on Machine Learning and Applications, pp. 900–905 (2008)
10. He, H., Garcia, E.A.: Learning from imbalanced data. IEEE Transactions on Knowledge and Data Engineering 21 (2009)
11. Japkowicz, N., Stephen, S.: The class imbalance problem: A systematic study. In: Intelligent Data Analysis, vol. 6, pp. 429–449 (2002)
12. Karakos, D., Eisner, J., Khudanpur, S., Priebe, C.E.: Cross-instance tuning of unsupervised document clustering algorithms. In: Proceedings of NAACL HLT, pp. 252–259 (2007)
13. Laurikkala, J.: Improving identification of difficult small classes by blancing class distribution. In: Proceedings of the 8th Conference of AI in Medicine in Europe: Artificial Intelligence Medicine, pp. 63–66 (2001)
14. Liu, A., Martin, C., Cour, B.L., Ghosh, J.: Effects of oversampling versus cost-sensitive learning for bayesian and svm classifiers. Annals of Information Systems 8, 159–192 (2010)
15. McCarthy, K., Zarbar, B., weiss, G.: Does cost-sensitive learning beat sampling for classifying rare classes? In: Proceedings of International Workshop Utility-Based Data Mining, pp. 69–77 (2005)
16. Sharkey, A.J. (ed.): Combining Artificial Neural Nets: Ensemble and Modular Multi-Net Systems. Springer, Heidelberg (1999)

17. Weiss, G., Provost, F.: The effect of class distribution on classifier learning: An empirical study. Tech. Rep. Dept. of Computer Science, Rutgers University (2001)
18. Weiss, G., Provost, F.: Learning when training data are costly: The effect of class distribution on tree induction. Journal of Artificial Intelligence Research 19, 315–354 (2003)
19. Zhu, H., Rohwer, R.: Information geometric measurements of generalization. Tech. Rep. 4350, Aston University (1995)

A   CART, DKM and α-Divergence

Gini. Assume a binary classification problem, $y \in \{0,1\}$ and binary feature $x \in \{0,1\}$. Since for choosing a best feature x the distribution of y is fixed, we can derive the following equation:

$$Gini = \sum_y p(y)(1-p(y)) - \sum_x p(x) \sum_y p(y|x)(1-p(y|x)). \tag{11}$$

$$= E_X[\frac{1}{2} - \sum_y p(y|x)(1-p(y|x))] + const \tag{12}$$

$$\propto E_X[D_2(p(y|x)||q(y))] \tag{13}$$

where $q(y) = (\frac{1}{2}, \frac{1}{2})$. Equation (13) follows from equation (4). Linear relation between the Gini splitting formula and α-divergence completes the proof.

DKM. Assuming the similar settings as in Appendix A, the splitting criterion function of DKM is:

$$DKM = \prod_y \sqrt{p(y)} - \prod_x p(x) \prod_y \sqrt{p(y|x)} \tag{14}$$

$$= E_X[\frac{1}{2} - \prod_y \sqrt{p(y|x)}] + const \tag{15}$$

$$\propto E_X[D_{\frac{1}{2}}(p(y|x)||q(y))] \tag{16}$$

where $q(y) = p(\bar{y}|x)$. Equation (16) follows from equation (2). Linear relation between the DKM formula and α-divergence completes the proof.

Pruned Random Subspace Method for One-Class Classifiers

Veronika Cheplygina and David M.J. Tax

Pattern Recognition Lab, Delft University of Technology
v.cheplygina@tudelft.nl, d.m.j.tax@tudelft.nl

Abstract. The goal of one-class classification is to distinguish the target class from all the other classes using only training data from the target class. Because it is difficult for a single one-class classifier to capture all the characteristics of the target class, combining several one-class classifiers may be required. Previous research has shown that the Random Subspace Method (RSM), in which classifiers are trained on different subsets of the feature space, can be effective for one-class classifiers. In this paper we show that the performance by the RSM can be noisy, and that pruning inaccurate classifiers from the ensemble can be more effective than using all available classifiers. We propose to apply pruning to RSM of one-class classifiers using a supervised area under the ROC curve (AUC) criterion or an unsupervised consistency criterion. It appears that when the AUC criterion is used, the performance may be increased dramatically, while for the consistency criterion results do not improve, but only become more predictable.

Keywords: One-class classification, Random Subspace Method, Ensemble learning, Pruning Ensembles.

1 Introduction

The goal of one-class classification is to create a description of one class of objects (called target objects) and distinguish this class from all other objects, not belonging to this class (called outliers) [21]. One-class classification is particularly suitable for situations where the outliers are not represented well in the training set. This is common in applications in which examples of one class are more difficult to obtain or expensive to sample, such as detecting credit card fraud, intrusions, or a rare disease [5].

For a single one-class classifier it may be hard to find a good model because of limited training data, high dimensionality of the feature space and/or the properties of the particular classifier. In one-class classification, a decision boundary should be fitted around the target class such that it distinguishes target objects from *all* potential outliers. That means that a decision boundary should be estimated in all directions in the feature space around the target class. Compared to the standard two-class classification problems, where we may expect objects from the other class predominantly in one direction, this requires more parameters to fit, and therefore more training data.

C. Sansone, J. Kittler, and F. Roli (Eds.): MCS 2011, LNCS 6713, pp. 96–105, 2011.

To avoid a too complex model and overfitting on the training target data, simpler models can be created that use less features. In the literature, approaches such as the Random Subspace Method (RSM) [10]) or feature bagging [12] are proposed. In RSM, several classifiers are trained on random feature subsets of the data and the decisions of these classifiers are combined. RSM is expected to benefit in problems suffering from the "curse of dimensionality" because of the improved object/feature ratio for each individual classifier. It has been demonstrated that combining classifiers can also be effective for one-class classifiers [21,20]. RSM is successfully applied to a range of one-class classifiers in [12,13,1].

Although it was originally believed that larger ensembles are more effective, it has been demonstrated that using a subset of classifiers might be better than the using the whole set [24]. A simple approach is to evaluate L classifiers individually according to a specified criterion (such as accuracy on the training set) and select the L_s ($L_s < L$) best classifiers (i.e. prune the inaccurate classifiers). Pruning in RSM has been shown to be effective for traditional classifiers [4,18], however, to apply this to one-class classifiers, one faces the problem of choosing a good selection criterion. Most criteria require data from all classes, but in the case of one-class classifiers one assumes a (very) poorly sampled outlier class. This paper evaluates two criteria for the pruned RSM: the area under the ROC curve (AUC) [2], which uses both target and outlier examples, and the consistency criterion, using only target data [22].

In Sect. 2, some background concerning one-class classifiers and RSM is given. The pruned random subspace method (PRSM) is proposed in Sect. 3. In Sect. 4, experiments are performed to analyze the behavior of the PRSM compared to the basic RSM and other popular combining methods. Conclusion and suggestions for further research are presented in Sect. 5.

2 Combining One-Class Classifiers

Supervised classification consists of approximating the true classification function $y = h(\mathbf{x})$ using a collection of object-label pairs $X = \{(\mathbf{x}_1, y_1)..., (\mathbf{x}_N, y_N)\}$ ($\mathbf{x} \in \mathbb{R}^d$) with an hypothesis h, and then using h to assign y values to previously unseen objects $\mathbf{x}$. Let us assume there are two classes, i.e. $y \in \{T, O\}$. A traditional classifier needs labeled objects of both classes in order to create h, and its performance suffers if one of the classes is absent from the training data. In these situations one-class classifiers can be used. A one-class classifier only needs objects of one class to create h. It is thus trained to accept objects of one class (target class) and reject objects of the other (outlier class).

A one-class classifier consists of two elements: the "similarity" of an object to the target class, expressed as a posterior probability $p(y = T|\mathbf{x})$ or a distance (from which $p(y = T|\mathbf{x})$ can be estimated), and a threshold θ on this measure, which is used to determine whether a new object belongs to the target class or not:

$$h(\mathbf{x}) = \delta(p(y = T|\mathbf{x}) > \theta) = \begin{cases} +1, & \text{when } p(y = T|\mathbf{x}) > \theta, \\ -1, & \text{otherwise,} \end{cases} \quad (1)$$

where δ is the indicator function. In practice, θ is set such that the classifier rejects a fraction f of the target class.

Estimating $p(y = T|\mathbf{x})$ is hard, in particular for high-dimensional feature spaces and low sample size situations. By making several (lower dimensional) approximations of h and combining these, a more robust model can be obtained. Assume that $\mathbf{s}(\mathbf{x})$ maps the original d-dimensional data to a d_s-dimensional subspace:

$$\mathbf{s}(\mathbf{x}; \mathbf{I}) = [x_{I_1}, x_{I_2}, ..., x_{I_{d_s}}]^T \tag{2}$$

where $\mathbf{I} = [I_1, I_2, ..., I_{d_s}]$ is the index vector indicating the features that are selected. Typically, the features are selected at random, resulting in RSM [10].

When in each of the subspaces a model $h_i(\mathbf{s}(\mathbf{x}; \mathbf{I}_i))$ (or actually $p_i(y = T|\mathbf{s}(\mathbf{x}; \mathbf{I}_i))$) is estimated, several combining methods can be used [11] to combine the subspace results. Two standard approaches are averaging of the posterior probabilities (also called mean combining):

$$\tilde{p}(y = T|\mathbf{x}) = \frac{1}{L} \sum_{i=1}^{L} p_i(y = T|\mathbf{s}(\mathbf{x}; \mathbf{I}_i)) \tag{3}$$

or voting, i.e.:

$$\tilde{p}(y = T|\mathbf{x}) = \frac{1}{L} \sum_{i=1}^{L} I(p_i(y = T|\mathbf{s}(\mathbf{x}; \mathbf{I}_i)) > \theta_i) \tag{4}$$

In Table 1, the average AUC performances are shown of a combination of Gaussian models (such that $p(y = T|\mathbf{x})$ in (1) is modelled as a normal distribution) for three datasets, which are described in more detail in Sect.4. The number of subspaces L is varied between 10 and 100, and the combining rule is varied between averaging (mean), product, voting and maximum of posterior probabilities. The subspace dimensionality d_s is fixed at 25% of the features.

As can be observed, RSM is able to significantly outperform the base classifiers, but this improvement is quite unpredictable, i.e. we cannot directly see what parameters are needed to achieve the best performance. In particular, a larger value of L does not always lead to increased performance. In fact, the optimal number of subspaces L differs across the datasets ($L = 25$ for Concordia, $L = 10$ for Imports, $L = 50$ for Sonar). The best combining method also depends on the dataset, however, we observe that mean produces better results than voting in most situations. Furthermore, the subspace dimensionality d_s also can affect performance of RSM [6], in fact, values between 20% and 80% can be found in literature [1,13]. In our experiments, 25% gave reasonable performances overall, so this value is fixed for the rest of this paper.

3   Pruned Random Subspaces for One-Class Classification

A successful classifier should have good performance *and* be predictable in terms of how its parameters influence its performance. It should be possible to choose at

Table 1. AUC performances ($\times 100$, averaged over 5 times 10-fold cross-validation) of the RSM combining Gaussian classifiers on three datasets. "Baseline" indicates performance of a single Gaussian classifier on all features. Two parameters of RSM are varied: combining rule (rows) and number of subspaces L (columns). Results in bold indicate performances not significantly worse ($\alpha = 0.05$ significance level) than the best performance per column.

Data	Baseline	Comb. rule	Number of subspaces L				
			10	25	50	75	100
Concordia 3	92.1	mean	**91.2**	**94.5**	**93.9**	**94.0**	**93.8**
		product	90.9	94.0	93.5	93.4	92.6
		vote	89.9	93.7	93.4	93.6	93.4
		maximum	90.2	93.7	93.5	93.4	93.4
Imports	74.0	mean	**78.0**	73.7	71.2	69.6	68.8
		product	**78.3**	73.5	71.6	67.5	65.6
		vote	**78.7**	**77.0**	**74.5**	72.8	72.7
		maximum	75.6	**76.3**	**76.0**	**75.2**	**74.3**
Sonar	63.0	mean	**65.8**	**65.1**	**65.9**	**65.6**	**65.5**
		product	**65.8**	**65.1**	64.5	62.6	61.7
		vote	62.6	63.2	64.5	**65.1**	64.7
		maximum	64.1	**65.1**	64.6	64.0	64.4

least some of the parameters beforehand, rather than trying all possible parameter combinations, as in Table 1. Based on our observations, the only parameter that can be chosen with some certainty is the combining rule, as it performs reasonably in most situations (although other combining methods might be comparable). However, it is more difficult to choose a value for the number of subspaces L because of its noisy effect on the performance.

It appears that RSM produces subspace classifiers of different quality. In some subspaces, the distinction between the target and outlier objects is much clearer than in others. Averaging is sensitive to such variations [16], thereby causing the performance of the whole ensemble to vary significantly. By excluding the less accurate classifiers, the performance of the ensemble should stabilize. In this case, the choice for the number of subspaces is less critical and less prone to noise in the subspace selection.

Therefore we propose PRSM, a pruned version of RSM. The implementation of PRSM is shown in Algorithm 1. The essential difference with the RSM is visible in the second to last line. Instead of combining all subspace results, PRSM only uses the best L_s outcomes. In order to find the best subspaces, the subspaces should be ranked according to an evaluation measure C.

We use two different evaluation measures particularly suitable for OCCs: AUC [2] and consistency [22]. The AUC is obtained by integrating the area under the Receiver-Operating-Characteristic curve, given by $(\varepsilon^t, 1 - \varepsilon^o)$ pairs, where ε^t is the error on the target class and ε^o is the error on the outlier class, as shown in (5). Because the true values of ε^t and ε^o are unknown, in practice, the AUC

Algorithm 1. PRSM Algorithm

Input: Training set X, base classifier h, number of classifiers L, number of selected
classifiers L_s, subspace dim. d_s, subspace criterion C, combining method M
Output: Ensemble E
   for $i = 1$ to L **do**
      $I_i \leftarrow RandomSelectFeatures(X, d_s)$
      $score_i \leftarrow CrossValidate(X, h, I_i, L_s, C)$
   end for
   $\mathbf{I} \leftarrow rankByScore(\mathbf{I}, score)$
   $\mathbf{h} \leftarrow Train(X, h, \mathbf{I}_{1:L_s})$
   return $E \leftarrow Combine(\mathbf{h}, M)$

---

is obtained by varying f, estimating the $(\hat{\varepsilon}^t, 1 - \hat{\varepsilon}^o)$ pairs using a validation set,
and integrating under this estimated curve.

$$AUC = 1 - \int_0^1 \varepsilon^o(\varepsilon^t)\, d\varepsilon^t. \tag{5}$$

The consistency measure indicates how consistent a classifier is in rejecting
fraction f of the target data. It is obtained by comparing f with $\hat{\varepsilon}^t$, an estimate
of the error on the target class:

$$Consistency = |\hat{\varepsilon}^t - f| \tag{6}$$

The AUC measure is chosen for its insensitiveness to class imbalance, which
may often be a problem in one-class classification tasks. However, to obtain
the AUC both $\hat{\varepsilon}^t$ and $\hat{\varepsilon}^o$ are needed, which means that the validation set must
contain outliers. Therefore, pruning using AUC is not strictly a one-class method
and the performance estimates might be more optimistic than in the pure one-
class setting. On the other hand, to obtain the consistency measure only $\hat{\varepsilon}^t$ is
required, which means no outlier information is used.

Using a validation set has another implication: there may be too little data
to use a separate validation set. An alternative is to do the evaluation using
the training set as in [4]. However, this is not possible for classifiers which have
100% performance on the training set, such as 1-NN. In our implementation, we
perform 10-fold cross-validation using the training set to evaluate the subspace
classifiers. After ranking the classifiers, they are retrained using the complete
training set.

4   Experiments

In this section, experiments are performed to analyze the behavior of the PRSM
compared to the basic RSM and other combining methods. First, the PRSM is
compared with RSM with respect to the absolute performance and the stability
or predictability. Next, the PRSM is compared to other classifiers across a range
of datasets. We use the base classifier and RSM, Bagging [3], and AdaBoost [19]

ensembles of the same base classifier. A separate comparison is performed for each base classifier. For the comparison, we use the Friedman test [9] and the post-hoc Nemenyi test [14], as recommended in [8].

Table 2. List of datasets. N_T and N_O represent the numbers of target and outlier objects, d stands for dimensionality.

Dataset	N_T	N_O	d	Dataset	N_T	N_O	d
Arrhythmia	237	183	278	Prime Vision	50	150	72
Cancer non-ret	151	47	33	Pump 2×2 noisy	64	176	64
Cancer ret	47	151	33	Pump 1×3 noisy	41	139	64
Concordia 2	400	3600	256	Sonar mines	111	97	60
Concordia 3	400	3600	256	Sonar rocks	97	111	60
Glass	17	197	9	Spambase	79	121	57
Housing	48	458	13	Spectf normal	95	254	44
Imports	71	88	25	Vowel 4	48	480	10
Ionosphere	225	126	34	Wine 2	71	107	13

All experiments are performed in Matlab using PRTools [17] and the Data Description toolbox [7]. In [21], several strategies for deriving one-class classifiers are described. In this paper we consider three typical examples: the Gaussian (Gauss), Nearest Neighbor (1-NN), and k-Means one-class classifiers. Gauss is a *density* method, which fits a normal distribution to the data and rejects objects on the tails of this distribution. 1-NN is a *boundary* method, which uses distances between objects to calculate the local densities of objects, and rejects new objects with a local density lower than that of its nearest neighbor. k-Means is a *reconstruction* method, which assumes that the data is clustered in k groups, finds k prototype objects for these groups, and creates a boundary out of the hyperspheres placed at these objects.

Several datasets are used from the UCI Machine Learning Repository [23], and have been modified in order to contain a target and an outlier class [15]. Furthermore, an additional dataset, the Prime Vision dataset, provided by a company called Prime Vision[1] in Delft, The Netherlands, is used. A summary of the datasets is included in Table 2. All the datasets have relatively low object-to-feature ratios, either originally or after downsampling (in case of Prime Vision and Spambase datasets).

In Fig. 1, a comparison between PRSM+AUC (PRSM using the AUC criterion), PRSM+Con (PRSM using the consistency criterion) and RSM for the Concordia digit 3 dataset is shown. For varying number of subspaces L the AUC of the combined classifier is presented. We use $d_s = 64$ (that is 25% of the features) and compare the performances of RSM with L_s classifiers to PRSM with L_s classifiers out of $L = 100$.

---

[1] http://www.primevision.com

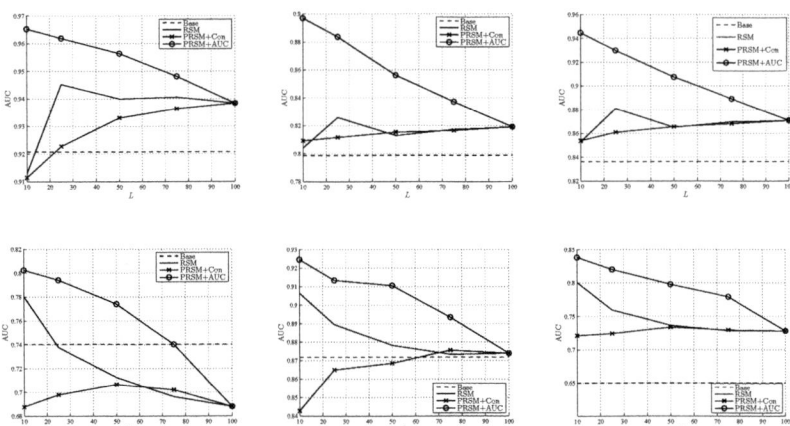

Fig. 1. The AUC (on test set) for varying number of subspaces for RSM, PRSM+AUC and PRSM+Con for the Concordia digit 3 dataset (top) and for the Imports dataset (bottom). From left to right: Gauss, 1-NN and k-Means base classifiers.

The results indicate that both PRSM+AUC and PRSM+Con are less noisy and their performances are more predictable with a varying L. For PRSM+AUC, optimal performances are achieved using just a small subset of $L_s = 10$ subspaces, whereas for PRSM+Con, larger values such as $L_s = 50$ are better. These values are fixed for the further experiments. However, in terms of absolute performance, only PRSM+AUC is able to significantly outperform the standard RSM. PRSM+Con sometimes has a lower performance than the baseline classifier, and is never performs significantly better than the unpruned RSM with $L = 100$ subspaces.

Next, RSM, PRSM+Con and PRSM+AUC are compared to the baseline classifier, and a few standard ensemble approaches, namely Bagging [3] and AdaBoost [19]. In Bagging, $L = 100$ bootstrapped versions of the training set X are generated, a base classifier is trained on these bootstrapped training sets, and the combined classifier is the average (as in (3)) of the L trained base classifiers. In AdaBoost, new training sets are iteratively generated by reweighing objects. Objects that are often misclassified get a higher weight, and are thus more likely to be selected in a training set. The combined classifier uses voting.

The results of the comparison are shown in Table 3. A key observation is that overall, the classifiers have comparable performance, however, there is no overall winning classifier. For some datasets, such as Glass, it is often best to just use the base classifier as no ensemble is able to improve on this performance. This can be explained because this dataset is very small (only 17 training target objects), and fitting a more complex classifier than the base classifier just overfits on the training data.

Surprisingly, also AdaBoost does not perform very well. AdaBoost is originally developed to boost two-class classifiers by reweighing training objects. In these experiments, the base classifier is a one-class classifier, trained to describe the

Table 3. AUC performances ($\times 100$) of the base classifier and ensemble methods for the Gauss, 1-NN and k-Means base classifiers. PRSc indicates the PRSM using the consistency criterion, PRSa indicates the PRSM using the AUC criterion. Bag means bagging, and AB means AdaBoost. Results in bold indicate the best (or not significantly worse than best) performances per dataset.

Data	Base	RSM	PRSc	PRSa	Bag	AB	Data	Base	RSM	PRSc	PRSa	Bag	AB
							Gauss						
Arr	75.6	78.0	77.8	**79.4**	75.5	50.0	PV	89.4	88.3	88.1	**89.8**	**90.2**	50.6
Cancl	50.2	**53.6**	**53.4**	52.8	51.1	**52.8**	Pump1	**94.8**	83.1	83.8	92.3	88.5	50.0
Canc2	62.4	60.9	59.6	**65.7**	61.0	52.8	Pump2	91.8	84.0	86.2	**98.2**	82.6	50.0
Conc2	83.6	87.0	86.3	**91.3**	82.5	51.5	Sonar1	65.7	64.1	63.5	**70.4**	65.2	54.6
Conc3	92.1	93.8	93.2	**96.4**	91.2	51.8	Sonar2	62.2	64.7	65.0	**69.1**	62.4	54.1
Glass	**86.2**	74.7	73.3	78.3	85.1	70.0	Spam	81.5	79.1	78.7	**85.5**	77.1	61.1
House	88.1	84.1	84.3	**91.3**	88.0	73.7	Spect	94.5	88.3	88.4	88.8	**94.9**	88.5
Impor	72.6	68.5	71.0	**80.3**	73.8	70.0	Vow4	**99.1**	95.6	94.5	96.2	**99.1**	96.7
Ion	96.5	96.9	97.0	**97.3**	96.4	87.4	Wine	94.4	92.9	90.4	**96.4**	94.3	90.1
							1-NN						
Arr	73.8	75.5	75.3	**78.1**	73.8	69.4	PV	88.0	88.7	88.5	**90.9**	87.9	86.2
Cancl	**52.6**	53.2	53.1	**53.5**	**53.4**	**51.8**	Pump1	77.3	76.0	76.3	**82.0**	76.9	77.2
Canc2	55.8	57.6	58.0	**68.3**	56.2	57.5	Pump2	78.5	77.2	76.9	**84.4**	77.9	72.1
Conc2	69.7	80.3	80.4	**88.9**	68.7	64.0	Sonar1	67.8	68.8	68.2	**78.5**	66.5	59.2
Conc3	83.7	87.2	86.6	**94.6**	82.8	78.1	Sonar2	72.2	72.3	72.4	**74.6**	70.7	63.3
Glass	**72.8**	**73.7**	72.5	**74.5**	69.9	62.5	Spam	56.7	60.5	60.8	**74.7**	53.4	53.2
House	87.5	**94.7**	**94.8**	**94.7**	83.7	83.4	Spect	95.4	**95.6**	**95.8**	**95.9**	95.2	87.3
Impor	85.6	86.5	86.3	**92.4**	80.2	77.3	Vow4	**99.4**	99.2	99.1	**99.5**	99.1	97.6
Ion	95.9	96.1	96.1	**97.3**	96.3	90.2	Wine	87.1	91.7	90.5	**94.0**	88.2	92.6
							k-means						
Arr	74.0	75.2	75.3	**77.1**	74.4	65.3	PV	86.5	88.2	88.1	**89.6**	86.8	84.4
Cancl	**51.1**	**51.7**	**51.9**	51.0	**52.0**	46.4	Pump1	71.6	72.6	72.3	**79.3**	71.0	63.6
Canc2	56.2	58.6	57.6	**63.9**	56.5	56.8	Pump2	66.8	69.5	69.4	**79.2**	68.8	66.2
Conc2	60.8	69.8	69.5	**80.9**	59.6	50.0	Sonar1	61.8	63.6	63.8	**67.3**	61.0	55.1
Conc3	79.9	81.9	81.5	**89.5**	79.9	69.8	Sonar2	65.9	67.4	67.4	**71.1**	65.7	61.2
Glass	70.1	**73.9**	72.5	**74.9**	69.6	69.0	Spam	50.6	53.7	54.0	**71.5**	50.2	52.7
House	83.4	**93.5**	92.5	**94.2**	82.5	87.2	Spect	84.5	87.2	**87.4**	**87.8**	85.8	84.1
Impor	65.0	72.3	73.1	**83.4**	63.2	68.2	Vow4	95.7	**98.5**	98.2	97.9	**98.3**	**98.1**
Ion	96.3	97.3	97.3	**97.7**	97.3	97.3	Wine	86.6	92.5	91.7	**94.7**	87.4	89.8

Table 4. Results of the Friedman/Nemenyi test. F stands for F-statistic, Signif. for any significant differences, CV for critical value and CD for critical difference.

Base clasf.	Tests				Ranks					
	F	CV	Signif.?	CD	Base	RSM	PRSc	PRSa	Bag	AB
Gauss	16.12	2.32	Yes	1.78	2.78	3.72	3.89	1.61	3.44	5.56
1-NN	32.99	2.32	Yes	1.78	3.94	2.89	3.17	1.06	4.44	5.50
k-Means	40.38	2.32	Yes	1.78	4.83	2.33	2.61	1.44	4.50	5.28

target data. Errors on the target class are generally weighted more heavily than errors on the outlier class. This results in a heavy imbalance in the sampling of the two classes, and consequently in poorly trained base classifiers.

In most situations, PRSM+AUC outperforms the other ensemble methods, and the base classifiers. The improvement is not always very large, but for some datasets it is quite significant, for instance, for the Arrhythmia, Imports, Spambase, and the Concordia datasets. These are often the high-dimensional datasets. A noticeable exception is the Ionosphere dataset. Here, the base classifiers already perform well, and the improvement achieved by the ensembles is modest.

The results of the statistical tests are shown in Table 4. For each base classifier, the F-statistic value indicates there are any significant differences between the classification methods. Significant differences are found if F is larger than a critical value (2.33 in this case) which depends on the number of datasets (18) and classifiers (6) in the experiment. If this is the case, the Nemenyi test can be used to compare any two classifiers. A significant difference between two classifiers occurs when the difference in classifier ranks is larger than the critical difference for the Nemenyi test, 1.77 in this case.

RSM is only significantly better than the base classifier for k-Means and even worse than the base classifier for Gauss. PRSM+Con produces worse (but not significantly worse) results than RSM for all three base classifiers. On the other hand, PRSM+AUC is significantly better than RSM for Gauss and 1-NN. Bagging and AdaBoost perform poorly in general, except for Gauss where Bagging is slightly better than RSM.

5   Conclusions

In this paper, we investigated the effect of pruning on Random Subspace ensembles on the Gaussian, Nearest Neighbor, and k-Means one-class classifiers. Experiments show that pruning improves the predictability of the outcomes, but does not always improve the classification performance. Pruning using the area under the ROC curve shows a significant improvement over the standard ensemble. The number of subspaces that is required for good performance is very low: 10 out of 100 subspaces is often already showing optimal performance. Pruning using the consistency criterion, based on only target objects, however, is not able to improve the results. This suggests that additional information in the form of extra outlier objects, has to be used in order to improve performance.

Our results suggest that combining a few accurate classifiers may be more beneficial than combining all available classifiers. However, pruning can only be successful if an appropriate evaluation criterion is selected. We demonstrated that the area under the ROC curve is a successful criterion, however, it requires the presence of outliers. A succesful criterion in the absence of outliers requires further investigation.

References

1. Biggio, B., Fumera, G., Roli, F.: Multiple classifier systems under attack. In: El Gayar, N., Kittler, J., Roli, F. (eds.) MCS 2010. LNCS, vol. 5997, pp. 74–83. Springer, Heidelberg (2010)
2. Bradley, A.: The use of the area under the ROC curve in the evaluation of machine learning algorithms. Pattern Recognition 30(7), 1145–1159 (1997)
3. Breiman, L.: Bagging predictors. Machine Learning 24(2), 123–140 (1996)
4. Bryll, R., Gutierrez-Osuna, R., Quek, F.: Attribute bagging: improving accuracy of classifier ensembles by using random feature subsets. Pattern Recognition 36(6), 1291–1302 (2003)
5. Chandola, V., Banerjee, A., Kumar, V.: Anomaly detection: A survey. ACM Computing Surveys 41(3), 1–58 (2009)
6. Cheplygina, V.: Random subspace method for one-class classifiers. Master's thesis, Delft University of Technology (2010)
7. DD_tools, the Data Description toolbox for Matlab, http://prlab.tudelft.nl/david-tax/dd_tools.html
8. Demšar, J.: Statistical comparisons of classifiers over multiple data sets. The Journal of Machine Learning Research 7, 1–30 (2006)
9. Friedman, M.: The use of ranks to avoid the assumption of normality implicit in the analysis of variance. Journal of the American Statistical Association 32(200), 675–701 (1937)
10. Ho, T.: The random subspace method for constructing decision forests. IEEE Transactions on Pattern Analysis and Machine Intelligence 20(8), 832–844 (1998)
11. Kittler, J., Hatef, M., Duin, R., Matas, J.: On combining classifiers. IEEE Transactions on Pattern Analysis and Machine Intelligence 20(3), 226–239 (2002)
12. Lazarevic, A., Kumar, V.: Feature bagging for outlier detection. In: 11th ACM SIGKDD Int. Conf. on Knowledge Discovery in Data Mining, pp. 157–166 (2005)
13. Nanni, L.: Experimental comparison of one-class classifiers for online signature verification. Neurocomputing 69(7-9), 869–873 (2006)
14. Nemenyi, P.: Distribution-free multiple comparisons. Ph.D. thesis, Princeton (1963)
15. OC classifier results, http://homepage.tudelft.nl/n9d04/occ/index.html
16. Oza, N., Tumer, K.: Classifier ensembles: Select real-world applications. Information Fusion 9(1), 4–20 (2008)
17. PRtools, Matlab toolbox for Pattern Recognition, http://www.prtools.org
18. Rokach, L.: Collective-agreement-based pruning of ensembles. Computational Statistics & Data Analysis 53(4), 1015–1026 (2009)
19. Schapire, R., Freund, Y.: Experiments with a new boosting algorithm. In: 13th Int. Conf. on Machine Learning, p. 148. Morgan Kaufmann, San Francisco (1996)
20. Tax, D., Duin, R.: Combining one-class classifiers. In: Kittler, J., Roli, F. (eds.) MCS 2001. LNCS, vol. 2096, pp. 299–308. Springer, Heidelberg (2001)
21. Tax, D.: One-class classification; Concept-learning in the absence of counter-examples. Ph.D. thesis, Delft University of Technology (June 2001)
22. Tax, D., Muller, K.: A consistency-based model selection for one-class classification. In: 17th Int. Conf. on Pattern Recognition, pp. 363–366. IEEE, Los Alamitos (2004)
23. UCI Machine Learning Repository, http://archive.ics.uci.edu/ml/
24. Zhou, Z., Wu, J., Tang, W.: Ensembling neural networks: Many could be better than all. Artificial Intelligence 137(1-2), 239–263 (2002)

Approximate Convex Hulls Family for One-Class Classification

Pierluigi Casale, Oriol Pujol, and Petia Radeva

Computer Vision Center, Bellaterra, Barcelona, Spain
Dept. Applied Mathematics and Analysis, University of Barcelona, Barcelona, Spain
{pierluigi,oriol,petia}@cvc.uab.es

Abstract. In this work, a new method for one-class classification based on the Convex Hull geometric structure is proposed. The new method creates a family of convex hulls able to fit the geometrical shape of the training points. The increased computational cost due to the creation of the convex hull in multiple dimensions is circumvented using random projections. This provides an approximation of the original structure with multiple bi-dimensional views. In the projection planes, a mechanism for noisy points rejection has also been elaborated and evaluated. Results show that the approach performs considerably well with respect to the state the art in one-class classification.

Keywords: Convex Hull, Random Projections, One-Class Classification.

1 Introduction

Many problems in Pattern Recognition can be solved efficiently by exploiting the geometrical structure of the problem itself, providing an intuitive understanding of the solution. In that framework, the convex hull geometric structure has always been considered a powerful tool in geometrical pattern recognition ([1],[2]). The convex hull of a set of points is defined as the smallest polytope containing the full set of points. In recent researches, Bennet et al.([3],[4]) show that there exists a geometrical interpretation of the Support Vector Machine (SVM) related to the convex hull. Finding the maximum margin between two classes is equivalent to find the nearest neighbors in the convex hull of each class when classes do not overlap. This intuitive explanation provides an immediate visualization of the main concepts of SVM from a geometrical point of view. Nevertheless, using the convex hull in real applications is limited by the fact that its computation in a high dimensional space has an extremely high cost. Many approximations circumventing this complexity have been proposed in the context of SVM related algorithms. For instance, Takahashi and Kudo [5] use the convex hull as a maximum margin classifier. They approximate the facets of the convex hull where the support planes are present by a set of reflexive support functions separating one class from the other ones. Pal and Bhattacharya [6] propose a self-evolving two-layer neural network model for computing the approximate convex hull of a

C. Sansone, J. Kittler, and F. Roli (Eds.): MCS 2011, LNCS 6713, pp. 106–115, 2011.

set of points in 3-D and spheres. The vertices of the convex hull are mapped with the neurons of the top layer. Mavroforakis and Theodoridis [7] introduce the notion for Reduced Convex Hull based on the geometric interpretation of SVM, formalizing the case when classes are not separable. Recently, convex hulls have been also used for enhancing the performance of classifiers. Zhou et al. [8], use a convex hull for reducing the number of training examples. A SVM trained on the boundary samples of each class convex hull maintains the same classification performance as if it has been trained using the entire training set.

On the other side, one-class classification problems arise where only data of a target class are available without counterexamples [9]. Effective one-class classification strategies are based on density estimation methods and boundary methods. Gaussian Model, Mixture of Gaussian Model and Parzen Density Estimation are density estimation methods widely used. Density estimation methods work well when there exists *a-priori* knowledge of the problem or a big load of data is available. Boundary methods only intend to model the boundary of the problem disregarding the underlying distribution. Well known approaches to boundary methods are k-Centers and Nearest Neighbors Method. The state of the art in one-class classification is represented by Support Vectors Data Description where the minimum hypersphere containing all the data is computed in a multidimensional space. For a complete review of one-class classification methods, refer to [10].

In this work, a parametrized family of functions based on the convex hull geometric structure is proposed for modeling one-class classification problems, following the approach of boundary methods. The computational limitation derived from high dimensional problems is bypassed by approximating the multidimensional convex hull by an ensemble of bi-dimensional convex hulls obtained by projecting data onto random planes. In those planes, computing the convex hull and establishing whether points belong to the geometric structure are both well known problems having very efficient solutions [11]. Additionally, a robust alternative is studied. In that approach, robustness to outliers[1] is provided computing the final bi-dimensional convex hull using a family of convex-hulls created using a subsampling strategy.

Random Projections are based on the idea that high dimensional data can be projected into a lower dimensional space without significantly losing the structure of the data [12]. Random projections and high dimensional geometry lie at the heart of several important approximation algorithms [13]. Blum [14] reports that projecting data down to a random subspace and solving the learning problem on that reduced space yield comparable results to solving the problem in the original space. Rahimi and Recht [15] use also random projections for building a weighted sum of linear separators. In their work, the authors show that under certain conditions using random projections is equivalent to use the kernel trick. At the same time, random projections provide a faster decaying of the testing error rate with respect to standard AdaBoost.

---

[1] In the present work, **no-target points** are defined as points that do not belong to the one-class classification target class. **Outliers points** are points that are numerically distant from the rest of the data in a class.

This paper is organized as follows. In Section 2, the proposed one-class classification method based on approximate convex hull is described in detail. In Section 3, validation process and parameters setting are described and experimental results reported. Finally, in Section 4, we discuss results and conclude.

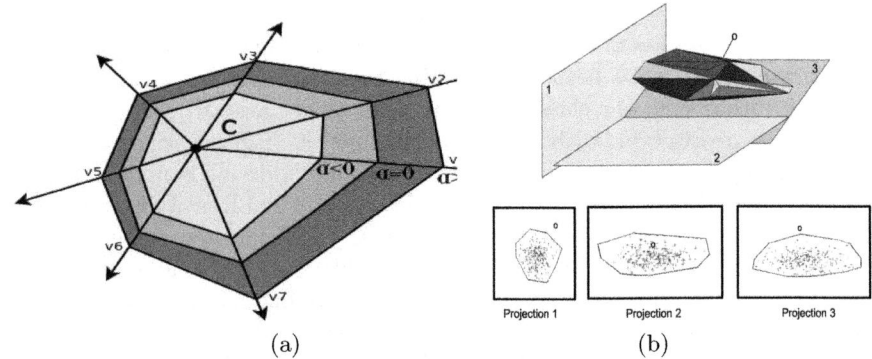

(a)   (b)

Fig. 1. (a) Expansion of the Approximate Convex Hull (ACH): starting from the center C of the original convex hull, vertices can be expanded using the parameter α. The innermost convex hull represents the expansion for a negative value of α –a contraction –, the outermost convex hull represents the expansion for a positive value of α; (b) Projections in ACH : the original tri-dimensional convex hull is approximated by several random projections. A point is outside of the convex hull if there exists at least one projection in which the point is outside of the projected convex hull.

2   Approximate Convex Hull Functions Family for One-Class Classification

One-class classification methods can be solved by modeling the boundary of the set of points defining the problem. The Convex Hull (CH) of a set of points is the minimal convex set containing all the points. The convex hull provides a good tool for delimiting the target class. In this setting, the one-class classification task is reduced to the problem of knowing if test data lie inside or outside the hull. This basic notion of convex hull is used for defining the parametric family as follows. Let $S = \{x_i\} \in \Re^d$ be the dataset representing the class to be approximated, and $V \subseteq S$ a set of k vertices defining a convex hull.

Definition 1. *Given $v_i \in CH(S), i \in \{1, .., k\}$, the parametric **family of convex hulls** depending on a scalar α and a center $C \in \Re^d$ is given as **a set of convex hulls** with vertices defined by $v_i^\alpha = v_i + \alpha(v_i - C)$.*

The parameter α allows to enlarge or contract the hull with respect to the center C. In the context of classification, α allows to adjust the operating point in the Receiver Operating Characteristic curve. It is a constant value for the multidimensional convex hull. In Figure 1(a), examples of positive and negative expansions of factor α are shown.

The creation of the convex hull is computationally intensive. In general, the computational cost of a d-dimensional convex hull over N data examples is $\mathcal{O}(N^{\lfloor d/2 \rfloor + 1})$. This cost is prohibitive in time and memory and, for the classification task, only checking if a point lies inside the convex hull is needed. Our approximation strategy consists in approximating the d-dimensional convex hull with a set of t bi-dimensional projections of the data. Convex hulls are computed on those planes and testing points are checked to belong to every projected bi-dimensional convex hull. If a projection exists where the testing point is found outside of the bi-dimensional convex hull, then the point does not lie inside the original d-dimensional convex hull. Figure 1(b) shows an example of a 3D convex hull with a test point outside the hull and three candidate projection planes randomly chosen. At the bottom of the figure it may be observed that in two of those projections the projected test point is outside the projected convex hull. Table 1 describes the pseudo-algorithm for creating the Approximate Convex Hull (ACH) and testing if a point lies inside the hull. For checking if a point lies inside a CH, the ray casting algorithm or many other methods can be used [11].

Table 1. Approximate Convex Hull Family

– **Create ACH(α,C):**
  Given a training set $S \in \mathcal{R}^d$, where d is the number of features
  Given a number of Projections t
      1.**For** $j = 1..t$
      2. $P^j = randmatrix()$ % Create a Normal Random Matrix
      3. $S_j = (P^j)^T S$ % Project data onto the Random Space
      4. $\{v_i\} = ConvexHull(S_j)$ % Return vertices V_i^j
      5. $\{V_i^j\} = \{v_i + \alpha(v_i - C)\}, \quad i \in \{1, \ldots, K\}$
  Return: $\{V, P\}$

– **Test:**
  Given a point $x \in \mathcal{R}^d$;
  Given a set of t Convex Hulls and Random Matrices $\{V, P\}$
      1.**For** $j = 1..t$
      2. $x_p^j = (P^j)^T x$
      3.   **If** $x_p^j \notin \{V_i^j\}$
      4.   **Return:** OUTSIDE
  Return: INSIDE

This approach has great advantages from both computational and memory storage point of view. On one hand, given a training set of N examples, the computational cost of building the convex hull in a bi-dimensional space is $\mathcal{O}(N \log N)$ [11]. Let K be the number of points defining the convex hull. The memory needed for storing the convex hull is $K << N$. The cost of testing if a point lies inside or outside the convex hull is $\mathcal{O}(K)$. Thus, if we use t projections the final computational cost for building the Approximate Convex Hull Family is $\mathcal{O}(tN \log N)$ and the test cost $\mathcal{O}(tK)$.

Table 2. Approximate Robust Hull Family

- **Create ARH(α,C)**:
  Given a training set $S \in \mathcal{R}^d$, where d is the number of features
  Given a number of Projections t, a number of sampling rounds N, $S_N = \varnothing$

      1. **For** $j = 1..t$
      2. $P^j =$ randmatrix() % Create a Normal Random Matrix
      3. $S_j = (P^j)^T S$ % Project data onto the Random Space
      4. **For** $k = 1..N$
      5. $S_k^j =$ Sample(S_j) % Get a sampling from X_p
      6. $S_N = S_N \cup S_k^j$
      4. $\{v_i\} = ConvexHull(S_N)$ % Return vertices v_i
      5. $\{V_i^j\} = \{v_i + \alpha(v_i - C)\}, \quad i \in \{1, \dots, K\}$
  Return: $\{V, P\}$

2.1 Approximate Robust Hull Family for One-Class Classification

A question that needs to be addressed when using a convex hull is its robustness with respect to outliers. Convex hulls are very sensitive to outliers. The presence of outliers can heavily influence the definition of the boundary and, consequently, the classification process. In order to reduce the influence of outliers, a robust convex hull built on repeated samplings of the original dataset is proposed. Similar to bagging [16], this sampling improves classification models in terms of stability and classification accuracy. It reduces variance and helps to avoid overfitting.

The modification resides in the training step, where data are sampled sequentially for N rounds of sampling. The final convex hull is built on the points given by the union of all the sampling rounds. The sampling strategy automatically reduces the influence of the elements on the boundary, allowing to define a core set of examples of the training set. The algorithm for the Approximate Robust Hull(ARH) is shown in Table 2. There are no changes in the testing part.

2.2 On the number of Projections in ACH/ARH

The number of projections is a critical parameter for both methods. In order to get an empirical estimation of the number of projections needed for the convex hull to be effective, the following experiment has been performed. A set of 100.000 random points are generated on a d-dimensional hypersphere surface of radius R. A test point outside the hypersphere is set at relative distance d/R. The experiment evaluates the number of projections needed for finding one in which the point is found outside of the projected convex hull, when varying the relative distance of the test point with respect to the sphere. The experiment has been repeated 100 times for different values of the dimensionality. The outcomes of the experiments are reported in Figure 2 for (a) dimension 5, (b) dimension 10, (c) dimension 50 and (d) dimension 100. In the figures, we see that as the relative distance of the test point increases, the number of random projections needed for

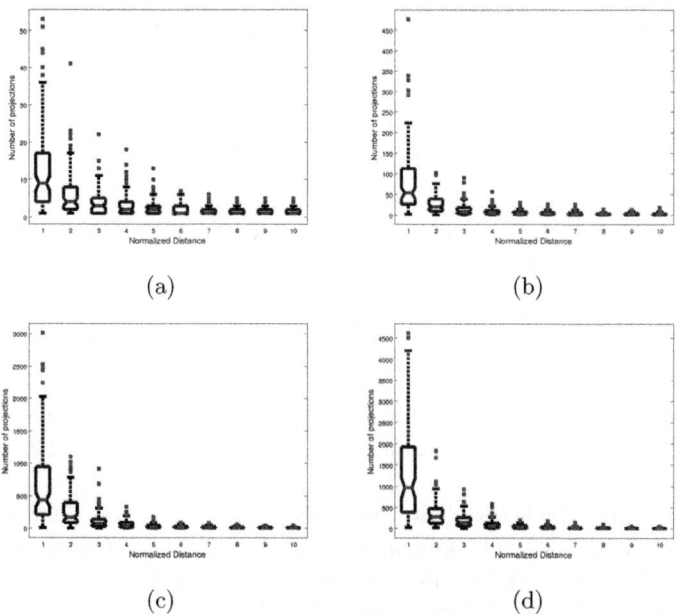

Fig. 2. Whisker plots for the distribution of the number of projections needed to find the test point outside of a hypersphere with dimension (a) 5, (b) 10, (c) 50, (d) 100

finding it outside of the projected hull decreases. Moreover, as the dimensionality increases, the number of random projections also increases. Observe that very close points require a large number of projections. This effect could increase the computational burden of our approach. However, there is a synergistic effect in the proposal that mitigates this initial drawback. Using a negative value for the parameter α, one can arbitrarily reduce the size of the convex hull. Observe that a small decrease in α increases the relative distance in $\frac{1+|\alpha|}{1-|\alpha|}$ and thus the number of projections needed can be drastically reduced without hindering the overall performance. This preliminary empirical results jointly with the exposed effect require a further study from theoretical point of view.

3   Validation and Results

In this section, ACH and ARH have been compared with different one-class models on several dataset from UCI machine learning repositories [17]. Datasets contain two or more classification problems, depending on the number of classes. One-class problems have been obtained using only one class as target class. Data from all the other classes have been considered no-target class and have not been used in the training step. For each target class, training and testing have been separated using a 2-folds split. The process is repeated 10 times and results averaged. The list of datasets used is shown in Table 3. In the table, classes used for one-class problems are listed with their cardinality. An identification number is also reported to reference datasets along the section. All the experiments have been performed using PR-Tools Toolbox [18] and its extension for one-class classification DD-Tools.

Table 3. List of One Class Problems

Id	Dataset	Targets	Elements	Id.	Dataset	Targets	Elements
1,2	Tic-Tac-Toe	1,2	626,332	17,18	Ionosphere	1,2	126,225
3,4	Sonar	1,2	97,111	19,20,21	Wine	1,2,3	59,71,48
5,6	Monks-1	1,2	272,284	22,23,24	Glass	1,2,3	70,76,68
7,8	Monks-2	1,2	300,301	25,26,27	Iris	1,2,3	50,50,50
9,10	Monks-3	1,2	275, 279	28,29,30	Ecoli	2,7,8	53,77,143
11,12	Haberman	1,2	225,81	31,32,33	Yeast	3,5,6	463,429,44
13,14	Pima	1,2	500,268	34,35,36	Yeast	7,8,10	163,51,244
15,16	Breast	1,2	238,445	37,..,47	Vowel	1,..,11	90 for each class

Comparison Methods. The proposed method has been compared with Gaussian model, Mixture of Gaussians (MoG), Parzen Density Estimation (PDE), k-Centers(kC), k-Nearest Neighbor (k-NN), k-Means(kM), Support Vector Data Descriptor (SVDD) and Minimum Spanning Trees (MST).

Evaluation Metrics. Classification methods can be evaluated on the base of ROC curve, representing the percentage of true positive accepted versus the false positive accepted and, in particular, using the Area Under ROC Curve (AUC). For evaluating the performance of ACH and ARH, the ROC curve has been computed using the expansion parameter α.

Parameters Setting. In the comparison methods, for Gaussian estimation, MoG, k-Centers, k-Means and SVDD, the optimal parameters have been chosen using 2-folds cross validation on the training set, varying the parameter between 1 and 10. For k-NN, the value of k has been found using its own optimization routine. For PDE and MST, no optimization parameters are needed. For ACH, the number of random projections is a parameter of the method. In the experiments, the number of random projections has been set to 75 and the number of subsamplings in the robust counterpart to 7. The sampling factor has been arbitrarily set to 50%. It is important to note that the method is very stable and small changes in these parameters slightly change the results.

Results on UCI Datasets.[2] Results obtained are shown in Figure 3. The numbers on the X-axis represents the Id as shown in Table 3. Whiter color represents higher AUC. Last two rows represent ACH and ARH performance. From the figure, it seems that there exists a group of problems where the performance of the proposed methods clearly improves the AUC with respect to the other methods. In Table 4, the counts of win, ties and losses are reported for each method. Both ACH and ARH are significantly better than the compared methods with probability $p > 0.02$, according to two-tailed signed test.

---

[2] Numerical results with mean value and standard deviation are available at http://www.maia.ub.es/~oriol/Personal/downloads.html

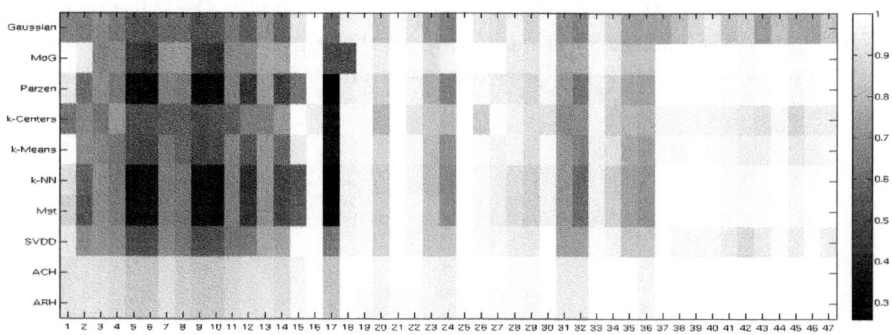

Fig. 3. AUC obtained with One-Class Classification Methods on UCI datasets. Whiter color represents bigger area. The last two lines represent the ACH and ARH methods.

Table 4. Counts of Wins,Ties,Losses for the Compared Methods

	Gauss	Mog	PDE	kC	kM	kNN	MST	SVDD	ACH	ARH
Gauss	–	9/2/36	14/2/31	19/0/28	8/3/36	16/1/30	16/1/30	6/1/40	1/1/45	2/0/45
Mog	36/2/9	–	44/1/2	39/0/8	37/1/9	44/1/2	44/1/2	27/2/18	8/4/35	13/2/32
PDE	31/2/14	2/1/44	–	26/0/21	18/1/28	20/14/13	19/14/14	13/1/33	3/0/44	3/1/43
kC	28/0/19	8/0/39	21/0/26	–	18/0/29	21/0/26	20/0/27	11/0/36	3/0/44	3/0/44
KM	36/3/8	9/1/37	28/1/18	29/0/18	–	29/1/17	29/1/17	16/2/29	3/1/43	4/0/43
kNN	30/1/16	2/1/44	13/14/20	26/0/21	17/1/29	–	8/23/16	12/1/34	3/0/44	3/1/43
MST	30/1/16	2/1/44	14/14/19	27/0/20	17/1/29	16/23/8	–	13/1/33	2/0/45	2/0/45
SVDD	40/1/6	18/2/27	33/1/13	36/0/11	29/2/16	34/1/12	33/1/13	–	5/0/42	5/0/42
ACH	45/1/1	35/4/8	44/0/3	44/0/3	43/1/3	44/0/3	45/0/2	42/0/5	–	36/9/2
ARH	45/0/2	32/2/13	43/1/3	44/0/3	43/0/4	43/1/3	45/0/2	42/0/5	2/9/36	–

Discussion. Results obtained on real problems show that there exists a significant difference between the performance of the proposed methods and the other ones. MoG and SVDD provide also very good performance with respect to the majority of the methods, except with ACH and ARH. In particular, results show that ACH performs much better than ARH. This fact could be surprising at first glance and it is possibly due to the characteristics of the datasets used. In order to show the effectiveness of ARH with respect to ACH, both methods have been compared on a specific toy problem. Both ACH and ARH have been trained on a bi-dimensional Gaussian distribution having two significant outliers. The number of projections has been set to 10 and the number of samplings to 7. No-target points have been created generating points outside a sphere of radius 3.5 built around the Gaussian distribution. An example of training set is shown in Figure 4(a). Both ACH and ARH have been tested on a separate testing set, generated in the same manner, without the two outliers. The experiment has been performed ten times and results averaged. ROC curves obtained are shown in Figure 4(b). For ACH, the mean AUC obtained is 0.961. For ARH, the mean AUC obtained is 0.993. This situation proves that the robust procedure adopted in the creation of ARH is really useful in the presence of significant outliers.

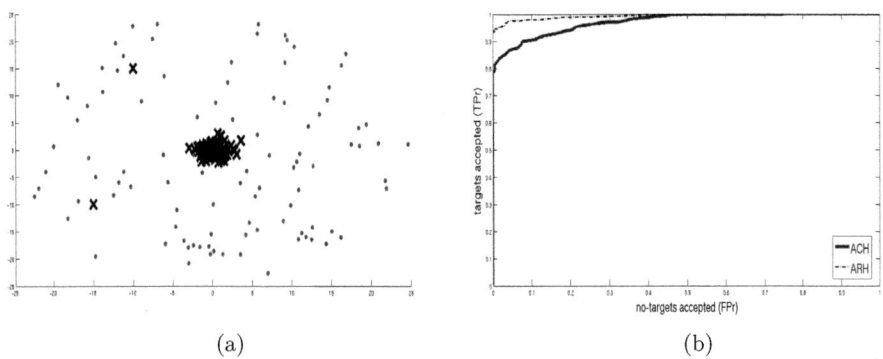

(a)   (b)

Fig. 4. (a) Artificial Gaussian dataset with two outliers. (b) Roc Curves for ACH and ARH.

4   Conclusions

In this work, a new method for one-class classification, the Approximated Convex Hull Family is proposed. The method is based on a family of parametrized convex hulls able to fit the target geometric shape by expanding the original convex hull created on the training set. The limitation due to the computation of high dimensional convex hulls is overcome using random projections. Indeed, the multidimensional convex hull is projected down to bi-dimensional planes, randomly selected. In those planes, creating a convex hull and testing if a point lies inside the hull are well known problems in computational geometry. Experiments show that the number of projections needed for ensuring if a point lies inside or outside the convex hull depends on the dimensionality of the problem and the distance of the point from the boundary. A slight modification of the method, the Approximated Robust Hull Family, allows to weaken the presence of outliers in the target distribution. The modified method uses a robust convex hull, built with a bagging-like strategy.

Both methods perform significantly better than most one-class classification methods on the wide majority of the problems considered. In addition, a toy problem proves the effectiveness of the Approximated Robust Hull Family in the presence of outliers.

Though both proposed methods perform well on the wide majority of the problems taken into account, future research lines are directed to extending the methods for non convex shapes, providing an even more general framework. Finally, the low computational and memory storage requirements allow both proposed methods to be used in devices having limited resources. Future research lines are also directed to exploit these potential capabilities with an implementation of the proposed methods in embedded systems and mobile phones with application to activity recognition and user verification.

Acknowledgments. This work is partially supported by a research grant from projects TIN2009-14404-C02, La Marato de TV3 082131 and CONSOLIDER (CSD2007-00018).

References

1. Bhattacharya, B.K.: Application of computational geometry to pattern recognition problems. PhD thesis (1982)
2. Toussaint, G.T.: The convex hull as a tool in pattern recognition. In: AFOSR Workshop in Communication Theory and Applications (1978)
3. Bennett, K.P., Bredensteiner, E.J.: Duality and geometry in svm classifiers. In: ICML, pp. 57–64 (2000)
4. Bi, J., Bennett, K.P.: Duality, geometry, and support vector regression. In: NIPS, pp. 593–600 (2002)
5. Takahashi, T., Kudo, M.: Margin preserved approximate convex hulls for classification. In: ICPR (2010)
6. Pal, S., Bhattacharya, S.: IEEE Transactions on Neural Networks 18(2), 600–605 (2007)
7. Mavroforakis, M.E., Theodoridis, S.: A geometric approach to support vector machine (svm) classification. Neural Networks, IEEE Transactions on 17, 671–682 (2006)
8. Zhou, X., Jiang, W., Tian, Y., Shi, Y.: Kernel subclass convex hull sample selection method for SVM on face recognition, vol.73 (2010)
9. Japkowicz, N.: Concept-learning in the absence of counter-examples: An autoassociation-based approach to classification. IJCAI, 518–523 (1995)
10. Tax, D.M.J.: One-class classification. PhD thesis (2001)
11. Preparata, F.P., Shamos, M.I.: Computational geometry: an introduction. Springer, New York.Inc (1985)
12. Johnson, W., Lindenstauss, J.: Extensions of lipschitz maps into a hilbert space. Contemporary Mathematics (1984)
13. Vempala, S.: The Random Projection Method. AMS (2004)
14. Blum, A.: Random projection, margins, kernels, and feature-selection. LNCS (2005)
15. Rahimi, A., Recht, B.: Weighted sums of random kitchen sinks: Replacing minimization with randomization in learning. In: NIPS, pp. 1313–1320 (2008)
16. Breiman, L.: Bagging predictors. Machine Learning 24, 123–140 (1996)
17. Frank, A., Asuncion, A.: UCI machine learning repository (2010)
18. Juszczak, P., Paclik, P., Pekalska, E., de Ridder, D., Tax, D., Verzakov, S., Duin, R.: A Matlab Toolbox for Pattern Recognition, PRTools4.1 (2007)

Generalized Augmentation of Multiple Kernels *

Wan-Jui Lee, Robert P.W. Duin, and Marco Loog

Pattern Recognition Laboratory,
Delft University of Technology, The Netherlands

Abstract. Kernel combination is meant to improve the performance of single kernels and avoid the difficulty of kernel selection. The most common way of combining kernels is to compute their weighted sum. Usually, the kernels are assumed to exist in independent empirical feature spaces and therefore were combined without considering their relationships.

To take these relationships into consideration in kernel combination, we propose the generalized augmentation kernel which is extended by all the single kernels considering their correlations. The generalized augmentation kernel, unlike the weighted sum kernel, does not need to find out the weight of each kernel, and also would not suffer from information loss due to the average of kernels.

In the experiments, we observe that the generalized augmentation kernel usually can achieve better performances than other combination methods that do not consider relationship between kernels.

1 Introduction

The selection of kernel functions, the model, and the parameters, is one of the most difficult problem of designing a kernel machine. Recently, an interesting development seeks to construct a good kernel from a series of kernels. Most approaches in the literature aim to derive a weighted sum kernel, and the main concern is to find out the weight of each kernel [2,4,6,8,9,10,13,18,19]. In [10], semi-definite programming (SDP) was used to optimize over the coefficients in a linear combination of different kernels with respect to a cost function. The optimization worked in a transductive setting, and therefore all the information of training and testing patterns was required in the process. To prevent over-fitting, the search space of possible combined kernel matrices is constrained by bounding the kernel matrices with a fixed trace. If kernel target alignment is used as the cost function, this method could be seen as a generalization of the kernel matrix learning method in [7]. Instead of learning the combined kernel matrix in a transductive setting, hyperkernel methods [18] directly learned the combined kernel function in the inductive setting by minimizing a regularized risk functional. In these optimization methods, not only the cost function and constraints were considered, much effort was also spent to speed up the optimization procedure. These methods mainly focus on finding the best weight of each kernel, and then perform the weighted sum of these kernels in order to derive a combined kernel. In these settings, local information is easily averaged out, and therefore these methods might suffer from information loss and the abilities of single

---

* We acknowledge financial support from the FET programme within the EU FP7, under the SIMBAD project (contract 213250).

C. Sansone, J. Kittler, and F. Roli (Eds.): MCS 2011, LNCS 6713, pp. 116–125, 2011.

kernels also become weaker. For example, if the dataset has varying local distributions, different kernels will be good for different areas. Averaging the kernels of such a dataset would lose some capability to describe these local distributions.

To unfold the local characteristics of data, [11,12,21] all proposed to augment s kernels of size $m \times m$ into a kernel of size $(s \times m) \times (s \times m)$. These methods all put the original kernel matrices on the diagonal. The main difference is on the off-diagonal. The methods proposed in [12,21], which we name as the augmented kernel in short, only put zeros on the off-diagonal due to lacking of knowledge about the cross terms. In this case, it implies that different kernels live in the different subspaces and there is no interaction between these subspaces. Also, the empirical feature functions of different kernels are unrelated. This is a rather constrained assumption, and does not provide much flexibility. These cross terms are however, defined in [11] as the inner product of square roots of kernel functions to meet the mercer condition [3,7,10,20]. Unfortunately, one can only derive the square root of a kernel function if the function is self-similar. Therefore, the only adoptable kernel for computing the composite kernel proposed in [11] is the Radial Basis function (RBF). Moreover, the cross terms of composite kernels are very similar to the product of two RBF kernels and therefore usually results in very small values. In most cases, it is very much the same as putting zeros on the off-diagonal.

In this work, we propose a method to augment single kernels into a generalized augmentation kernel by considering the cross terms on the off-diagonal and these cross terms can be derived from any type of kernel. It duplicates empirical feature functions on the off-diagonal with scaling parameters which indicate how related the empirical feature functions from two different kernels should be. This scaling parameter also allows us to smoothly vary between the original augmented kernel and the direct sum kernel, and therefore we can have a generalized description of multiple kernels and enlarge the searching space of the optimal solution. The experimental results suggest that the generalized augmentation kernel usually can find a better solution than the sum kernel, the composite kernel and the augmented kernel.

The rest of the paper is organized as follows. In Section 2, the overview of support vector machine is recaped. The direct sum of kernels and the augmented kernel and their empirical feature spaces are described in Section 3, respectively. Our proposal for constructing the generalized augmentation kernel from single kernels is given in Section 4. Simulation results are presented in Section 5. Finally, a conclusion is given in Section 6.

2 Overview of Support Vector Machine

For convenience, we introduce the support vector classifier with d input variables x_{i1}, $x_{i2}, \ldots, x_{id}$ for 2-class problem with class labels $+1$ and -1 in this section. $\mathbf{x}_i$ and y_i represent i^{th} input datum (a vector) and its corresponding class label [20,3]. Extension to multi-class problems can be achieved by training multiple support vector machines.

To control both training error and model complexity, the optimization problem of SVM is formalized as follows:

$$\text{minimize} \ \frac{1}{2} < \mathbf{w}, \mathbf{w} > + C \sum_{i=1}^{n} \xi_i,$$

$$\text{subject to} \ < \mathbf{w} \cdot \mathbf{x}_i > + b \geq +1 - \xi_i, \ \text{for} \ y_i = +1$$

$$< \mathbf{w} \cdot \mathbf{x}_i > + b \leq -1 + \xi_i, \ \text{for} \ y_i = -1$$

$$\xi_i \geq 0, \forall i. \tag{1}$$

By using Lagrange multiplier techniques, Eq.(1) could lead to the following dual optimization problem:

$$\text{maximize} \ \sum_{i=1}^{n} \alpha_i - \sum_{i=1}^{n} \sum_{j=1}^{n} \alpha_i \alpha_j y_i y_j < \mathbf{x}_i, \mathbf{x}_j >,$$

$$\text{subject to} \ \sum_{i=1}^{n} \alpha_i y_i = 0, \alpha_i \in [0, C]. \tag{2}$$

Using Lagrange multipliers, the optimal desired weight vector of the discriminant hyperplane is $\mathbf{w} = \sum_{i=1}^{n} \alpha_i y_i \mathbf{x}_i$. Therefore the best discriminant hyperplane can be derived as

$$f(\mathbf{x}) = < \sum_{i=1}^{n} \alpha_i y_i \mathbf{x}_i, \mathbf{x} > + b = (\sum_{i=1}^{n} \alpha_i y_i < \mathbf{x}_i, \mathbf{x} >) + b, \tag{3}$$

where b is the bias of the discriminant hyperplane.

2.1 Empirical Feature Function of Kernels

In Eq.(3), the only way in which the data appears is in the form of dot products, that is $< \mathbf{x}_i, \mathbf{x} >$. The discriminant hyperplane is thereby linear and can only solve a linearly separable classification problem. If the problem is nonlinear, instead of trying to fit a nonlinear model, the problem can be mapped to a new space by a nonlinear transformation using a suitably chosen kernel function. The linear model used in the new space corresponds to a nonlinear model in the original space. To make the above model nonlinear, consider a mapping $\phi(\mathbf{x})$ from the input space into some feature space as

$$\phi : \mathbb{R}^d \rightarrow \mathcal{H}. \tag{4}$$

The training algorithm only depends on the data through dot products in $\mathcal{H}$, i.e. on functions of the form $< \phi(\mathbf{x}_i), \phi(\mathbf{x}_j) >$. Suppose a kernel function K defined by

$$K(\mathbf{x}_i, \mathbf{x}_j) = < \phi(\mathbf{x}_i), \phi(\mathbf{x}_j) >, \tag{5}$$

is used in the training algorithm. Explicit knowledge of ϕ is thereby avoided. The dot product in the feature space can be expressed as a kernel function. Similar to Eq.(3) in linear problems, for a nonlinear problem, we will have the following discriminant function

$$f(\mathbf{x}) = \sum_{i=1}^{n} \alpha_i y_i K(\mathbf{x}_i, \mathbf{x}) + b. \tag{6}$$

3   Sum Kernel and Augmented Kernel

Most kernel combination methods try to average out the kernel matrices in one way or another [10,1,9,8,14,18]. Suppose s original kernels are given as $K_1, K_2, ...,$ and K_s with size $m \times m$ and the empirical feature functions of these kernels are $\phi_1(\mathbf{x}), \phi_2(\mathbf{x})...,$ and $\phi_s(\mathbf{x})$. Combining kernels by summing up all the kernels is equivalent to taking the Cartesian product of their respect empirical feature spaces. And weighing the kernels is the same as scaling the empirical feature spaces. To make use of all the local characteristics of each single kernel, the augmented kernel is proposed in [12,21], which is of the form

$$
K_1 \bigoplus K_2 \bigoplus \cdots \bigoplus K_s = \begin{pmatrix} K_1 & 0 & \cdots & 0 \\ 0 & K_2 & \cdots & 0 \\ \vdots & \vdots & \ddots & \vdots \\ 0 & 0 & \cdots & K_s \end{pmatrix}_{s \times m, s \times m.} \tag{7}
$$

A direct sum kernel is with m coefficients while the augmented kernels is with $s \times m$ coefficients. Therefore, every object in every kernel can contribute during the training for augmented kernel.

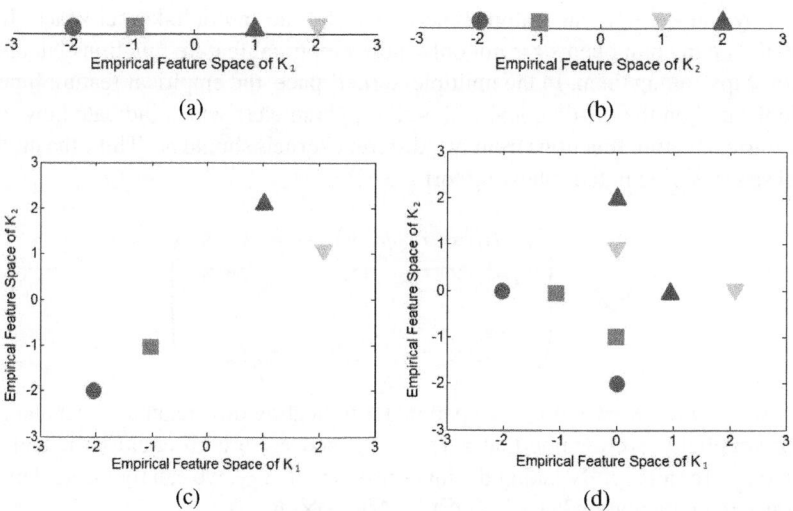

Fig. 1. Geometrical interpretation of (a) the empirical feature space of K_1 (b) the empirical feature space of K_2 (c) the empirical feature space of $K_1 + K_2$ (d) the empirical feature space of $K_1 \bigoplus K_2$

By augmenting the empirical feature functions of kernels into one matrix $X_\bigoplus$ as

$$
X_\bigoplus = \begin{pmatrix} \phi_1(\mathbf{x}) & 0 & \cdots & 0 \\ 0 & \phi_2(\mathbf{x}) & \cdots & 0 \\ \vdots & \vdots & \ddots & \vdots \\ 0 & 0 & \cdots & \phi_s(\mathbf{x}) \end{pmatrix}_{s \times m, s \times m,} \tag{8}
$$

one can show that the augmented kernel obeys the mercer theorem by taking the inner product of $X_\oplus$, that is $X_\oplus^T X_\oplus = K_1 \oplus K_2 \oplus \cdots \oplus K_s$.

Figure 1 is the illustration of the empirical feature spaces of direct sum kernel and the augmented kernel. From Figure 1(c) and Figure 1(d), we can see that instead of 4 objects as in the empirical feature spaces of the individual kernels and the direct sum kernel, there are duplicated objects (in total 8) in the empirical feature space of the augmented kernel. However, the augmented kernel assumes that all the empirical feature functions $\phi_1(\mathbf{x}), \phi_2(\mathbf{x})...,$ and $\phi_s(\mathbf{x})$ are independent and uncorrelated. Therefore objects only live in their subspaces, and that is why we see two clear vertical and horizontal sets of objects in Figure 1(d). All the objects are therefore only allowed to be on the coordinates and the rest of the space is completely empty.

4   Generalized Augmentation Kernel in Multiple-Kernel Spaces

The augmented kernel assumes that the empirical feature spaces of kernels are all independent and uncorrelated. However, this assumption is not necessarily true, especially if kernels of the same type but different shapes are selected to combine. To combine kernels which are not necessarily uncorrelated, we first define the space which is expanded by these empirical feature functions to be the multiple-kernel space. In this space, the main components are not only these empirical feature functions but also the relationships among them. In the multiple-kernel space, the empirical feature functions are duplicated on the off-diagonal with scaling parameters which indicate how related the empirical feature functions from two different kernels should be. Thus, the multiple-kernel space $X_\otimes$ is in the following form:

$$
X_\otimes = \begin{pmatrix} r_{11}\phi_1(\mathbf{x}) & r_{12}\phi_1(\mathbf{x}) & \cdots & r_{1s}\phi_1(\mathbf{x}) \\ r_{21}\phi_2(\mathbf{x}) & r_{22}\phi_2(\mathbf{x}) & \cdots & r_{2s}\phi_2(\mathbf{x}) \\ \vdots & \vdots & \ddots & \vdots \\ r_{s1}\phi_s(\mathbf{x}) & r_{s2}\phi_s(\mathbf{x}) & \cdots & r_{ss}\phi_s(\mathbf{x}) \end{pmatrix} \tag{9}
$$

with size $s \times m, s \times m$ and r_{ij} is a parameter indicating how related $\phi_i(\mathbf{x})$ and $\phi_j(\mathbf{x})$ is. For simplicity, we assume that $r_{11}, r_{22}, ...,$ and r_{ss} are all equal to 1, and r_{ij} is equal to r_{ji}, for all i, j. By taking the inner product of $X_\otimes$, we can therefore define the generalized augmentation kernel $K_1 \otimes K_2 \otimes \cdots \otimes K_s$ as

$$
\begin{pmatrix} r_{11}r_{11}K_1 + \cdots + r_{s1}r_{s1}K_s & r_{12}r_{11}K_1 + \cdots + r_{s2}r_{s1}K_s & \cdots & r_{1s}r_{11}K_1 + \cdots + r_{ss}r_{s1}K_s \\ r_{11}r_{12}K_1 + \cdots + r_{s1}r_{s2}K_s & r_{12}r_{12}K_1 + \cdots + r_{s2}r_{s2}K_s & \cdots & r_{1s}r_{12}K_1 + \cdots + r_{ss}r_{s2}K_s \\ \vdots & \vdots & \ddots & \vdots \\ r_{11}r_{1s}K_1 + \cdots + r_{s1}r_{ss}K_s & r_{12}r_{1s}K_1 + \cdots + r_{s2}r_{ss}K_s & \cdots & r_{1s}r_{1s}K_1 + \cdots + r_{ss}r_{ss}K_s \end{pmatrix}
$$

with size $s \times m, s \times m$. Therefore the size of the generalized augmentation kernel matrix is $(s \times m) \times (s \times m)$ while the sizes of the original kernel matrices are $m \times m$. After the construction of the generalized augmentation kernel matrix, the support vector machine can proceed the learning of support vectors and their corresponding coefficients.

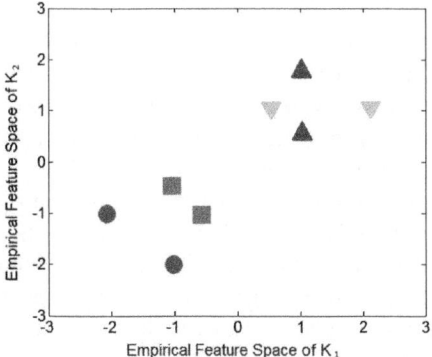

Fig. 2. Geometrical interpretation of the empirical feature space of $K_1 \otimes K_2$ with $r_{11} = 1$, $r_{22} = 1$, $r_{12} = 0.5$, and $r_{21} = 0.5$ given the empirical feature spaces of K_1 and K_2 as in Figure 1(a) and Figure 1(b)

Both training and testing objects have to be replicated as the generalized augmentation kernel matrix is s times larger than the base kernels.

The geometrical interpretation of the generalized augmentation kernel is given in Figure 2. By setting the degree of correlation between the empirical feature functions of two kernels as 0.5, the duplicated objects, unlike in the augmented kernel, can now position in different parts of the space as well. This obviously enlarges the searching space of optimal solutions for support vector machines. Moreover, the parameter r_{ij} allows us to smoothly vary between the original augmented kernel ($r = 0$) and the direct sum kernel ($r = 1$). Note that the generalized augmentation kernel is not necessarily the same as the direct sum kernel when $r = 1$. The generalized augmentation kernel will be composed of duplicated direct sum kernels and therefore it might find a different optimal SVM classifier than the direct sum kernel.

5   Experimental Results

In this section, we compare the experimental results obtained by our generalized augmentation kernel with those of other kernel combination methods and a classifier combination method. The kernel combination methods include the augmented kernel [12,21], the direct sum kernel and the composite kernel [11]. The Fisher learning rule is used to derive the classifier combiner. One synthetic dataset and 7 benchmark datasets [15,16,17] are used in the experiments. To test whether the generalized augmentation kernel is more capable of describing data with different local distributions than the other kernel or classifier combination methods, two of the 8 datasets used in the experiments are with different local distributions, and the other 6 datasets are regular real datasets. The single kernel and the combined kernel SVM classifiers in the experiments are implemented with LIBSVM [5] and the classifier combiners are built with PRTOOLS [17]. In every experiment, two RBF kernels are constructed, and kernel combination and classifier combination methods were used to combine these kernels or classifiers.

5.1   Experiment 1: Data with Varying Local Distributions

Spiral and sonar datasets are used in experiment 1. The SVM parameter C is set to 1 in all experiments to obtain a reasonable number of support vectors. The spiral dataset is a synthetic 2-dimensional dataset with 400 data patterns in 2 classes as shown in Figure 3. The sonar dataset contains information of 208 objects, 60 attributes, and two classes, rock and mine. The attributes represent the energy within a particular frequency band integrated over a certain period of time. In all the experiments, the dataset is randomly split into training and testing datasets with 80% and 20% ratio. For both datasets, two RBF kernels are built and different methods are used to combine these two kernels in each experiment. The results are averaged over 500 experiments. The sigma's of these single RBF kernels are assigned heuristically in the following way. The smallest sigma is the average distance of each data pattern to its nearest neighbor. The largest sigma is the average distance of each data pattern to its furthest neighbor.

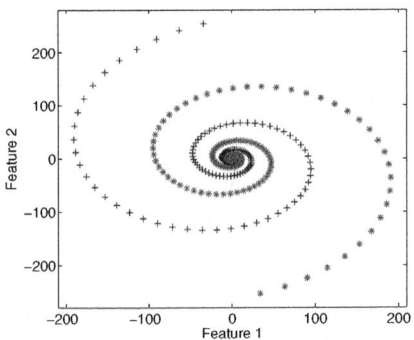

Fig. 3. Distribution of the spiral dataset

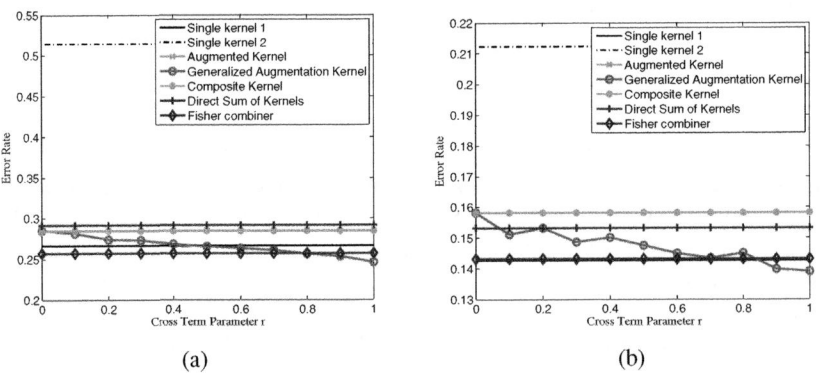

Fig. 4. Experiment results of (a) spiral dataset, and (b) sonar dataset with single kernels, generalized augmentation kernel, augmented kernel, composite kernel and the Fisher combiner

The results for both single kernel classifiers, kernel combination methods and classifier combination methods with spiral dataset and sonar dataset are in Figure 4(a) and Figure 4(b), respectively. For these two datasets, the sum kernel, the augmented kernel and the composite kernel all have very similar performances. The Fisher classifier combiner is much better than kernel combination methods except for the generalized augmentation kernel under the situation that a high correlation factor is assigned during the construction of the generalized augmentation kernel. This suggests that the classifier combiner is more preferable than kernel combination methods when data is with varying local distributions, especially if the empirical feature spaces of the combined kernel is assumed to be uncorrelated.

5.2 Experiment 2: Benchmark Data

Six real world datasets [15,16,17] as shown in Table 1 with the number of features, objects and classes, are used to have a more general investigation in experiment 2. The SVM parameter C is set to 1 in all experiments to obtain a reasonable number of support vectors. In all the experiments, the dataset is randomly split into training and testing datasets with 80% and 20% ratio. For all datasets, two RBF kernels are built with the heuristic mentioned above and different methods are used to combine these two kernels in each experiment, and the results are the averages of 500 repeated experiments. The results of all single kernel classifiers, kernel combination methods and classifier combination methods with all datasets are shown in Table 2. The results of the generalized augmentation kernel with different values of the cross term parameter r are given in Table 3.

Table 1. Benchmark datasets

dataset	biomed	IMOX	auto-mpg	heart	iris	wine
# features	5	8	6	13	4	13
# classes	2	4	2	2	3	3
# objects	194	192	398	297	150	178

Table 2. Results of single kernels, augmented kernel, direct sum kernel, composite kernel and the Fisher combiner

method	datasets					
	biomed	IMOX	auto-mpg	heart	iris	wine
	average error rate of 500 experiments					
single kernel 1	0.1245	0.0788	0.1704	0.3828	0.0692	0.2610
single kernel 2	0.1336	0.0794	0.1497	0.3041	0.0365	0.3147
direct sum kernel	**0.1206**	**0.0587**	0.1418	0.3346	0.0423	0.2803
augmented kernel	0.1362	0.0607	**0.1396**	**0.3258**	**0.0402**	0.2684
composite kernel	0.1362	0.0608	**0.1396**	**0.3259**	**0.0402**	0.2684
fisher combiner	0.1318	0.0893	0.1737	0.4242	0.0750	**0.2622**

Table 3. Results of the generalized augmentation kernel with different values of the cross term parameter r

cross term parameter r	datasets					
	biomed	IMOX	auto-mpg	heart	iris	wine
	average error rate of 500 experiments					
0	0.1362	0.0607	0.1396	**0.3258**	0.0402	0.2684
0.2	0.1263	0.0576	0.1375	0.3370	0.0396	0.2491
0.4	0.1231	0.0524	**0.1365**	0.3388	0.0388	0.2435
0.6	0.1196	0.0458	0.1435	0.3339	0.0381	0.2377
0.8	0.1222	0.0449	0.1456	0.3416	0.0423	0.2390
1	**0.1189**	**0.0443**	0.1513	0.3580	**0.0367**	**0.2316**

From Table 2 and Table 3, we can see that given the right value to the parameter r, the generalized augmentation kernel can perform better than the best kernel combination (in bold) in all the datasets. Nevertheless, these r values seem to be dependant on the dataset and therefore there is a need to choose a good value beforehand. Also in these datasets, the classifier combiner is in general worse than the kernel combination methods.

6   Conclusions

In this study we proposed a method to construct the generalized augmentation kernel with multiple kernels. With the replication of empirical feature functions and adding a scaling factor to influence the correlation between kernels, we give more flexibility in positioning the duplicated objects in the combined empirical feature space.

However, this scaling factor, also called the cross term parameter, seems to be dependant on the dataset, and therefore it is more beneficial if this parameter can be well chosen before the process of learning. So far, we only choose the optimal parameter based on testing results and this is not really a practical solution. Moreover, our experiments only use two kernels so far in order to have a better insight on the cross term parameter. It is necessary, however, to include experiments with more kernels later on.

References

1. Bennett, K.P., Momma, M., Embrechts, M.J.: MARK: A Boosting Algorithm for Heterogeneous Kernel Models. In: Proc. 8th ACMSIGKDD Int. Conf. Knowledge Discovery and Data Mining, pp. 24–31 (2002)
2. Bousquet, O., Herrmann, D.: On the Complexity of Learning the Kernel Matrix. In: Proc. Advances in Neural Information Processing Systems, pp. 415–422 (2003)
3. Burges, C.J.C.: A Tutorial on Support Vector Machines for Pattern Recognition. Knowledge Discovery and Data Mining 2(2), 1–43 (1998)
4. Camps-Valls, G., Gomez-Chova, L., Muñoz-Marí, J., Vila-Francés, J., Calpe-Maravilla, J.: Composite Kernels for Hyperspectral Image Classification. IEEE Geoscience and Remote Sensing Letters 3(1), 93–97 (2006)
5. Chang, C.C., Lin, C.J.: LIBSVM: A Library for Support Vector Machines, Taiwan, National Taiwan University (2001), http://www.csie.ntu.edu.tw/cjlin/libsvm

6. Crammer, K., Keshet, J., Singer, Y.: Kernel Design Using Boosting. In: Proc. of the Fifteenth Annual Conference on Neural Information Processing Systems (2002)
7. Cristianini, N., Kandola, J., Elisseeff, A., Shawe-Taylor, J.: On Kernel Target Alignment. Technical Report NeuroColt, pp. 2001–2099. Royal Holloway University, London (2001)
8. de Diego, I.M., Moguerza, J.M., Mu noz, A.: Combining Kernel Information for Support Vector Classification. In: Proc. Multiple Classifier Systems, pp. 102–111 (2004)
9. Fung, G., Dundar, M., Bi, J., Rao, B.: A Fast Iterative Algorithm for Fisher Discriminant Using Heterogeneous Kernels. In: Proc. 21st Int. Conf. Machine Learning (2004)
10. Lanckriet, G.R.G., Cristianini, N., Bartlett, P., Ghaoui, L.E., Jordan, M.I.: Learning the Kernel Matrix with Semidefinite Programming. Journal of Machine Learning Research 5, 27–72 (2004)
11. Lee, W.-J., Verzakov, S.A., Duin, R.P.W.: Kernel Combination Versus Classifier Combination. In: Proc. Multiple Classifier Systems, pp. 22–31 (2007)
12. Lin, C.T., Yeh, C.M., Liang, S.F., Chung, J.F., Kumar, N.: Support-Vector-Based Fuzzy Neural Network for Pattern Classification. IEEE Trans. on Fuzzy Systems 14(1), 31–41 (2006)
13. Micchelli, C.A., Pontil, M.: Learning the Kernel Function via Regularization. Journal of Machine Learning Research 6, 1099–1125 (2005)
14. Moguerza, J.M., Munoz, A., de Diego, I.M.: Improving Support Vector Classification via the Combination of Multiple Sources of Information. In: SSPR/SPR, pp. 592–600 (2004)
15. Asuncion, A., Newman, D.J.: UCI Machine Learning Repository. University of California, Department of Information and Computer Science, Irvine, CA (2007),
   http://www.ics.uci.edu/mlearn/MLRepository.html
16. Jain, A.K., Ramaswami, M.D.: Classifier design with Parzen window. Pattern Recogition and Artificial Intelligence (1988)
17. Duin, R.P.W., Juszczak, P., Paclik, P., Pękalska, E., de Ridder, D., Tax, D.M.J.: PRTOOLS4, A Matlab Toolbox for Pattern Recognition, Delft University of Technology, Pattern Recognition Laboratory, The Netherlands (2004), http://www.prtools.org
18. Ong, C.S., Smola, A.J., Williamson, R.C.: Learning the Kernel with Hyperkernels. Journal of Machine Learning Research 6, 1043–1071 (2005)
19. Tsang, I.W.H., Kwok, J.T.Y.: Efficient Hyperkernel Learning Using Second-Order Cone Programming. IEEE Trans. on Neural Networks 17(1), 48–58 (2006)
20. Vapnik, V.: The Nature of Statistical Learning Theory. Springer, Heidelberg (1995)
21. Yan, F., Mikolajczyk, K., Kittler, J., Tahir, M.A.: Combining Multiple Kernels by Augmenting the Kernel Matrix. In: Proc. Multiple Classifier Systems, pp. 175–184 (2010)

A Modified Neutral Point Method for Kernel-Based Fusion of Pattern-Recognition Modalities with Incomplete Data Sets

Maxim Panov[1], Alexander Tatarchuk[2], Vadim Mottl[2], and David Windridge[3]

[1] Moscow Institute of Physics and Technology,
Institutsky Per. 9, Dolgoprudny, Moscow Region, 141700, Russia
panov.maxim@gmail.com
[2] Computing Center of the Russian Academy of Sciences,
Vavilov St. 40, Moscow, 119991, Russia
aitech@yandex.ru, vmottl@yandex.ru
[3] Center for Vision, Speech and Signal Processing, University of Surrey,
The Stag Hill, Guildford, GU2 7XH, UK
D.Windridge@surrey.ac.uk

Abstract. It is commonly the case in multi-modal pattern recognition that certain modality-specific object features are missing in the training set. We address here the missing data problem for kernel-based Support Vector Machines, in which each modality is represented by the respective kernel matrix over the set of training objects, such that the omission of a modality for some object manifests itself as a blank in the modality-specific kernel matrix at the relevant position. We propose to fill the blank positions in the collection of training kernel matrices via a variant of the Neutral Point Substitution (NPS) method, where the term "neutral point" stands for the locus of points defined by the "neutral hyperplane" in the hypothetical linear space produced by the respective kernel. The current method crucially differs from the previously developed neutral point approach in that it is capable of treating missing data in the training set on the same basis as missing data in the test set. It is therefore of potentially much wider applicability. We evaluate the method on the Biosecure DS2 data set.

1 Introduction

It is well-established that the classification performance of modality-specific classifiers can be improved by combining several different object-representation modalities within a single pattern-recognition procedure. This fusion may be performed at the early or late stage. In the former case of early fusion [1,2], the growing overall dimensionality of the object representation with increasing the number of modalities can be reduced by incorporating some form of modality-selection within the final classification procedure [3,4], thereby eliminating the danger of over-fitting. Such modality-selectivity is correlated with the generalization performance of the training process, so that, if performed ideally, the

C. Sansone, J. Kittler, and F. Roli (Eds.): MCS 2011, LNCS 6713, pp. 126–136, 2011.

recognition system user is free to include object-representation modalities without constraint.

This freedom creates a new difficulty – the greater the number of modalities employed for comprehensive object representation, the more likely is the omission of some modality-specific feature in the available data.

The problem of missing features has been intensively studied in the pattern recognition literature. However, the aspect of combining diverse pattern-recognition modalities makes special demands on the method of handling blanks in object information.

In [1], three levels of fusing several biometric modalities are compared:

- sensor level, when what is fused are signals acquired immediately from sensors forming different initial object representations;
- classifier score level, that presupposes fusion of scores of multiple classifiers as preliminary decisions made from different modalities to be combined;
- decision level, implying fusion of final decisions made separately by single classifiers on the basis of each modality.

Practically all known methods of compensating for missing data tacitly address the latter two levels of combining modalities [5,6] and boil down to replacing the missing features via some surrogate values or designing a fusion classifier for all possible combinations of observable features.

At the same time, it is noted in [1] that the sensor level of fusing modalities can potentially yield better results if it is possible to find an appropriate algorithm for combining signals of incomparable physical type. One such algorithm is given in [2] under the assumption that a kernel-based methodology is utilized to obtain a recognition rule for each particular modality, for which a discriminant hyperplane is specified in the linear space associated with each modality. In this case, the kernel trick [11,12] transforms the problem of combining diverse modalities with missing data into that of appropriately treating blanks in the modality-specific kernel matrices when fusing them into a unified matrix.

Two types of incomplete data samples are to be distinguished – those in the training set, during the classifier learning stage, and those in the test set, when the classifier is already operational.

For the latter case, it was proposed in [8] to adopt the neutral point substitution (NPS) method originally developed in [7] as a means of kernel-based combining of disjoint multi-modal training data, i.e., when only one feature is known for each object. Important advantages of the NPS method are that it is implicitly incorporated into the SVM training framework and it is free from the necessity of inventing any heuristic surrogates for replacing the missing data. It is shown in [8] that the omission of features of the given object at the testing stage is theoretically equivalent, in the case of completely disjoint data sets, to the sum-rule fusion of classifiers within the available modalities [9].

However, the NPS method of treating missing object representations does not lend itself, in its original version, to immediate extension to the training stage (except in the degenerate case of completely disjoint data). The purpose of this paper is to fill in this gap while retaining the advantages of a strictly mathematical approach to the missing-data problem for the case of training sets with a more typical density of blanks.

As the data source for experiments, we use the publicly available biometric database Biosecure DS2 [10].

2   Inferring a Modality-Specific Kernel-Based Classifier from an Unbalanced Training Set

2.1   Modality-Specific Kernel Functions

Let each real world object $\omega \in \Omega$ be represented by several characteristics (features) measured by respective sensors in sensor-specific scales $x_i(\omega)\colon \Omega \to \mathbb{X}_i$, $i \in I$, where $I = \{1, \ldots, n\}$ is the set of sensors. It is typical in the practice of data analysis that the signals of the initial sensors are of different physical natures and hardly lend themselves to joint treatment. We keep in this paper to the kernel-based approach to combining arbitrary object-representation modalities under the basic assumption that a modality-specific kernel function $K_i(x_i', x_i'')$ is defined in the output scale of each particular sensor [2].

A kernel is a symmetric two-argument function $K_i(x_i', x_i'')\colon \mathbb{X}_i \times \mathbb{X}_i \to \mathbb{R}$, which forms a positive semidefinite matrix $\left[K_i(x_i(\omega_j), x_i(\omega_l)); j, l = 1, ..., m \right]$ for each finite collection of objects $\{\omega_j, j = 1, ..., m\}$ [11]. Any kernel $K_i(x_i', x_i'')$ embeds the scale of the respective sensor $\mathbb{X}_i$ into a hypothetical linear space $\mathbb{X}_i \subseteq \tilde{\mathbb{X}}_i$, in which the null element and linear operations are defined in a particular way [12]:

$$\phi_i \in \tilde{\mathbb{X}}_i, \quad x_i' + x_i'' \colon \tilde{\mathbb{X}}_i \times \tilde{\mathbb{X}}_i \to \tilde{\mathbb{X}}_i, \quad c x_i \colon \mathbb{R} \times \tilde{\mathbb{X}}_i \to \tilde{\mathbb{X}}_i.$$

The role of inner product is played by the symmetric kernel function itself, which is inevitably bilinear $K_i(\alpha' x_i' + \alpha'' x_i'', x_i) = \alpha' K_i(x_i', x_i) + \alpha'' K_i(x_i'', x_i)$.

The major convenience factor of the kernel-based approach to data analysis is its ability to provide the constructor of a data-analysis system with the possibility of working with objects of arbitrary nature in unified terms of linear functions $f(\omega) = f(x_i(\omega))\colon \Omega \to \mathbb{X}_i \to \mathbb{Y}$, where $\mathbb{Y}$ is any desired linear space. More strictly, the carrier of kernel-specific linear functions is not the feature scale $\mathbb{X}_i$ itself, but rather its linear closure $\mathbb{X}_i \subseteq \tilde{\mathbb{X}}_i \to \mathbb{Y}$.

However, it should be kept in mind that $\tilde{\mathbb{X}}_i$ is thus a hypothetical linear space deriving from the kernel trick, in contrast to its observable subset $\mathbb{X}_i \subseteq \tilde{\mathbb{X}}_i$ which is the output scale of a particular sensor associated with its respective feature $x_i(\omega) \in \mathbb{X}_i$ relating to the set of real-world objects $\omega \in \Omega$.

In particular, to determine a scalar linear function $f_i(x) \colon \tilde{\mathbb{X}}_i \to \mathbb{R}$, it is enough to specify a direction element (vector, in linear-space terms) $a_i \in \tilde{\mathbb{X}}_i$ and a numerical threshold $b_i \in \mathbb{R}$, then the function will be expressed by the formula $f_i(x|a_i, b_i) = K_i(a_i, x) + b_i$. The equation $f_i(x|a_i, b_i) = K_i(a_i, x) + b_i = 0$ defines a hyperplane which dichotomizes the hypothetical linear space $\tilde{\mathbb{X}}_i$ and, as a consequence, the feature scale $\mathbb{X}_i \subseteq \tilde{\mathbb{X}}_i$ along with the original set of objects:

$$f_i\big(x_i(\omega)|a_i, b_i\big) = K_i\big(a_i, x_i(\omega)\big) + b_i \gtrless 0. \tag{1}$$

The inequality (1) plays the role of a modality-specific kernel-based linear two-class classifier in the set of real-world objects of arbitrary kind.

Before discussing methods of combining diverse modalities of objects represented in a training set with missing measurements, we consider in the next Section the structure of a modality-specific classifier, and introduce the notion of neutral points in the linear closure of the feature scale $\tilde{\mathbb{X}}_i \supseteq \mathbb{X}_i$. This notion will be the main mathematical instrument for filling blanks in the training set.

2.2   A Single Modality-Specific Kernel-Based Classifier Inferred from an Incomplete Training Set

Let $\Omega^* = \{(\omega_j, y_j), j = 1, ..., N\}$ be the training set of real-world objects allocated by the trainer between two classes $y_j = y(\omega_j) = \pm 1$. In the case of training incompleteness, the partial set of training information for the subset of objects $\Omega_i^* \subset \Omega^*$, at which the ith modality $i \in I$ is acquired $x_{ij} = x_i(\omega_j) \in \mathbb{X}_i$, will consist of the matrix of available kernel values and class-indices:

$$\Omega_i^* \Rightarrow \{K_i(x_{ij}, x_{il}), y_j; \omega_j, \omega_l \in \Omega_i^*\}. \tag{2}$$

Perhaps the most widely adopted technique for finding a discriminant hyperplane (1) for a given training set of classified objects represented by a single kernel is the Support Vector Machine (SVM) [11]. The idea underlying the classical SVM for linearly separable training sets is that of finding the discriminant hyperplane which provides the maximum margin between the closest training points of both classes:

$$\begin{cases} K_i(a_i, x_{ij}) + b_i \geqslant \varepsilon_i, \ y_j = 1, \\ K_i(a_i, x_{ij}) + b_i \leqslant -\varepsilon_i, \ y_j = -1, \end{cases} \omega_j \in \Omega_i^*, \ 2\varepsilon_i \to \max_{K_i(a_i, a_i) = 1} (a_i \in \tilde{\mathbb{X}}_i, b_i \in \mathbb{R}). \tag{3}$$

The attempt to maximize the overall margin between the classes $2\varepsilon_i \to \max$ is what has given rise to the terminology "Support Vector Machine", because the direction vector of the optimal discriminant hyperplane $\hat{a}_i$ obtained as the solution of the optimization problem (3) is completely determined (supported) by the projections of a few number of objects into the modality-specific feature space $\mathbb{X}_i \subseteq \tilde{\mathbb{X}}_i$.

In the more realistic case of a linearly inseparable training set, the normalized form of criterion (3) can be put as

$$\begin{cases} K_i(a_i, a_i) + C_i \sum_{\omega_j \in \Omega_i^*} \delta_{ij} \to \min(a_i \in \tilde{\mathbb{X}}_i, \ b_i \in \mathbb{R}, \delta_{ij} \in \mathbb{R}), \\ y_j\big(K_i(a_i, x_{ij}) + b_i\big) \geqslant 1 - \delta_{ij}, \delta_{ij} \geqslant 0, \ \omega_j \in \Omega_i^*, \end{cases} \tag{4}$$

where coefficient $C_i > 0$ penalizes the shifts δ_{ij} of objects breaking the linear separability of classes [11]. The dual form of this criterion is a quadratic programming problem with respect to modality-specific Lagrange multipliers λ_{ij} at the inequality constraints:

$$\begin{cases} \sum_{\omega_j \in \Omega_i^*} \lambda_{ij} - (1/2) \sum_{\omega_j \in \Omega_i^*} \sum_{\omega_l \in \Omega_i^*} y_j y_l K_i(x_{ij}, x_{il}) \lambda_{ij} \lambda_{il} \to \max, \\ \sum_{\omega_j \in \Omega_i^*} y_j \lambda_{ij} = 0, \ 0 \leqslant \lambda_{ij} \leqslant C_i/2, \ \omega_j \in \Omega_i^*. \end{cases} \tag{5}$$

As the most essential result of training, the solution of the dual problem ($\hat{\lambda}_{ij} \geqslant 0, \omega_j \in \Omega_i^*$) picks out a subset of *support objects* within the modality-specific training set (2):

$$\hat{\Omega}_i = \{\omega_j \in \Omega_i^* : \hat{\lambda}_{ij} > 0\} \subseteq \Omega_i^*. \tag{6}$$

The positive Lagrange multipliers at the support objects ($\hat{\lambda}_{ij} > 0, \omega_j \in \hat{\Omega}_i$) completely determine the values of the variables which optimize (4), first of all, the direction vector and position of the hyperplane:

$$\hat{a}_i = \sum\nolimits_{\omega_j \in \hat{\Omega}_i} y_j \hat{\lambda}_{ij} x_{ij} \in \tilde{\mathbb{X}}_i, \tag{7}$$

$$\hat{b}_i = \frac{\sum_{\omega_j \in \Omega_i^*, 0 < \hat{\lambda}_{ij} < C/2} \hat{\lambda}_{ij} K_i(\hat{a}_i, x_{ij}) + (C/2) \sum_{\omega_j \in \Omega_i^*, \hat{\lambda}_{ij} = C/2} y_j}{\sum_{\omega_j \in \Omega_i^*, 0 < \hat{\lambda}_{ij} < C/2} \hat{\lambda}_{ij}}. \tag{8}$$

As collateral solutions, the training problem (4) yields also the forced shifts $\hat{\delta}_{ij} \geqslant 0$ of the training objects, but for our purpose there is no need to compute these values.

The direction vector $\hat{a}_i \in \tilde{\mathbb{X}}_i$ of the modality-specific discriminant hyperplane is expressed in (7) as the sum in terms of the hypothetical linear operations defined in $\tilde{\mathbb{X}}_i$ by the modality-specific kernel by virtue of the kernel trick. However, there is no need to compute it explicitly. Substitution of the formal equality (7) into (1) and (8) yields the family of recognition rules immediately applicable to any new object $\omega \in \Omega$ under the only condition that the ith modality $x_i(\omega) \in \mathbb{X}_i$ is completely defined for it, i.e., kernel values $K_i(x_{ij}, x_i(\omega))$ are known for all the objects of the training set:

$$\hat{f}_i(\omega | \Omega_i^*, C_i, b_i) = \sum_{\omega_j \in \hat{\Omega}_i} y_j \hat{\lambda}_{ij} K_i(x_{ij}, x_i(\omega)) + \hat{b}_i \geqslant 0,$$

$$\hat{b}_i = \frac{\sum_{\omega_j \in \Omega_i^*, 0 < \hat{\lambda}_{ij} < C/2} \hat{\lambda}_{ij} \sum_{\omega_k \in \Omega_i^*, \hat{\lambda}_{ik} > 0} y_k \hat{\lambda}_{ik} K_i(x_{ij}, x_{ik}) + (C/2) \sum_{\omega_j \in \Omega_i^*, \hat{\lambda}_{ij} = C/2} y_j}{\sum_{\omega_j \in \Omega_i^*, 0 < \hat{\lambda}_{ij} < C/2} \hat{\lambda}_{ij}}. \tag{9}$$

2.3   Neutral Points in the Modality-Specific Linear Space of Object Representation

Let the training set of object representations in terms of the ith modality Ω_i^* (2) be fixed. Suppose the training problem in terms of the ith modality (4)-(5) has been solved, namely, the Lagrange multipliers are known ($\hat{\lambda}_{ij}, \omega_j \in \Omega_i^*$).

This solution determines the optimal discriminant hyperplane in the hypothetical linear closure $\tilde{\mathbb{X}}_i$ of the modality-specific feature scale $\mathbb{X}_i$. Depending on the sign of the decision function (9), the ith modality votes for assigning a new object $\omega \in \Omega$ to the positive or negative class, but a firm decision will be impossible if the object maps exactly to the discriminant hyperplane. For this reason, we call the points of the discriminant hyperplane *neutral points*, using the special symbols $x_{\phi,i} \in \tilde{\mathbb{X}}_{\phi,i}$ to denote them:

$$\tilde{\mathbb{X}}_{\phi,i} = \left\{ x_{\phi,i} \in \tilde{\mathbb{X}}_i : \sum\nolimits_{\omega_j \in \hat{\Omega}_i} y_j \hat{\lambda}_{ij} K_i(x_{ij}, x_{\phi,i}) + \hat{b}_i = 0 \right\} \subset \tilde{\mathbb{X}}_i. \tag{10}$$

It is clear that $\tilde{\mathbb{X}}_{\phi,i}$ is a set of continuum cardinality. All the neutral points $x_{\phi,i} \in \tilde{\mathbb{X}}_{\phi,i}$ possess the same property of ambiguous class membership (10), but, in what follows, it will be convenient for us to distinguish one of them having the minimum norm:

$$\hat{x}_{\phi,i} = \arg\min_{x_{\phi,i} \in \tilde{\mathbb{X}}_{\phi,i}} K_i(x_{\phi,i}, x_{\phi,i}). \tag{11}$$

In terms of the linear operations in $\tilde{\mathbb{X}}_i$, this point is proportional to the direction vector of the optimal discriminant hyperplane (7) $\hat{x}_{\phi,i} = c_i \hat{a}_i = c_i \sum_{\omega_j \in \hat{\Omega}_i} y_j \hat{\lambda}_{ij} x_{ij}$. The coefficient $c_i \in \mathbb{R}$ is given by the equation $K_i(\hat{a}_i, c_i \hat{a}_i) + \hat{b}_i = c_i K_i(\hat{a}_i, \hat{a}_i) + \hat{b}_i = 0$, whence it follows that $c_i = -\hat{b}_i / K_i(\hat{a}_i, \hat{a}_i)$, and, with respect to (7),

$$\hat{x}_{\phi,i} = \frac{\hat{b}_i}{\sum_{\omega_j \in \hat{\Omega}_i} \sum_{\omega_k \in \hat{\Omega}_i} y_j y_k K_i(x_{ij}, x_{ik}) \hat{\lambda}_{ij} \hat{\lambda}_{ik}} \sum_{\omega_j \in \hat{\Omega}_i} y_j \hat{\lambda}_{ij} x_{ij} \in \tilde{\mathbb{X}}_i. \tag{12}$$

The neutral points (12) (or more exactly, the coefficients of their representation as linear combinations of object features in the hypothetical linear spaces $\tilde{\mathbb{X}}_i$), are additional results of training from the incomplete modality-specific training sets ($\Omega_i^*, i \in I$), that contain only those objects in the entire training set Ω^* for which the respective modality is defined. The central idea behind harnessing such values for joint training with respect to all of the modalities is using $\hat{x}_{\phi,i}$ instead of missed actual values of the respective modality-specific features for incompletely represented objects. Such a strategy of replacing missed feature values is then free of arbitrary assumptions regarding the nature of the original natural data set.

3   Fusing Pattern-Recognition Modalities at the Training Stage for Incomplete Data

3.1   The Principle of Additive Kernel Fusion

We will call the union of all modality-specific training sets $\Omega^* = \bigcup_{i \in I} \Omega_i^*$ (2) over all the available modalities $I = \{1, \ldots, n\}$ the *unified training set*. We shall say the unified training set Ω^* is full if each object $\omega_j \in \Omega^*$ is represented by all modality-specific signals $(x_{ij} = x_i(\omega_j) \in \mathbb{X}_i, i \in I)$, i.e., all the kernel-specific training sets coincide $\Omega_1^* = \ldots = \Omega_n^*$.

A full training set Ω^* allows for immediate combination of the various modalities by kernel fusion. It is enough to define an appropriate combined kernel (inner product) $K(x', x'')$, $x = (x_i, i \in I) \in \tilde{\mathbb{X}}$, in the Cartesian product $\tilde{\mathbb{X}} = \tilde{\mathbb{X}}_1 \times \ldots \times \tilde{\mathbb{X}}_{n=|I|}$ of the linear spaces $\tilde{\mathbb{X}}_i \supseteq \mathbb{X}_i$ defined by the respective kernels. In particular, the sum of the initial kernels $K(x', x'') = \sum_{i \in I} K_i(x_i', x_i'')$ will be a kernel in $\tilde{\mathbb{X}}$. From this point of view, any choice of a point $a = (a_i \in \tilde{\mathbb{X}}_i, i \in I) \in \tilde{\mathbb{X}}$ and real number $b \in \mathbb{R}$ yields a discriminant hyperplane $\hat{f}(\omega|\Omega^*) = K(a, x(\omega)) + b = \sum_{i \in I} K_i(a_i, x_i(\omega)) + b \gtrless 0$ with direction vector a in the Cartesian product $\tilde{\mathbb{X}}$, and produces, thereby, a kernel fusion technique. However, just as in the case of a single kernel, there is no need to implicitly evaluate the hypothetical direction vector which exists only in terms of the kernel trick.

The straightforward application of the SVM training principle (4)-(12) to the Cartesian product of the particular linear spaces $\boldsymbol{x}_j = (x_{ij}, i \in I) \in \tilde{\mathbb{X}} = \tilde{\mathbb{X}}_1 \times \ldots \times \tilde{\mathbb{X}}_n$, $\omega_j \in \Omega^*$, results in the dual training problem, in which $C > 0$ is the penalty coefficient on the shifts of objects that break the linear separability of the training set in $\tilde{\mathbb{X}}$:

$$\begin{cases} \sum_{\omega_j \in \Omega^*} \lambda_j - (1/2) \sum_{\omega_j \in \Omega^*} \sum_{\omega_l \in \Omega^*} y_j y_l \left(\sum_{i \in I} K_i(x_{ij}, x_{il}) \right) \lambda_j \lambda_l \to \max, \\ \sum_{\omega_j \in \Omega^*} y_j \lambda_j = 0,\ 0 \leqslant \lambda_j \leqslant C/2,\ \omega_j \in \Omega_i^*. \end{cases} \quad (13)$$

This quadratic programming problem over Lagrange multipliers $(\lambda_j, \omega_j \in \Omega^*)$ has the same structure as that for a single modality (5). The only difference is that the training set occurs in (13) through kernels $K(\boldsymbol{x}_j, \boldsymbol{x}_l) = \sum_{i \in I} K_i(x_{ij}, x_{il})$ in the unified linear space $\tilde{\mathbb{X}}$ instead of single modality-specific kernels $K_i(x_{ij}, x_{il})$ (5) in the particular spaces $\tilde{\mathbb{X}}_i$.

Let the training set be not full, i.e., such that each object ω_j is, in general, represented by only a fraction of the modalities $x_{ij} \in \mathbb{X}_i$, $i \in I_j \subseteq I$. Then, there are objects ω_j not represented by certain of the modalities $(x_{ij}, i \notin I_j)$, and the respective kernel values adjacent to these objects are unknown in the dual criterion (13) $\left(K_i(x_{ij}, x_{il}) = ?,\ i \notin I_j,\ \omega_l \in \Omega^* \right)$.

The central idea for overcoming the problem of incomplete data at the training stage outlined in the next Section, is that of substituting the neutral values $\hat{x}_{\phi,i}$ for the missing modalities x_{ij}.

3.2   Neutral Point Substitution for Missing Representations of Training Objects

Let the SVM be applied within each modality-specific partial training set Ω_i^* (Section 2.2). Then, the sets of support objects $\hat{\Omega}_i$ along with Lagrange multipliers $(\hat{\lambda}_{ij}, \omega_j \in \Omega_i^*)$ and biases of discriminant hyperplanes $\hat{b}_i$ are found for all the modalities $i \in I$ in accordance with (5), (6) and (12). As a result, the hypothetical neutral points $\hat{x}_{\phi,i} \in \tilde{\mathbb{X}}_i$ are defined by (12) as linear combinations of modality-specific object features.

Thus, it is possible to compute the neutral-point substitutes for missing values of kernels in (13):

$$\begin{aligned} K_i(x_{ij}, x_{il}) &\Leftarrow K_i(\hat{x}_{\phi,i}, x_{il}),\ i \notin I_j,\ \omega_l \in \Omega^*; \\ K_i(\hat{x}_{\phi,i}, x_{il}) &= \frac{\hat{b}_i \sum_{\omega_k \in \Omega_i^*} y_k \hat{\lambda}_{ik} K_i(x_{ik}, x_{il})}{\sum_{\omega_k \in \hat{\Omega}_i} \sum_{\omega_q \in \hat{\Omega}_i} y_k y_q K_i(x_{ik}, x_{iq}) \hat{\lambda}_{ik} \hat{\lambda}_{iq}}. \end{aligned} \quad (14)$$

The Lagrange multipliers $(\lambda_j, \omega_j \in \Omega^*)$, found as a solution of the dual problem (13) after such a substitution, determine, on the full application of the sequence of computations (6)-(12), first, the set of support objects $\hat{\Omega} = \{\omega_j \in \Omega^* : \hat{\lambda}_j > 0\}$, then the bias of the discriminant hyperplane $\hat{b}$, and finally yield the decision rule

$$\hat{f}(\omega | \Omega^*, C) = \sum_{\omega_j \in \hat{\Omega}} y_j \hat{\lambda}_j \sum_{i \in I} K_i(x_{ij}, x_i(\omega)) + \hat{b} \geqslant 0, \qquad (15)$$

which is the result of fusing the available pattern-recognition modalities taking into account the final imbalance of the incomplete training set Ω^*.

4 Experiments: Biometric-Based Identity Authentication from Incomplete Data

To demonstrate the above principle experimentally, we employ the Biosecure database [10], derived from a European project whose aim is to integrate multi-disciplinary research efforts in biometric-based identity authentication.

We randomly chose a total of 333 different individuals from the database, with distinct identities $Z = \{z = 1, ..., 333\}$. Each of them is represented by four time-spaced measurements $v_i^t(z) \in \mathbb{V}_i$, $t = 0, 1, 2, 3$, of eight modalities $i \in I = \{1, ..., n\}$, $n = 8$, where $\mathbb{V}_i$ is the scale for measuring the ith modality:

- two versions of the frontal face image (high- and low-resolution ones from, respectively, professional and web camera), $v_i^t(z) \in \mathbb{V}_i$, $i = 1, 2$,
- six fingerprints of the right hand (optical and thermal imprints of the index finger, middle finger and thumb) $v_i^t(z) \in \mathbb{V}_i$, $i = 3, ..., 8$.

The Cartesian product of all the modality-specific measurement scales will be denoted as $\mathbb{V} = \mathbb{V}_1 \times ... \times \mathbb{V}_n$. However, not all of potential measurements $\{v_i^t(z), z \in Z, i \in I, t = 0, 1, 2, 3\}$ are available in the data base. Approximately one fourth of them have missing constituents, but for each of the chosen persons $z \in Z$ at least one of the measurement sets, let it be $t = 0$, is full, i.e., all the modalities $(v_i^0(z), i \in I)$ are properly represented in the data base, and neither of the remaining sets $(v_i^{1,2,3}(z), i \in I)$ is completely missed.

In the experiments, we used this full set $\boldsymbol{v}^0(z) = (v_i^0(z), i \in I) \in \mathbb{V}$ as the personal template of person $z \in Z$, whereas the remaining three sets

$$\boldsymbol{v}^{1,2,3}(z) = (v_i^{1,2,3}(z), i \in I) \in \mathbb{V}, \qquad (16)$$

some of whose elements may be missing, served as his/her independent representation in the experiments. Let symbols $V_i(z) = \{v_i^t(z), t = 1, 2, 3\} \subset \mathbb{V}_i$ and $V_i = \bigcup_{z \in Z} V_i(z) \subset \mathbb{V}_i$ stand, respectively, for the set of the representations of person z in terms of the ith modality and the total set of such representations of all the persons involved in the experiments.

Thus, we distinguish here between the people's identities $z \in Z$ and the three times greater number of their multi-modal computer representations $\boldsymbol{v}^t(z) = (v_i^t(z), i \in I) \in V = \bigcup_{z \in Z} V(z) \subset \mathbb{V}$, $V(z) = V_1(z) \times ... \times V_n(z) \subset \mathbb{V}$, $t = 1, 2, 3$.

To constitute the total set of real-world pattern-recognition objects $\omega \in \Omega$, we choose the set of pairs

$$\Omega = \{\omega = (\boldsymbol{v}^t(z), \tilde{z})\} = \mathbb{V} \times \mathbb{V} \times \mathbb{V} \times Z, \qquad (17)$$

where $\boldsymbol{v}^t(z) = (v_i^t(z), i \in I)$ is one of the three received representations of a person $t = 1, ..., 3$, and $\tilde{z}$ is its claimed identity, which may be true or false. We shall say that object $\omega = (\boldsymbol{v}(z), \tilde{z})$ belongs to the class of clients $y = 1$ if the identity claim is correct $z = \tilde{z}$, and to the class of impostors $y = -1$ in the case of a fraudulent claim $z \neq \tilde{z}$.

For each modality $i \in I$, a real-valued similarity measure $S_i(v_i', v_i'') : \mathbb{V}_i \times \mathbb{V}_i \to \mathbb{R}$ is defined in the Biosecure database. It appears natural to measure the credibility of the identity claim $\tilde{z}$ in the received pair $\omega = (\boldsymbol{v}(z), \tilde{z}) = ((v_i, i \in I), z)$ from the viewpoints of different modalities $i \in I$ as the real-valued modality-specific features $x_i(\omega) = S_i(v_i(z), v_i^0(\tilde{z})) \in \mathbb{R}$. In this case, the natural modality-specific kernel is dot product of feature values $K_i(\omega', \omega'') = K_i(x_i(\omega'), x_i(\omega'')) = K_i(x_i', x_i'') = x_i' x_i''$.

We thus used information on 333 persons (person identities) $Z = \{z = 1, ..., 333\}$, each of which is represented by three independent sets of multi-modal measurements $\boldsymbol{v}^{1,2,3}(z)$ in accordance with (16). All in all, we have $999 \times 333 = 332667$ pairs of person representations and person identities $\omega = (\boldsymbol{v}(z), \tilde{z})$, which is the size of the full set of objects $|\Omega| = 332667$ (17) in the experiments.

From the part of the full data set Ω, which contains only complete person representations $(v_i^t(z), i \in I)$, we chose the fixed test set consisting of 20962 objects, namely, pairs <complete person representation/claimed identity>. From the rest of Ω, containing complete as well as incomplete person representations, we further randomly chose 500 training sets each consisting of 200 pairs with the correct claimed identity $y = 1$ and 800 incorrectly claimed pairs $y = -1$. On average, one fourth of 1000 objects in each of the random training sets were incompletely represented, i.e., about 250 of them had at least one missing value in the feature vector.

The goal of the experiment is to show that filling-in blanks in the multi-modal training sets by the generalized neutral-point technique improves generalization performance of the inferred recognition rule in comparison with other methods of imputation. We compared the SVM-NPS technique outlined in this paper with the following five SVM-based methods of treating blanks in the training data:

- SVM handling only objects represented by all the features, in our case, about 3/4 of the training set (SVM-Full);
- sum-rule of combining single SVM-based modality-specific classifiers, inferred each from the partial training subset containing only objects for which the respective modality is known (SVM-SumRule);
- SVM handling all the objects with replacing the unknown features by their averaged known values over the entire training set (SVM-OverallMean);
- the same with replacing the unknown features by their averaged known values over the objects of the same class (SVM-ClassSpecificMean);
- the same with replacing the unknown features by their averaged known values over 5 nearest neighboring objects in the feature space (SVM-5NN).

The interpretation of training results was based on computing the Equal Error Rate of the direction vector of the discriminant hyperplane inferred by each of the six techniques under comparison from each of the 500 random training sets and applied to the test set. The EER value of the respective technique was further averaged over all the training sets.

The following table summarizes the averaged EERs in percentages for the imputation methods under comparison starting with our SVM-NPS:

SVM-NPS	SVM-Full	SVM-SumRule	SVM-OverallMean	SVM-ClassSpecificMean	SVM-5NN
1.07	1.93	1.93	1.38	1.92	1.85

As we can see, the SVM-NPS approach shows almost two times better performance than the SVM-based learning from the training set consisting only of objects with the complete set of features and the SVM-based sum-rule of combining modality-specific classifiers. All other imputation methods are also far outperformed.

5   Conclusions

In this paper, we have set out to generalize the previous neutral point method for accommodating missing data within multi-modal kernel fusion problems in order to accommodate arbitrary amounts of missing *training* (as opposed to test) data. By using imbalance-sensitive SVM methods, we have shown that the SVM-NPS approach to multi-modal pattern-recognition with incomplete data displays exceptionally good generalization performance as compared to the sum rule fusion of modality-specific classifiers, and to the known SVM-based methods of missing-data imputation. Future experimental study will set out to determine the full bounds of its practical applicability.

Acknowledgements

The research leading to these results has received funding from the Russian Foundation for Basic Research, Grant No. 08-01-00695-a. We also gratefully acknowledge the support of EPSRC through grant EP/F069626/1.

References

1. Ross, A., Jain, A.K.: Multimodal biometrics: An overview. In: Proc. of the 12th European Signal Processing Conference (EUSIPCO), Austria, pp. 1221–1224 (2004)
2. Mottl, V., Tatarchuk, A., Sulimova, V., Krasotkina, O., Seredin, O.: Combining pattern recognition modalities at the sensor level via kernel fusion. In: Haindl, M., Kittler, J., Roli, F. (eds.) MCS 2007. LNCS, vol. 4472, pp. 1–12. Springer, Heidelberg (2007)
3. Tatarchuk, A., Sulimova, V., Windridge, D., Mottl, V., Lange, M.: Supervised selective combining pattern recognition modalities and its application to signature verification by fusing on-line and off-line kernels. In: Benediktsson, J.A., Kittler, J., Roli, F. (eds.) MCS 2009. LNCS, vol. 5519, pp. 324–334. Springer, Heidelberg (2009)

4. Tatarchuk, A., Urlov, E., Mottl, V., Windridge, D.: A support kernel machine for supervised selective combining of diverse pattern-recognition modalities. In: El Gayar, N., Kittler, J., Roli, F. (eds.) MCS 2010. LNCS, vol. 5997, pp. 165–174. Springer, Heidelberg (2010)
5. Nandakumar, K., Jain, A.K., Ross, A.: Fusion in multibiometric identification systems: What about the missing data? In: Proc. of the 3rd IAPR/IEEE International Conference on Biometrics, Italy (June 2009)
6. Poh, N., Bourlai, T., Kittler, J., et al.: Benchmarking quality-dependent and cost-sensitive score-level multimodal biometric fusion algorithms. IEEE Trans. on Information Forensics and Security 4(4), 849–866 (2009)
7. Windridge, D., Mottl, V., Tatarchuk, A., Eliseyev, A.: The neutral point method for kernel-based combination of disjoint training data in multi-modal pattern recognition. In: Haindl, M., Kittler, J., Roli, F. (eds.) MCS 2007. LNCS, vol. 4472, pp. 13–21. Springer, Heidelberg (2007)
8. Poh, N., Windridge, D., Mottl, V., Tatarchuk, A., Eliseyev, A.: Addressing missing values in kernel-based multimodal biometric fusion using neutral point substitution. IEEE Trans. on Information Forensics and Security 5(3), 461–469 (2010)
9. Kittler, J., Hatef, M., Duin, R.P.W., Matas, J.: On combining classifiers. IEEE Trans. on Pattern Analysis and Machine Intelligence 20, 226–239 (1998)
10. Poh, N., Bourlai, T., Kittler, J.: A multimodal biometric test bed for quality-dependent, cost-sensitive and client-specific score-level fusion algorithms. Pattern Recognition 43(3), 1094–1105 (2010), http://www.biosecure.info/
11. Vapnik, V.: Statistical Learning Theory. John-Wiley & Sons, Inc, New York (1998)
12. Mottl, V.: Metric spaces admitting linear operations and inner product. Doklady Mathematics 67(1), 140–143 (2003)

Two-Stage Augmented Kernel Matrix for Object Recognition

Muhammad Awais, Fei Yan, Krystian Mikolajczyk, and Josef Kittler

Centre for Vision, Speech and Signal Processing (CVSSP),
University of Surrey, UK
{m.rana,f.yan,k.mikolajczyk,j.kittler}@surrey.ac.uk

Abstract. Multiple Kernel Learning (MKL) has become a preferred choice for information fusion in image recognition problem. Aim of MKL is to learn optimal combination of kernels formed from different features, thus, to learn importance of different feature spaces for classification. Augmented Kernel Matrix (AKM) has recently been proposed to accommodate for the fact that a single training example may have different importance in different feature spaces, in contrast to MKL that assigns same weight to all examples in one feature space. However, AKM approach is limited to small datasets due to its memory requirements.

We propose a novel two stage technique to make AKM applicable to large data problems. In first stage various kernels are combined into different groups automatically using kernel alignment. Next, most influential training examples are identified within each group and used to construct an AKM of significantly reduced size. This reduced size AKM leads to same results as the original AKM. We demonstrate that proposed two stage approach is memory efficient and leads to better performance than original AKM and is robust to noise. Results are compared with other state-of-the art MKL techniques, and show improvement on challenging object recognition benchmarks.

1 Introduction

Object and image recognition has undergone a rapid progress in last decade due to advances in both features design and kernel methods [1] in machine learning. In particular, recent introduction of multiple kernel learning methods set a new direction of research. The state-of-the-art object and image recognition algorithms use multiple kernel learning based methods for classification, dimensionality reduction and clustering in a wide range of applications [1], [2]. Due to importance of complementary information in MKL, much research was done in field of feature design [3], [4] to diversify kernels, leading to large number of kernels in typical visual classification tasks. Kernels are often computed independently of each others thus may be highly informative, noisy or redundant. Proper selection and fusion of kernels is therefore crucial to maximize performance and to address the efficiency issues in large scale visual recognition applications.

MKL was first proposed by Lancriet et al. [5] using semi-definite programming, where kernel weights were learned by maximizing soft margin between

C. Sansone, J. Kittler, and F. Roli (Eds.): MCS 2011, LNCS 6713, pp. 137–146, 2011.

two classes. Since algorithm proposed in [5] was limited to small kernel sizes and low number of kernels, a number of other methods were proposed to address these problems [6], [7]. All these MKL methods focus on linear combination of kernels, in which a single kernel corresponding to a particular feature space is attributed a single weight. This is a strong constraint as it does not exploit information from individual samples in different feature spaces, e.g., in context of object recognition, some samples can carry more shape information while others may carry more texture information for same object category. To address this problem AKM was proposed [8] in which different features extracted form same sample are treated as different samples of same class. Fundamental problem with AKM is its large augmented matrix which requires a lot of memory and makes it inapplicable to large datasets. In this paper we derive primal and dual of AKM, discuss its empirical feature space and address its issues with a two stage architecture. In the first stage, groups are formed from a set of base kernels based on similarity between kernels. Next, a representative kernel for each group is learned by a linear combination of within group kernels. These representative kernels are highly informative containing most of information from each group. Our grouping approach is also useful for methods proposed in [9], [10], which assumed that kernel groups are available. We further reduce complexity of AKM by exploiting independence of empirical feature spaces of representative kernels in augmented kernel matrix. Due to independence, only most influential training examples from the representative kernels can be used to build an AKM of a reduced size without compromising its performance. In second stage, AKM scheme is used to include contribution of most influential samples from all representative kernels in final classifier. Our experiments show that proposed strategy of grouping kernels and selecting subsets of training examples makes approach efficient and improves classifier performance. AKM results are compared to other MKL techniques, using different regularization, ℓ_1, ℓ_2, and ℓ_∞ norms. We demonstrate significant improvement on challenging object recognition benchmark Pascal VOC 2007 [11] and multiclass flower datasets [12], [13]. Moreover, proposed memory efficient learning strategy is also applicable in other MKL techniques which is particularly important in large scale data scenario.

Rest of paper is organized as follows. In section 2 we discuss the structure of AKM matrix and derive its primal and dual for SVM. We then compare empirical feature spaces of a linear combination MKL and AKM schemes. Our proposed two stage multiple kernel learning for AKM is presented in section 3. In section 4 we present the result and compare with other state-of-art MKL methods for object recognition.

2   Linear Combination vs Augmented Kernel Matrix

We first present structure of AKM and give primal formulations for a binary classification. We then present concept of empirical feature space for AKM scheme.

Consider we are given m training samples (x_i, y_i), where x_i is a sample in input space and $y_i \in \pm 1$ is its label. Feature extraction results in n training kernels (K_p) of size $m \times m$ and corresponding n test kernels $(\dot{K}_p)$ of size $m \times l$. Each kernel $K_p = \langle \Phi_p(x_i), \Phi_p(x_j) \rangle$ implicitly maps samples x_i from input space to feature space with mapping function $(\Phi_p(x_i)_{p=1,...,n})$. In MKL aim is to find linear combination $\sum_{p=1}^{n} \beta_p K_p$, normal vector $\mathbf{w}$ and bias b of separating hyperplane simultaneously such that soft margin between two classes is maximized. Primal and its corresponding dual for a linear combination of kernels are derived for various formulations in [14], [5], [6], [7]. The decision function is then $f(x) = sign(\sum_{i=1}^{m} \alpha_i y_i k(x_i, x) + b)$, where $k(x_i, x)$ is dot product of test sample x with i^{th} training sample in feature space, $\alpha \in \mathbb{R}^m$, and b are Lagrange multiplier and bias. Contribution of a given feature channel is fixed by β_p, which may be suboptimal, as in a particular feature channel one example can carry more shape information than texture or vice versa. In contrast, in AKM [8], given the set of base training kernels augmented kernel is defined as follows:

$$K = K_1 \oplus \cdots \oplus K_n = \begin{bmatrix} K_1 & \cdots & 0 \\ \vdots & \ddots & \vdots \\ 0 & \cdots & K_n \end{bmatrix} \tag{1}$$

where base kernels are on diagonal. Zeros on off diagonal reflect that there is no cross terms between different kernel matrices. Note that all base kernels are of size $m \times m$ while AKM is of size $(n \times m) \times (n \times m)$, thus it uses $n \times m$ training samples instead of m. The SVM primal of AKM scheme is then given:

$$\min_{\mathbf{w},\boldsymbol{\xi},b} \frac{1}{2} \sum_{p=1}^{n} \langle \mathbf{w_p}, \mathbf{w_p} \rangle + C \sum_{i=1}^{n \times m} \xi_i \tag{2}$$

$$s.t. \ y_i(\sum_{p=1}^{n} \langle \mathbf{w_p}, \Phi_p(x_i) \rangle + b) \geq 1 - \xi_{pi}, \ \xi_{pi} \geq 0, \ i = 1,...,m, \ p = 1,...,n$$

The dual of Eq. (2) can be derived using Lagrange multiplier techniques:

$$\max_{\alpha} \sum_{p=1}^{n} \sum_{i=1}^{m} \alpha_{pi} - \frac{1}{2} \sum_{p=1}^{n} \sum_{i,j=1}^{m} \alpha_{pi} \alpha_{pj} y_i y_j k_p(x_i, x_j) \tag{3}$$

$$s.t. \sum_{p=1}^{n} \sum_{i=1}^{m} \alpha_{pi} y_i = 0, \ 0 \leq \alpha \leq C,$$

Decision function of AKM is $f(x) = sign(\sum_{p=1}^{n} \sum_{i=1}^{m} \alpha_{pi} y_i k_p(x_i, x) + b)$, where α_{pi} are Lagrange multipliers and x is test sample. Note that same samples from different feature channels are added as separate examples of same class, therefore one Lagrange multiplier is learnt for each sample from each feature channel.

The concept of empirical feature space is crucial to analyze spread and shape of data. Kernel matrices consist of dot products between samples in some feature spaces. These feature spaces are usually very high or even infinite dimensional. However, in [15] it is shown that there exists an empirical feature space in which the intrinsic geometry of data is identical to true feature space, thus, in many problems it is sufficient to study empirical feature space. Empirical feature spaces X and $\dot{X}$ for training kernel K of size $m \times m$ and test kernel $\dot{K}$ of size $m \times l$ can be derived by eigen value decomposition as shown in [8].

Consider a linear combination of two training kernels K_1, K_2 with sample points in r_1, r_2 dimensional empirical feature space given by matrices X_1, X_2 of sizes $r_1 \times m$ and $r_2 \times m$, respectively. By definition of a dot product, computing weighted sum of base kernels is equivalent to computing cartesian product of associated empirical feature spaces, after scaling them with $\sqrt{\beta_p}$, $p = 1, ..n$. An illustration of empirical feature space is given in figure 1. K_1, K_2 are two base kernels with rank $r_1 = r_2 = 1$ i.e., the samples live in one dimensional empirical feature space as shown in figure 1(a) and (b). Note, this toy example is for illustration purpose, whereas, in practice the empirical feature spaces can be up to m dimensional. Figure 1(c) shows the empirical feature space of a sum of two kernels. Note that the number of samples in figure 1(c) is equal to m which is the same as the number of samples in K_1 and K_2.

Let K be AKM of two training kernels K_1, K_2. The matrix X of training vectors in empirical feature space associated with K can be computed by eigen value decomposition [8]. However, by exploiting property of block diagonal augmented matrix K, its associated matrix X is directly given by:

$$X = \begin{pmatrix} X_1 & 0 \\ 0 & X_2 \end{pmatrix} \tag{4}$$

where X is a block diagonal matrix of size $(r_1 + r_2) \times 2m$, with matrix X_1 and X_2 on its diagonal. The empirical feature space for augmented kernel matrix from two one-dimensional kernels K_1 and K_2 is shown in figure 1(d). Note that there are now total of $2m$ training examples in the empirical feature space of AKM.

3   Two-Stage Multiple Kernel Learning

In this section we present a two stage architecture for multiple kernel learning which combines the MKL and AKM schemes. Kernel matrix of AKM needs large amount of memory and is very slow in training of classifier. For example, the extra memory required by cross terms in a large augmented kernel matrix of n base kernels is $n(n - 1)$ times larger than linear combination of these base kernels. This makes AKM less inapplicable to large datasets especially when n is large. We address this problem by introducing grouping of base kernels followed by a selection of training samples. Two stage approach serves two goals. It addresses the memory problems of AKM but also filters out noisy and redundant feature channels. Adding redundant feature channels as separate examples increases the memory requirements in AKM and adding noisy feature channel as separate examples leads to a significant performance loss. These two problems are alleviated by applying the grouping stage.

3.1 Kernel Grouping

We define multiple groups of base kernels using a similarity criterion. One such grouping criterion can be based on the modality of features or their extraction technique. For example, feature channels based on colour can belong to one group, texture based feature channels to another group and shape based ones to yet another group. However, this kind of grouping is not automatic and needs prior information about input spaces of kernel which may not be available. We exploit Kernel Alignment [16] as a measure of similarity between kernels to group then in unsupervised manner. Given an unlabeled sample set $S = \{x_i\}_{i=1}^m$, we use the Frobenius inner product between kernel matrices i.e., $\langle K_1, K_2 \rangle_F = \sum_{i,j=1}^m K_1(x_i, x_j) K_2(x_i, x_j)$. The empirical alignment between kernels with respect to the set S is defined as:

$$\hat{A}(S, K_1, K_2) = \frac{\langle K_1, K_2 \rangle_F}{\sqrt{\langle K_1, K_1 \rangle_F \langle K_2, K_2 \rangle_F}} \tag{5}$$

where K_i is the kernel matrix for the sample S. In [16] concentration and generalization of kernel alignment was introduced and proved. Concentration means that the probability of an empirical estimate deviating from its mean can be bounded as an exponentially decaying function of that deviation. In other words, the alignment is little dependent on the training set S as shown by theorem 3 in [16]. Generalization (test error) of a simple classification function is related to the value of the alignment as shown by theorem 4 in [16].

Using kernel alignment $\hat{A}(S, k_1, k_2)$ defined in Eq. (5) as a similarity measure we preform agglomerative clustering to find g groups of kernels. We initialize all kernels as clusters and merge two most similar clusters at a time. Similarity between two clusters is defined as largest distance between all possible pairs of clusters members. This continues until g groups re obtained. We used agglomerative as opposed to k-means to make it independent to initialization. Kullback-Leibler divergence can also be used as a similarity criterion between kernels [17].

Learning a linear combination of kernels within a group can discard or downweight redundant or noisy kernels thus result in a better kernel. Moreover, linear combination leads to more compact representation without loss of information. Therefore, for each group, MKL-SVM methods using ℓ_1, ℓ_2 and ℓ_∞ norms are applied to obtain the representative kernels. The kernel that obtains the highest score on the validation data is used as group representant. Thus, the grouping and within group combination results in a set of representative kernels containing most of the information from various feature channels.

3.2 Selection of Training Samples

Kernel grouping partially addresses the issue of large AKM matrix. However, the matrix can be further reduced without compromising the performance by selecting only the samples from representative kernels which are crucial for classification. The decision function of SVM is determined by the α_i, one for each

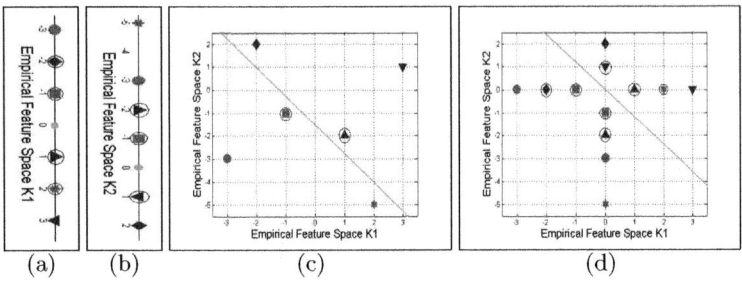

Fig. 1. Empirical feature spaces for Multiple kernels: (a) empirical feature space for K_1; (b) empirical feature space for K_2; (c) empirical feature space for $K_1 + K_2$; (d) empirical feature space for $K_1 \oplus K_2$

training sample. The α_i are non-zero for the support vectors only. Hence, for a single kernel, support vectors are sufficient for classification and all other samples can be discarded without performance degradation. This is supported by the fact that the feature spaces do not interfere with each other due to the structure of the augmented kernel matrix (cf. Eq. (1)). It can be proved by considering the dual of AKM, Eq. (3), which can be rewritten as follows:

$$\max_{\boldsymbol{\alpha}} \sum_{i=1}^{m} \alpha_{1i} - \frac{1}{2} \sum_{i,j=1}^{m} \alpha_{1i}\alpha_{1j}y_iy_j\langle \Phi_1(x_i), \Phi_1(x_j)\rangle + ...+ \qquad (6)$$

$$\sum_{i=1}^{m} \alpha_{ni} - \frac{1}{2} \sum_{i,j=1}^{m} \alpha_{ni}\alpha_{nj}y_iy_j\langle \Phi_n(x_i), \Phi_n(x_j)\rangle$$

$$s.t. \sum_{i=1}^{m} \alpha_{1i}y_i + ... + \sum_{i=1}^{m} \alpha_{ni}y_i = 0, \ 0 \le \alpha \le C,$$

The first constraint in Eq. (6) is the sum of constraints for n kernels. The support vectors for all individual kernels together satisfy this constraint and thus lie in the feasible set of the optimization problem in Eq. (6). This is also illustrated by a toy example of a binary classification in figure 1. All the support vectors in empirical feature space for base kernels K_1 and K_2 are shown by the enclosing black circles and the hyperplane is represented in green at origin in figure 1(a) and (b), respectively. Figure 1(c) shows the empirical feature space of unweighted linear combination of base kernel. There are only two support vectors in figure 1(c), and the classes are separated by the hyperplane. However, the separability of the training set does not necessarily guarantee better performance as it depends upon the generalization to the test set [1]. Figure 1(d) is the empirical feature space of AKM combination of base kernels. Feature spaces of two base kernels are orthogonal to each other. There are $2m$ training samples and all the support vectors of kernel K_1 and K_2 are support vectors of AKM due to the orthogonality of their feature space. It is clear from Eq. (6) and figure 1, that the support vectors of representative kernels from each group are sufficient to construct the

AKM matrix as the Lagrange multipliers of support vectors lie in the feasible set of Eq. (6). The use of support vectors only for different combinations of kernels is validated empirically in section 4.

4   Experiments and Discussion

This section presents experimental results on challenging binary and multiclass object recognition datasets Pascal VOC 2007 [11], Oxford Flower 17 [12] and Oxford Flower 102 [13].

Pascal VOC 2007 [11] consists of 20 object classes with 9963 image examples. The classification of 20 object categories is handled as 20 independent binary classification problems (as recommended by organizers of Pascal challenge). We present results using average precision (AP) [11], which is proportional to area under precision recall curve. Mean average precision (MAP) is computed by averaging scores for all 20 classes.

We compute 20 kernels by combining features introduced in [3], [4] with 2 sampling strategies (dense, interest points) and spatial location grids [18]: whole image (1x1), horizontal bars (1x3), vertical bars (3x1) and image quarters (2x2). In experiments we use SVM to compare several kernel combination schemes and two stage AKM scheme proposed in this paper. The multiple kernel SVM (MK-SVM) schemes differ by regularization norms used during learning, which include ℓ_1 [7], ℓ_2 [14], and ℓ_∞ (equal weights). We divide 20 kernels into 4 groups as discussed in section 3.1. For each group, MKL-SVM methods using ℓ_1, ℓ_2 and ℓ_∞ norms are applied to obtain representative kernels. Results for various learning techniques are presented in table 1.

Consistently lower performance of ℓ_1-norm, which typically leads to sparsely selected kernels, indicates that most of base kernels carry complementary information. Therefore, non-sparse multiple kernel methods, ℓ_2-norm and ℓ_∞-norm, give better results. Proposed two stage AKM scheme outperforms other MKL combination schemes. In case of ℓ_2 within group and AKM between groups, (AKM, ℓ_2), we obtain an improvement of 0.6%, and in case of ℓ_∞ within group and AKM between groups, an improvement of 0.7% over all linear combinations of MKL-SVM. In case of informative kernels, use of kernel grouping achieves comparable performance to corresponding non-grouping schemes. The best performance of state-of-the-art multiple kernel learning for these kernels is 62.1% , as shown in table 1 while performance of winning method for this challenge is 59.4% [11]. We beat winning method by 3.4%, moreover, 0.7% improvement by proposed two stage AKM over state-of-the-art MKL is still significant given that all kernels are highly informative due to carefully designed features. For example, leading methods in PASCAL VOC often differ by a fraction of a percent in MAP. It is important to note that AKM on its own is giving 61.0%, however, when it is used together with grouping stage it is performing 1.8% better. It is because linear combination within grouping stage gives good representative kernel with less noisy or redundant data. These highly informative representative kernel should be combined with AKM scheme so that information in each example of these kernels is exploited. We expect grouping scheme to show better

Table 1. MAP of PASCAL VOC 2007 with various MKL and AKM approaches

between groups \ within group	no grouping	linear ℓ_1	linear ℓ_2	linear ℓ_∞
linear ℓ_1	56.0	55.3	56.5	56.6
linear ℓ_2	61.4	60.8	61.3	56.5
linear ℓ_∞	62.1	61.1	62.1	62.0
AKM	61.0	60.8	**62.7**	**62.8**

performance if there are noisy or redundant kernels in set as shown by noisy feature channels experiment in next section.

We have also validated empirically the selection of support vectors for AKM on 20 binary classification problems of the Pascal 2007 [11] dataset. Only 0.3% to 0.5% of the support vectors of AKM differs from the union of individual support vectors of representative kernels, while the MAP results are same up to sixth decimal place. However, due to use of the significant examples only, we are using 3 to 4 times less samples per base kernel. Hence, size of AKM matrix is 60% to 70% less than original size without compromising performance. It is important to note that it is not possible to apply AKM without selection of significant examples in this benchmark due to memory requirements. We have used 4 groups of kernels thus AKM kernel is even smaller than original kernel of size 5011×5011. Note that for each group a classifier has to be trained i.e. 4 in this experiment. This is however done efficiently on small kernels and acceptable considering performance gain achieved over other multiple kernel learning methods. Moreover, in α-step of alternative MKL techniques [7], [14] we have to train linear combination of base kernels for different regularization norms several times before obtaining optimal weights values β for base kernels. All results presented for AKM in this paper are obtained using "support vectors only scheme".

Oxford Flower 17 [12] dataset consists of 17 categories with 80 images in each category. Dataset is split into training, validation and test using 3 predefined random splits. We have used used seven distance matrices provided online. Features used to compute these distance matrices include different types of shape, texture and color based descriptors [12]. We have used SVM as classifier and follow one-vs-all setup for multiclass classification [12]. We train an AKM classifier for each category and use the maximum response of the classifiers for each example to obtain the label and score for evaluation. Regularization parameter for the SVM is in the range $C \in \{10^{(-2,-1,...,3)}\}$.

Results are given in figure 2(a). For comparison we use recent evaluation results from [19] of state-of-the-art feature fusion techniques including MKL and boosting based classifier fusion. There are two baseline techniques, MKL-prod-SVM and MKL-avg-SVM, which are obtained from element wise product and averaging of base kernels and classifying with SVM. MKL baseline for kernel product gives the highest score of 85.5%. Moreover, it is very simple and fast in comparison to other MKL methods in figure 2(a). Our proposed scheme based on AKM gives 86.7%, which is better than all MKL and Boosting based methods.

We also investigate effect of adding random feature channels on different fusion schemes. In addition to 7 informative kernels of Flower17 dataset we have

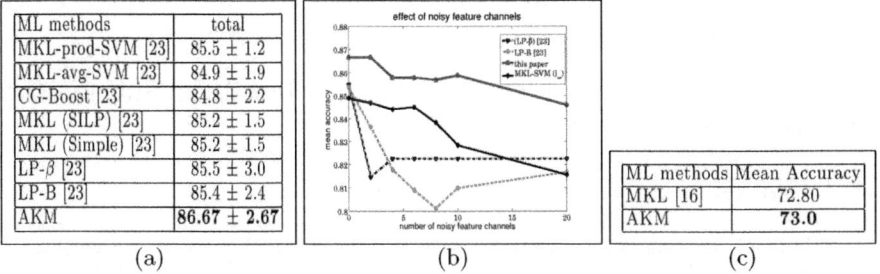

ML methods	total
MKL-prod-SVM [23]	85.5 ± 1.2
MKL-avg-SVM [23]	84.9 ± 1.9
CG-Boost [23]	84.8 ± 2.2
MKL (SILP) [23]	85.2 ± 1.5
MKL (Simple) [23]	85.2 ± 1.5
LP-β [23]	85.5 ± 3.0
LP-B [23]	85.4 ± 2.4
AKM	**86.67 ± 2.67**

ML methods	Mean Accuracy
MKL [16]	72.80
AKM	73.0

(a)   (b)   (c)

Fig. 2. (a) Mean accuracy on Oxford Flower17 and comparision with different machine learning methods; (b) Oxford Flower17. Mean Accuracy of diffierent fusion methods under noisy feature channels; (c) Mean accuracy on Oxford Flower 102 dataset.

generated 20 RBF kernels from 20 sets of random vectors. We started with all informative kernels, i.e., zero noisy kernels, then we added different number of noisy kernel. Mean accuracy of different state-of-the-art methods under noisy channels is presented in figure 2(b). MKL baseline drops down significantly with the number of noisy kernels while two-stage AKM is robust to noisy feature channels and perform significantly better than MKL or boosting based approaches.

Oxford Flower 102 [13] is an extended multi-class dataset containing 102 flower categories. The dataset is split into training, validation and test using predefined splits. For experiments we have used 4 χ^2 distance matrices provided online. The details of the features used to compute these distance matrices can be found in [13]. The experimental setup is the same as for Oxford Flower 17. AKM is performing comparable to MKL as shown in figure 2(c).

5 Conclusions

In this paper we have presented a novel two stage multiple kernel learning approach for augmented kernel matrix. The proposed method addresses the complexity problems of AKM and makes it robust to redundant and noisy kernels. We propose automatic grouping of kernels based on kernel alignment by agglomerative clustering of kernels. Learning representative kernels for each group results in a small set of highly informative kernels. Learning a combination within a group discards or downweights redundant and noisy kernels thus results in an optimal kernel from a set of informative base kernels. The complexity is further reduced by exploiting the property of independence of empirical feature spaces in the AKM scheme. It allows to use only the most influential examples from each representative kernel to construct the AKM matrix. We perform experiments on challenging object recognition datasets and the results validate our technique. The proposed approach makes it possible to use the AKM method for 20 kernels with several thousands of training examples. A performance increase is observed compared to MKL based on a linear combination of all base kernels. This observation is significant as it suggests that the information in the kernels can be exploited more effectively and the classification rate increases without using additional features.

Acknowledgements. This research was supported by UK EPSRC EP/F0034 20/1 and the BBC R&D grants.

References

1. Schölkopf, B., Smola, A.: Learning with Kernels. MIT Press, Cambridge (2002)
2. Vapnik, V.: The Nature of Statistical Learning Theory. Springer, Heidelberg (2000)
3. Mikolajczyk, K., Schmid, C.: A performance evaluation of local descriptors. PAMI 27(10), 1615–1630 (2005)
4. van de Sande, K., Gevers, T., Snoek, C.: Evaluation of color descriptors for object and scene recognition. In: CVPR (2008)
5. Lanckriet, G., Cristianini, N., Bartlett, P., Ghaoui, L., Jordan, M.: Learning the Kernel Matrix with Semidefinite Programming. JMLR 5, 27–72 (2004)
6. Bach, F., Lanckriet, G., Jordan, M.: Multiple Kernel Learning, Conic Duality, and the SMO Algorithm. In: ICML (2004)
7. Sonnenburg, S., Rätsch, G., Schafer, C., Schölkopf, B.: Large Scale Multiple Kernel Learning. JMLR 7, 1531–1565 (2006)
8. Yan, F., Mikolajczyk, K., Kittler, J., Tahir, M.A.: Combining multiple kernels by augmenting the kernel matrix. In: El Gayar, N., Kittler, J., Roli, F. (eds.) MCS 2010. LNCS, vol. 5997, pp. 175–184. Springer, Heidelberg (2010)
9. Szafranski, M., Grandvalet, Y., Rakotomamonjy, A.: Composite Kernel Learning. ML 79(1), 73–103 (2010)
10. Nath, J., Dinesh, G., Raman, S., Bhattacharyya, C., Ben-Tal, A., Ramakrishnan, K.: On the Algorithmics and Applications of a Mixed-norm Based Kernel Learning Formulation. In: NIPS (2009)
11. Everingham, M., Van Gool, L., Williams, C.K.I., Winn, J., Zisserman, A.: The PASCALVisual Object Classes Challenge (VOC 2007)Results (2007) http://www.pascal-network.org/challenges/VOC/voc2007/workshop/index.html
12. M. Nilsback A. Zisserman.: A visual Vocabulary for Flower Classification. In: CVPR, 2006.
13. Nilsback, M.-E., Zisserman, A.: Automated Flower Classification over a Large Number of Classes. In: ICCVGIP (2008)
14. Kloft, M., Brefeld, U., Laskov, P., Sonnenburg, S.: Nonsparse Multiple Kernel Learning. In: NIPS Workshop on Kernel Learning: Automatic Selection of Optimal Kernels (2008)
15. Schölkopf, B., Mika, S., Burges, C., Knirsch, P., Müller, K., Rätsch, G., Smola, A.: Input Space Versus Feature Space in Kernel-Based Methods. NN 10(5), 1000–1017 (1999)
16. Cristianini, N., Shawe-Taylor, J., Elisseeff, A., Kandola, J.: On Kernel-Target Alignment. In: NIPS (2001)
17. Lawrence, N., Sanguinetti, G.: Matching Kernel through Kullback-Leibler Divergence Minimisation. Technical Report CS-04-12, Department of Computer Science, University of Sheffield (2005)
18. Lazebnik, S., Schmid, C., Ponce, J.: Beyond Bags of Features: Spatial Pyramid Matching for Recognizing Natural Scene Categories. In: CVPR (2006)
19. Gehler, P., Nowozin, S.: On Feature Combination for Multiclass Object Classification. In: ICCV (2009)

Multiple Kernel Learning via Distance Metric Learning for Interactive Image Retrieval

Fei Yan, Krystian Mikolajczyk, and Josef Kittler

Centre for Vision, Speech, and Signal Processing
University of Surrey
Guildford, Surrey, GU2 7XH, UK
{f.yan,k.mikolajczyk,j.kittler}@surrey.ac.uk

Abstract. In this paper we formulate multiple kernel learning (MKL) as a distance metric learning (DML) problem. More specifically, we learn a linear combination of a set of base kernels by optimising two objective functions that are commonly used in distance metric learning. We first propose a global version of such an MKL via DML scheme, then a localised version. We argue that the localised version not only yields better performance than the global version, but also fits naturally into the framework of example based retrieval and relevance feedback. Finally the usefulness of the proposed schemes are verified through experiments on two image retrieval datasets.

1 Introduction

Kernel methods [1] have enjoyed considerable success in a wide variety of learning tasks since their introduction in the mid-1990s. In the past few years, an extension of the kernel methods, multiple kernel learning (MKL) [2,3,4], has drawn great attention in the machine learning community. The goal of MKL is to learn an "optimal" (and often linear) combination of a given set of base kernels. On the other hand, distance metric learning (DML) [5,6,7] is another very active area of machine learning in recent years. In supervised and linear DML, the objective is to learn a Mahalanobis distance in the original space, such that the distance between similarly labelled samples is reduced and that between differently labelled samples is increased.

In this paper, we combine MKL and DML by formulating MKL as a DML problem. More specifically, we learn a linear combination of a set of base kernels, or equivalently a composite feature space, by considering several DML objectives in the concatenation of the feature spaces induced by the base kernels. Such a scheme is of particular interest to applications with heterogeneous data types (e.g. strings, graphs, vectors). In such a situation, it is not straightforward to learn a distance function by combining the features in the input spaces. On the other hand, by mapping into feature spaces, different types of features are unified and standard DML methods can be applied. The learnt feature space can be considered optimal for distance based classifiers such as nearest neighbour (NN), which makes our scheme particularly attractive for image retrieval. We

C. Sansone, J. Kittler, and F. Roli (Eds.): MCS 2011, LNCS 6713, pp. 147–156, 2011.

demonstrate that by learning a composite feature space using DML objectives, the performance of an image retrieval system can be improved over a single kernel or the uniform weighting scheme.

The formulation above learns a composite feature space globally. We then further propose to learn a feature space locally, that is, for each query image. Such a formulation fits naturally into the framework of interactive retrieval. For each query image, we start with a uniform weighting of the base kernels, and ask the user to annotate a small number of retrieved images. Training triplets are then generated from these annotated images and used for learning a set of kernel weights for this particular query image. We show on two datasets that this local learning approach further boosts the performance of an image retrieval system.

The rest of this paper is organised as follows. In Section 2, we introduce previous work that is related to this paper. We then present our MKL via DML approach, first the global setting then the local setting, in Section 3. Experimental evidence showing the usefulness of our approach is provided in Section 4. Finally, conclusions are given in Section 5.

2   Related Work

In this section we discuss the approaches in multiple kernel learning and distance metric learning that we combine within an active learning scenario.

2.1   Multiple Kernel Learning

The goal of multiple kernel learning (MKL) is to learn an "optimal" (and often linear) combination of a set of base kernels, or equivalently, an "optimal" composite feature space. Suppose one is given n $m \times m$ training kernel matrices $K_h, h = 1, \cdots, n$ and m class labels $y_i \in \{1, -1\}, i = 1, \cdots, m$, where m is the number of training samples. The original formulation of MKL [2] considers a linear convex combination of these n base kernels: $K = \sum_{h=1}^{n} \beta_h K_h, \beta_h \geq 0, ||\beta||_1 = 1$. In [2] the soft margin of SVM is used as a measure of optimality, and the kernel weights are regularised with an ℓ_1 norm. The efficiency of this first MKL formulation was improved significantly in later works [3,4]. Various other norms have also been proposed to regularise the kernel weights [8]. In parallel to MK-SVM, another line of research focuses on MKL for Fisher Discriminant Analysis (FDA) [9,10], where the FDA type of class separation criterion is considered instead of the soft margin.

2.2   Distance Metric Learning

Supervised linear distance metric learning (DML) [5,6,7] has a strong connection to supervised dimensionality reduction [11]. Suppose we have a set of samples $x_i \in \mathbb{R}^D, i = 1, \cdots, m$. The goal of supervised linear DML is to learn a squared Mahalanobis distance $d_M(x_i, x_j) = (x_i - x_j)^T M(x_i - x_j)$, where M is a positive semi-definite (PSD) matrix, such that the "compactness" of similarly labelled samples and the "scattereness" of differently labelled samples are maximised simultaneously. The DML and dimensionality reduction techniques in [5,6,7,11] differ mainly in the definition of compactness and scattereness. Among them, SVM with relative comparison (SVM-RC) [6] and large margin nearest neighbour (LMNN) [7] are two representative techniques. SVM-RC assumes weak supervision is available in the form of relative comparison, such as "i is closer to j than i is to k". It learns a weighted Euclidean distance by minimising the violation of the supervision information. SVM-RC assumes for any sample, all samples with the same label should be closer to it than any sample with a different label. By contrast, LMNN only assumes that the g nearest neighbours with the same label should be closer than any sample with a different label. LMNN then learns a Mahalanobis distance by minimising the violation of this assumption.

3   Multiple Kernel Learning via Distance Metric Learning

In this section, we formulate multiple kernel learning as a distance metric learning problem. We first present the global version of this MKL via DML approach, and then describe the local version and its application to relevance feedback.

3.1   MKL via DML: The Global Version

Assume we are given n $m \times m$ PSD kernel matrices $K_h, h = 1, \cdots, n$. Each kernel induces a feature space and the hth kernel K_h can be considered as the pairwise dot product of m points in the feature space induced by K_h: $K_h^{i,j} = <\mathbf{x}_h^i, \mathbf{x}_h^j>$, where $\mathbf{x}_h^i, \mathbf{x}_h^j \in \mathbb{R}^{r_h}$ and r_h is the rank of K_h. It directly follows that the squared Euclidean distance between the ith and j^{th} samples in the hth feature space is given by $d_h(\mathbf{x}_h^i, \mathbf{x}_h^j) = K_h^{i,i} + K_h^{j,j} - 2K_h^{i,j}$, and this distance can be used in distance based applications such as information retrieval.

Now consider a weighted linear combination of the n kernels $K = \sum_{h=1}^{n} \beta_h K_h$, $\beta_h \geq 0$. The squared Euclidean distance between the ith and j^{th} samples in the composite feature space induced by K is given by:

$$d(\mathbf{x}^i, \mathbf{x}^j) = \sum_{h=1}^{n} \beta_h d_h(\mathbf{x}_h^i, \mathbf{x}_h^j) \tag{1}$$

The problem of learning a linear combination of the n kernel matrices can then be cast as one of learning a distance metric.

SVM-RC Formulation. We first consider the setting in SVM-RC [6]. Suppose we have a set of triplets of indices of the training samples, and for each triplet $\{i, j, k\}$ we have weak supervision information in the form of relative comparison: we know that samples i and j share the same label and i and k have different labels. As a result the distance between samples i and j should be smaller than that between i and k. However, in practice this cannot be satisfied by all triplets. As in SVM, we introduce a slack variable for each triplet and learn the kernel weights $\boldsymbol{\beta} = (\beta_1, \cdots, \beta_n)^T$ by minimising the violation of the relative comparison:

$$\min_{\boldsymbol{\beta},\boldsymbol{\xi}} \sum_{i,j,k} \xi_{ijk} \tag{2}$$
$$\text{s.t.} \quad \forall \{i,j,k\} : \quad d(\mathbf{x}^i, \mathbf{x}^k) - d(\mathbf{x}^i, \mathbf{x}^j) \geq 1 - \xi_{ijk}, \quad \boldsymbol{\xi} \geq \mathbf{0}, \quad \boldsymbol{\beta} \geq \mathbf{0}$$

where $d(\cdot, \cdot)$ is defined as in Eq. (1).

To avoid the trivial solution of an arbitrarily large $\boldsymbol{\beta}$, we put an ℓ_2 constraint on $\boldsymbol{\beta}$. Incorporating this regularisation and substituting Eq. (1) into Eq. (2), we arrive at the MKL via SVM-RC optimisation problem:

$$\min_{\boldsymbol{\beta},\boldsymbol{\xi}} \tfrac{1}{2}\boldsymbol{\beta}^T\boldsymbol{\beta} + C\sum_{i,j,k} \xi_{ijk} \tag{3}$$
$$\text{s.t.} \forall \{i,j,k\} : \sum_{h=1}^{n} \beta_h d_h(\mathbf{x}_h^i, \mathbf{x}_h^k) - \sum_{h=1}^{n} \beta_h d_h(\mathbf{x}_h^i, \mathbf{x}_h^j) \geq 1 - \xi_{ijk}, \boldsymbol{\xi} \geq \mathbf{0}, \boldsymbol{\beta} \geq \mathbf{0}$$

where C is a parameter controlling the trade-off between the ℓ_2 norm of $\boldsymbol{\beta}$ and the empirical error. The main difference between SVM-RC and our formulation in Eq. (3) is that SVM-RC assigns weights to different dimensions of a vector space, while Eq. (3) assigns weights to several vector spaces. In this light Eq. (3) can be thought of as a block version of SVM-RC, where each block corresponds to the feature space of a base kernel. Eq. (3) is recognised as a linearly constrained quadratic program (LCQP), and can be solved with off-the-shelf optimisation toolboxes such as Mosek [1].

LMNN Formulation. The formulation above assumes that for any sample all similarly labelled samples should be closer to it than any differently labelled sample. By contrast, LMNN [7] only assumes the similarly labelled g nearest neighbours should be closer than any differently labelled sample. We introduce a variable $\eta_{i,j}$ to indicate whether sample j is one of the g nearest neighbours of sample i that share the same label with i: $\eta_{ij} = 1$ if it is and $\eta_{ij} = 0$ otherwise. Ignoring the regularisation on $\boldsymbol{\beta}$ for the moment, we have:

$$\min_{\boldsymbol{\beta},\boldsymbol{\xi}} \sum_{i,j,k} \eta_{ij}\xi_{ijk} \tag{4}$$
$$\text{s.t.} \quad \forall \{i,j,k\} : \quad d(\mathbf{x}^i, \mathbf{x}^k) - d(\mathbf{x}^i, \mathbf{x}^j) \geq 1 - \xi_{ijk}, \quad \boldsymbol{\xi} \geq \mathbf{0}, \quad \boldsymbol{\beta} \geq \mathbf{0}$$

where $d(\cdot, \cdot)$ is defined as in Eq. (1). Note that the only difference between Eq. (2) and Eq. (4) is the η_{ij} term in the objective function.

Similarly as in the SVM-RC formulation, $\boldsymbol{\beta}$ must be regularised in order to get a meaningful solution. However, following LMNN, we regularise $\boldsymbol{\beta}$ slightly differently. Instead of minimising the ℓ_2 norm of $\boldsymbol{\beta}$, we minimise the sum of the distances between all samples and their g same labelled nearest neighbours. Incorporating this regularisation and substituting Eq. (1) into Eq. (4) we arrive at the MKL via LMNN optimisation problem:

$$\min_{\boldsymbol{\beta},\boldsymbol{\xi}} \sum_{ij} \eta_{ij} \sum_{h=1}^{n} \beta_h d_h(\mathbf{x}_h^i, \mathbf{x}_h^j) + C\sum_{i,j,k} \eta_{ij}\xi_{ijk} \tag{5}$$
$$\text{s.t.} \forall \{i,j,k\} : \sum_{h=1}^{n} \beta_h d_h(\mathbf{x}_h^i, \mathbf{x}_h^k) - \sum_{h=1}^{n} \beta_h d_h(\mathbf{x}_h^i, \mathbf{x}_h^j) \geq 1 - \xi_{ijk}, \boldsymbol{\xi} \geq \mathbf{0}, \boldsymbol{\beta} \geq \mathbf{0}$$

---

[1] http://www.mosek.com

where C is the trade-off parameter. As in the SVM-RC formulation, Eq. (5) can be seen as a block version of LMNN. Another difference between LMNN and Eq. (5) is that LMNN is a semidefinite program (SDP) while Eq. (5) is a linear program (LP), which can be solved again using the Mosek optimisation toolbox.

3.2   MKL via DML: The Localised Version

Given a set of base kernels (and the associated base feature spaces), the formulations in Eq. (3) and Eq. (5) learn distance metrics by weighting the base feature spaces, hence they can also be considered as multiple kernel learning methods. Both formulations require a set of triplets $\{i, j, k\}$, which can be drawn randomly from, or by considering all valid combinations in a set of (weakly) labelled samples. The learnt metrics are expected to be more discriminative than the squared Euclidean distance in the feature space associated with the uniformly weighted sum of the base kernels, and as a result expected to perform better in distance based applications such as image retrieval.

However, such schemes are global in the sense that the distance metrics are learnt from a fixed training set and applied universally ignoring the locations of a sample in the base feature spaces. Arguably, localised learning may be advantageous over global learning since it captures better the local shapes in the base feature spaces. Moreover, localised distance metric learning fits naturally into the framework of example based retrieval and relevance feedback. The MKL via localised DML scheme for relevance feedback can be summarised as follows:

1. User submits an example image as query and machine provides initial retrieval results using the Euclidean distance in the uniformly weighted sum of the n base feature spaces;
2. User labels the top m retrieved images as to whether they are relevant or not;
3. Triplets are drawn from the set of $m + 1$ labelled images including the query image: the query image is used as sample i; sample j is drawn from images labelled as relevant; and sample k drawn from the remaining images.
4. **A new distance metric is learnt using either Eq. (3) or Eq. (5)**, with the drawn triplets in step 3. The list of relevant images is recalculated with the new distance metric. Go to step 2 if desired.

Essentially, this localised learning scheme learns an optimal distance metric for each query image online, by capturing the local structures around the query image in the base feature spaces. In the next section, we will show experimental evidence that locally learnt metrics outperform globally learnt metrics.

4   Experiments

In this section we show experimental results of the proposed global and local MKL via DML methods, in an example based image retrieval setting. We first described the datasets used, and then present the results.

4.1 Datasets

Oxford Flower17 dataset [12] consists of 17 categories of flowers with 80 images per category. It comes with three predefined splits into train (17×40 images), validation (17×20 images) and test (17×20 images) sets. For each split, we use the $17 \times 60 = 1020$ images in the training and validation sets as images to be retrieved, and the $17 \times 20 = 340$ images in the test set as queries. For each query image, a ranking of the 1020 images in the database is given based on their distances to the query image according to some distance metric. An average precision is computed from this ranking. The mean average precision (MAP) of the 340 query images can then be used as the performance measure. We repeat this process for all three predefined splits and report the mean of the MAPs. The authors of [12] precomputed 7 distance matrices using various features [2], from which we computed 7 radial basis function (RBF) kernels and used them as base kernels.

Caltech101 [13] is a multiclass object recognition benchmark with 101 object categories. We randomly select 15 images from each class and use the $101 \times 15 = 1515$ images as images to be retrieved, and use up to 50 randomly selected images per class, that is, 3999 in total, as query images. We repeat this process of randomly selecting samples three times. Similarly as in Flower17 experiments, we compute an MAP for each random sampling, and report the mean of the three MAPs. 21 base kernels are generated by combining the colour based local descriptors in [14] and three kernel functions, namely, pyramid match kernel (PMK) [15], spatial pyramid match kernel (SPMK) [16], and RBF kernel with χ^2 distance.

4.2 Results

We show first in Fig. 1 left and Fig. 2 left the baseline performance. The first 4 bars in both plots show the minimum, maximum, median, and mean of the performance of the base kernels; while the last bar indicates the performance of the uniformly weighted sum of the base kernels. For the Oxford Flower17 dataset, the uniform weighting scheme outperforms the best single kernel by a large margin (0.3680 vs. 0.3022); while for the Caltech101 dataset, its advantage is only marginal (0.2303 vs. 0.2294).

In Fig. 1 right and Fig. 2 right we show the performance of the global version of the proposed MKL via SVM-RC and MKL via LMNN schemes. For the global version, triplets of relative comparison are drawn randomly from the 1020 images in the database. Note that this is not realistic since in a retrieval scenario the labels of the images in the database are not available. Nevertheless, we present the results of the global learning schemes to show the advantage of localised learning.

For both global schemes we use approximately the same number of triplets for training. For MKL via SVM-RC, we randomly draw 2×10^4 triplets of samples

---

[2] http://www.robots.ox.ac.uk/~vgg/research/flowers/index.html

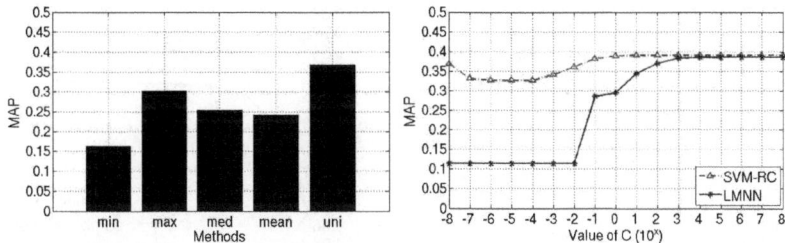

Fig. 1. Oxford Flower17. Left: baseline. Right: performance of global learning.

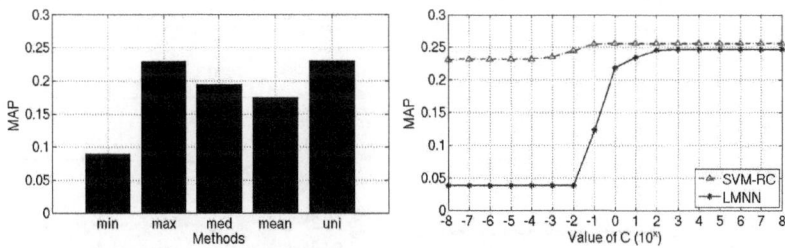

Fig. 2. Caltech101. Left: baseline. Right: performance of global learning.

such that two samples share the same label and the third one has a different label. For the MKL via LMNN scheme, we first randomly draw 100 samples as sample i. We then identify for each of them the nearest 3 samples with the same label, which form sample j. Finally, 70 samples are randomly drawn for each of the 100 "i samples" from those with different labels, and are used as sample k. This process results in $100 \times 3 \times 70 = 2.1 \times 10^4$ triplets. With $\sim 2 \times 10^4$ triplets, both methods use several GB of memory, and take ~ 15 seconds to learn a set of kernel weights on a single core processor.

We vary the value of the trade-off parameter C in both schemes from 10^{-8} to 10^8, and show in Fig. 1 right and Fig. 2 right how the performance varies accordingly. Results on both datasets show that for SVM-RC, when C is sufficiently small, the learnt kernel weights are uniform, leading to an MAP that is same as the uniform weighting scheme. For LMNN, when C is sufficiently small, the learnt kernel weights are all zeros, which means its performance becomes that of a random distance metric. For both methods, the optimal performance is reached

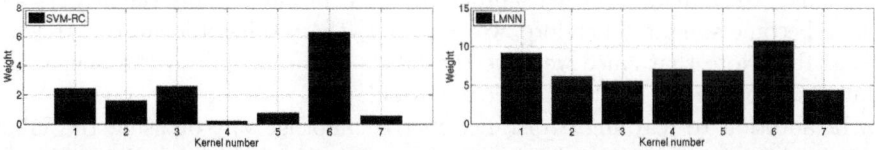

Fig. 3. Oxford Flower17. Learnt kernel weights in the global version of MKL via SVM-RC and MKL via LMNN. $C = 10^8$.

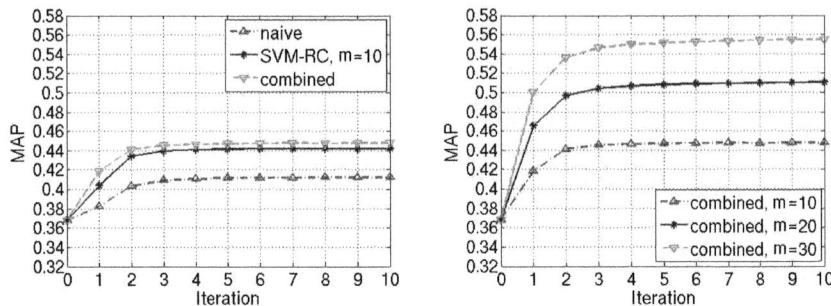

Fig. 4. Oxford Flower17. Localised learning for relevance feedback. Left: naive, learning based, and combined schemes, $m = 10$. Right: combined scheme, $m = 10, 20, 30$.

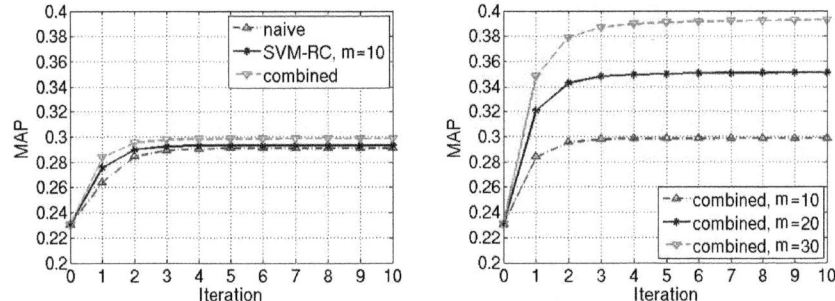

Fig. 5. Caltech101. Localised learning for relevance feedback. Left: naive, learning based, and combined schemes, $m = 10$. Right: combined scheme, $m = 10, 20, 30$.

when C is large enough. When $C = 10^8$, the MAPs of SVM-RC and LMNN formulations on the Oxford Flower17 dataset are 0.3892 and 0.3845 respectively, as compared to 0.3680, which is the MAP achieved by uniform weighting. The learnt kernel weights in both global schemes when $C = 10^8$ are shown in Fig. 3. On the Caltech101 dataset, similar improvements are observed (0.2553/0.2462 vs. 0.2303).

In the following experiment we turn to localised learning for relevance feedback. We draw triplets following the scheme outlined in Section 3.2. These "local triplets" are then used for learning a distance metric for this particular query image. Since the number of images a user labels, m, is typically small, the effect of pulling all similarly labelled samples and that of pulling only the nearest g of them become similar. Therefore, we show only the results of localised MKL via SVM-RC. Note that since we have the labels of the images in the benchmark dataset, the manual labelling process is simulated.

In addition to learning a distance metric, another way of using the labels provided by the user is simply to rank the positively labelled samples at the top of the list of retrieved images. We shall call this the naive scheme. Furthermore, this naive scheme can be combined with the learning based scheme: we learn a new distance metric using the labelled samples, retrieve again with the new metric, and then rank the positively labelled samples at the top of the new list. We shall call this the combined scheme in the following experiment.

The relevance feedback procedure can be applied iteratively. In each round, the newly labelled images are pooled with the labelled images in the previous rounds for triplet sampling, and a new distance metric is learnt, which will be used for retrieval in the next round. This allows a user to actively explore the database and improve the metric used for retrieval.

The performance of the three schemes: naive, learning based (localised MKL via SVM-RC), and combined, is plotted in Fig. 4 left and Fig. 5 left, where iteration 0 corresponds to uniform weighting of kernels. In the learning based scheme, the trade-off parameter C is set to 10^8, and the number of randomly sampled triplets is set to 10^3. It is clear from the figures that both the combined scheme and the learning based scheme outperform the naive scheme. This means that significant improvements are indeed from learning the kernel weights.

In Fig. 4 right and Fig. 5 right we show the performance of the combined scheme with various numbers of labelled images. As expected, the performance improves significantly as the number of manually labelled samples increases. Labelling even 30 images is fast as the user needs to indicate either the relevant or irrelevant images only. Finally, the MAPs of all methods under comparison on both datasets are summarised in Table 1. Note that the MAPs of localised learning are achieved without combining with the naive scheme. We can see from the table that the localised learning scheme not only outperforms the baseline methods, but also outperforms global learning. With 10^3 triplets at each iteration, it takes on average 0.061 seconds to learn the kernel weights.

Table 1. Performance on both datasets: a summary

	baseline		global		SVM-RC localised		
	single max.	uniform	SVM-RC	LMNN	$m = 10$, iter=1	$m = 30$, iter=1	$m = 30$, iter=4
Flower17	0.3022	0.3680	0.3892	0.3845	0.4036	0.4612	0.5046
Caltech101	0.2294	0.2303	0.2553	0.2462	0.2756	0.3120	0.3263

5   Conclusions

In this paper we have formulated multiple kernel learning as a distance metric learning problem. We consider two objective functions that are commonly used in distance metric learning, and optimise them under constraints based on relevance comparisons. We have proposed both global version and localised version of such a MKL via DML scheme. We argue that the localised version not only yields better performance than the global version, but also fits naturally into the framework of example based retrieval and relevance feedback. This claim is verified through experiments on two image retrieval datasets.

Acknowledgement

We would like to acknowledge support for this work from the EPSRC/UK grant EP/F069626/1 ACASVA Project.

References

1. Shawe-Taylor, J., Cristianini, N.: Kernel Methods for Pattern Analysis. Cambridge University Press, Cambridge (2004)
2. Lanckriet, G., Cristianini, N., Bartlett, P., Ghaoui, L.E., Jordan, M.: Learning the kernel matrix with semidefinite programming. JMLR 5, 27–72 (2004)
3. Bach, F., Lanckriet, G.: Multiple kernel learning, conic duality, and the smo algorithm. In: ICML (2004)
4. Sonnenburg, S., Ratsch, G., Schafer, C., Scholkopf, B.: Large scale multiple kernel learning. JMLR 7, 1531–1565 (2006)
5. Xing, E., Ng, A., Jordan, M., Russell, S.: Distance metric learning, with application to clustering with side-information. In: NIPS (2002)
6. Schultz, M., Joachims, T.: Learning a distance metric from relative comparisons. In: NIPS (2004)
7. Weinberger, K., Saul, L.: Distance metric learning for large margin nearest neighbor classification. JMLR 10, 207–244 (2009)
8. Kloft, M., Brefeld, U., Sonnenburg, S., Zien, A.: Efficient and accurate. In: NIPS (2009)
9. Ye, J., Ji, S., Chen, J.: Multi-class discriminant kernel learning via convex programming. JMLR 9, 719–758 (2008)
10. Yan, F., Mikolajczyk, K., Barnard, M., Cai, H., Kittler, J.: Lp norm multiple kernel fisher discriminant analysis for object and image categorisation. In: CVPR (2010)
11. Yan, S., Xu, D., Zhang, B., Zhang, H., Yang, Q., Lin, S.: Graph embedding and extensions: A general framework for dimensionality reduction. In: PAMI (2007)
12. Nilsback, M., Zisserman, A.: Automated flower classification over a large number of classes. In: Indian Conference on Computer Vision, Graphics and Image Processing (2008)
13. Fei-Fei, L., Fergus, R., Perona, P.: One-shot learning of object categories. PAMI 28(4), 594–611 (2006)
14. Sande, K., Gevers, T., Snoek, C.: Evaluation of color descriptors for object and scene recognition. In: CVPR (2008)
15. Grauman, K., Darrell, T.: The pyramid match kernel: Discriminative classification with sets of image features. In: ICCV (2005)
16. Lazebnik, S., Schmid, C., Ponce, J.: Beyond bags of features: Spatial pyramid matching for recognizing natural scene categories. In: CVPR (2006)

Dynamic Ensemble Selection for Off-Line Signature Verification[*]

Luana Batista, Eric Granger, and Robert Sabourin

Laboratoire d'imagerie, de vision et d'intelligence artificielle
École de technologie supérieure
1100, rue Notre-Dame Ouest, Montréal, QC, H3C 1K3, Canada
lbatista@livia.etsmtl.ca, {eric.granger,robert.sabourin}@etsmtl.ca

Abstract. Although not in widespread use in Signature Verification (SV), the performance of SV systems may be improved by using ensemble of classifiers (EoC). Given a diversified pool of classifiers, the selection of a subset to form an EoC may be performed either statically or dynamically. In this paper, two new dynamic selection (DS) strategies are proposed, namely OP-UNION and OP-ELIMINATE, both based on the K-nearest-oracles. To compare ensemble selection strategies, a hybrid generative-discriminative system for off-line SV system is considered. Experiments performed by using real-world SV data, comprised of genuine samples, and random, simple and skilled forgeries, indicate that the proposed DS strategies achieve a significantly higher level of performance in off-line SV than other well-known DS and static selection (SS) strategies. Improvements are most notable in problems where a significant level of uncertainty emerges due a considerable amount of intra-class variability.

1 Introduction

Signature Verification (SV) systems are relevant in many real-world applications, such as check cashing, credit card transactions and document authentication. In off-line SV, handwritten signatures are transcribed on sheets of paper, and at some later time scanned in order to obtain a digital representation. Given a digitized signature, an off-line SV system typically performs preprocessing, feature extraction and classification to authenticate the signature of an individual.

Handwritten signatures are behavioural biometric traits that are known to incorporate a considerable amount of intra-class variability. Although not in widespread use in off-line SV, a promising way to improving system performance is through ensemble of classifiers (EoC) [1,4]. The motivation of using EoCs stems from the fact that a diverse set of classifiers usually make different errors on input samples. Indeed, when the response of a set of $\mathcal{C}$ classifiers is averaged, the variance contribution in the bias-variance decomposition decreases by $1/\mathcal{C}$, resulting in a smaller expected classification error [9].

---

[*] This research has been supported by the Natural Sciences and Engineering Research Council of Canada.

C. Sansone, J. Kittler, and F. Roli (Eds.): MCS 2011, LNCS 6713, pp. 157–166, 2011.

Bagging, boosting and random subspaces are well-known methods for creating diversity among classifiers. While bagging and boosting use different samples subsets to train different classifiers, the random subspace method use different subspaces of the original input feature space. Given a diversified pool of classifiers, an important issue is the selection of a subset to form an EoC, such that the recognition rates are maximized during operations [6]. EoC selection may be performed either statically or dynamically. Based on a set of reference samples not used during training, static selection (SS) strategies select the EoC that provides the best classification rates on that set. Then, this EoC is employed during operations to classify any input sample. Dynamic selection (DS) strategies also need a reference set to select the best EoC, although this task is performed during operations, by taking into account the specific characteristics of a given sample to be classified.

In a pattern recognition system that starts with a limited number of reference samples, it is difficult to define *a priori* a single best EoC for the application. Ideally, the EoC should be continuously adapted whenever new reference samples become available. With DS, this new data can be incorporated to the reference set (after being classified by the pool of classifiers) without any additional step.

KNORA (*K*-nearest-oracles) is a DS strategy that has been successfully applied to handwritten numeral recognition [6]. For each input sample, the KNORA strategy finds its *K*-nearest neighbors in the reference set, and then selects the classifiers that have correctly classified those neighbors. Finally, the selected classifiers are combined in order to classify the input sample. The main drawback of KNORA is that a robust set of features must be defined in order to compute similarity between the input sample and the samples in the DS database.

As an alternative, this paper propose two new DS strategies, namely OP-UNION and OP-ELIMINATE, that use the classifier outputs (i.e., the output profile) to find the *K*-nearest neighbors. To validate the proposed and other reference dynamic and static selection strategies, a multi-classifier generative-discriminative system is considered. In this system, Hidden Markov Models (HMMs) are employed as feature extractors followed by Support Vector Machines (SVMs) as two-class classifiers. Proof-of-concept experiments are carried out on a real-world signature database [4,5,8], comprised of genuine samples, and random, simple and skilled forgeries. The rest of this paper is organized as follows. The next section presents the hybrid generative-discriminative system for off-line SV. Then, Section 3 proposes two new DS strategies. Finally, Section 4 describes the experimental methodology, and Section 5 presents and discusses the experiments.

2   A Hybrid System for Off-Line SV

Let $T^i = I^i_{trn(l)}$, $1 \leq l \leq N$, be the training set used to design a SV system for writer i. The set T^i contains genuine signature samples supplied by writer i, as well as random forgery samples supplied by other writers not enrolled to the system. For each signature $I^i_{trn(l)}$ in the training set T^i, a set of features is generated (see Figure 1). First, $I^i_{trn(l)}$ is described by means of pixel densities, which are extracted through a grid composed of retangular cells. Each column of cells j is converted into a low-level feature vector $\mathbf{F}^i_j = \{f^i_{j1}, f^i_{j2}, ...\}$, where each vector component $f^i_{jh} \in [0,1]$. These components correspond to the number of black pixels in a cell divided by the total number of pixels of this cell. The signature $I^i_{trn(l)}$ is therefore represented by a set of low-level feature vectors $\mathcal{F}^i_{trn(l)} = \{\mathbf{F}^i_j\}$, $1 \le j \le col$, where col is the number of columns in the grid.

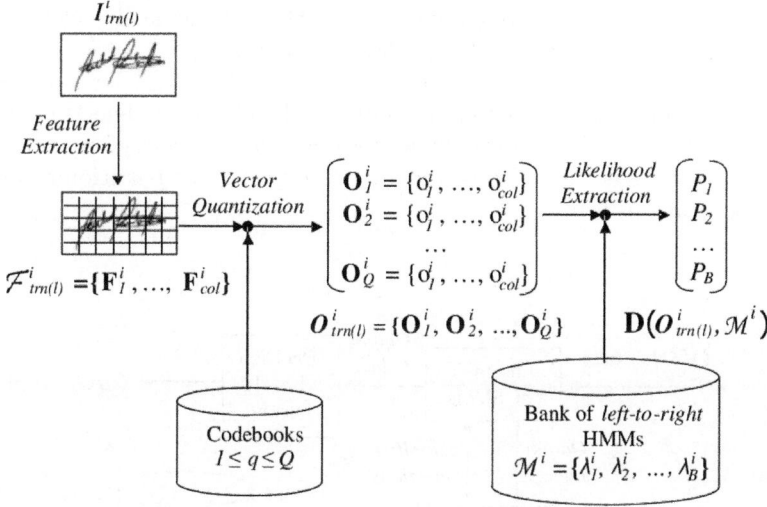

Fig. 1. Design of the generative stage for a specific writer i. After feature extraction, the signature is quantized into $\mathcal{Q}$ different sequences, from which $B > \mathcal{Q}$ likelihoods are obtained.

Then, $\mathcal{F}^i_{trn(l)}$ is quantized into a sequence of discrete observations $\mathbf{O}^i_q = \{o^i_j\}$, for $1 \le j \le col$. Each observation o^i_j is a symbol provided by the codebook q (generated using the K-$means$ algorithm). Since $\mathcal{Q}$ different codebooks are employed per writer i, each training signature $I^i_{trn(l)}$ yields a set of observation sequences $\mathbf{O}^i_{trn(l)} = \{\mathbf{O}^i_q\}$, for $1 \le q \le \mathcal{Q}$. The set of observation sequences, $\mathbf{O}^i_{trn(l)}$, is then input to the bank of $left$-to-$right$ HMMs $\mathcal{M}^i = \{\lambda^i_b\}$, $1 \le b \le B$, from which a high-level feature vector $\mathbf{D}\left(\mathbf{O}^i_{trn(l)}, \mathcal{M}^i\right) = \{P_1, ..., P_B\}$ is extracted. Each component P_b is a likelihood computed between an observation sequence $\mathbf{O}^i_q$ and a HMM λ^i_b, where λ^i_b can either correspond to the genuine class (i.e., trained with genuine samples from writer i), or to the impostor class (i.e., trained with random forgery samples). It is worth noting that the same sequences $\mathbf{O}^i_{trn(l)}$, $1 \le l \le N$, used to obtain the HMM likelihood vectors are also used to train the HMMs in $\mathcal{M}^i$. Appart from the different codebooks, a different number of states is employed to produce a bank of HMMs.

As long HMM likelihood vectors are produced during the design of the generative stage, the random subspace method (RSM) is used to select the input space in which multiple SVMs are trained. For each random subspace r, $1 \leq r \leq \mathcal{R}$, a smaller subset of likelihoods is randomly selected, with replacement, from $\mathbf{D}\left(O^i_{trn(l)}, \mathcal{M}^i\right)$, $1 \leq l \leq N$, and used to train a different SVM. During operations, a given input signature I^i_{tst} follows the same steps of feature extraction, vector quantization and likelihood extraction as performed with a training signature, resulting in the HMM likelihood vector $\mathbf{D}\left(O^i_{tst}, \mathcal{M}^i\right)$. Then, based on previously-classified signature samples – stored in the dynamic selection (DS) database –, the most accurate ensemble of SVMs is dynamically selected from the pool and used to classify $\mathbf{D}\left(O^i_{tst}, \mathcal{M}^i\right)$ (see Figure 2).

As described in Section 3, signature samples selected from the DS database are the K-nearest neighbors of the input sample to be classified. The DS database contains genuine samples supplied by writer i, as well as random forgery samples taken from writers not enrolled to the system. explains the partitioning of each dataset used in this work.

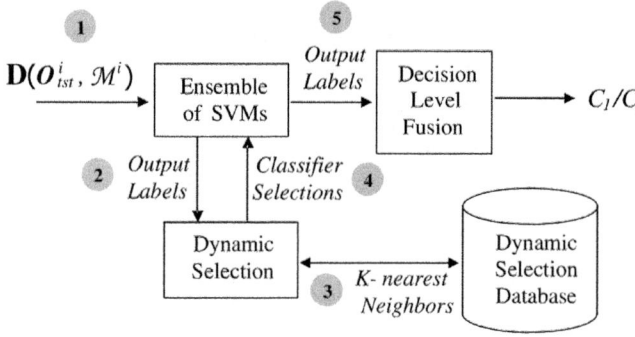

Fig. 2. System architecture of the discriminative stage for a specific writer i

Bank of HMMs. Let $\mathcal{M}^i = \{w_1 \cup w_2\}$ be the bank of HMMs, where $w_1 = \{\lambda_1^{(C_1)}, \lambda_2^{(C_1)}, ..., \lambda_R^{(C_1)}\}$ is the set of R HMMs of the genuine class C_1, and $w_2 = \{\lambda_1^{(C_2)}, \lambda_2^{(C_2)}, ..., \lambda_S^{(C_2)}\}$ is the set of S HMMs of the impostor's class C_2. Given the set of observation sequences $O^i_{trn(l)} = \{O^i_1, O^i_2, ..., O^i_Q\}$ extracted from a training signature $I^i_{trn(l)}$, the vector $\mathbf{D}\left(O^i_{trn(l)}, \mathcal{M}^i\right)$ is obtained by computing the likelihoods of $O^i_{trn(l)}$ for each HMM in $\mathcal{M}^i$, that is,

$$\mathbf{D}\left(O^i_{trn(l)}, \mathcal{M}^i\right) = \begin{bmatrix} P(O^i_q/\lambda_1^{(C_1)}) \\ P(O^i_q/\lambda_2^{(C_1)}) \\ ... \\ P(O^i_q/\lambda_S^{(C_2)}) \end{bmatrix} \tag{1}$$

If, for instance, $\lambda_1^{(C_1)}$ and $\lambda_S^{(C_2)}$ are trained with observation sequences extracted from the codebook $q = 10$, a compatible sequence from $O^i_{trn(l)}$, that is, $\mathbf{O}^i_{q=10}$, must be sent to both. Finally, the vector $\mathbf{D}\left(O^i_{trn(l)}, \mathcal{M}^i\right)$ is labeled according to the class of $O^i_{trn(l)}$. It is worth noting that, if $O^i_{trn(l)}$ belongs to class C_1, $\mathbf{D}\left(O^i_{trn(l)}, \mathcal{M}^i\right)$ should contain higher values in the first R positions and smaller values in the remaining S positions, allowing a two-class classifier to discriminate samples of class C_1 from class C_2.

3   New Strategies for Dynamic Ensemble Selection

Let $O^i_{ds(j)}$, $1 \le j \le M$, be the sequences of observations extracted from the DS database of writer i, and $\mathbf{D}\left(O^i_{ds(j)}, \mathcal{M}^i\right)$ be their corresponding likelihood vectors [1], for $1 \le j \le M$. For each DS vector $\mathbf{D}\left(O^i_{ds(j)}, \mathcal{M}^i\right)$, an output profile (OP) is calculated as follows. First, $\mathbf{D}\left(O^i_{ds(j)}, \mathcal{M}^i\right)$ is input to all SVM classifiers c_r, $r = 1, 2, ..., \mathcal{R}$, in the pool of classifiers $\mathcal{C}$. Each c_r receives as input only the vector positions related to its respective subspace. Then, the resulting output labels are stored as a vector to form a DS output profile, $\mathbf{OP}\left(\mathbf{D}\left(O^i_{ds(j)}, \mathcal{M}^i\right)\right)$. This procedure is repeated for all DS vectors, resulting in a set of DS-OPs. For simplicity, it is assumed that the DS-OPs are also stored in the DS database.

During operations, when a test vector $\mathbf{D}\left(O^i_{tst}, \mathcal{M}^i\right)$ is presented to the off-line SV system, four main steps are performed. First, the output profile $\mathbf{OP}\left(\mathbf{D}\left(O^i_{tst}, \mathcal{M}^i\right)\right)$ is calculated, as performed for the DS vectors. Second, the Euclidean distance is computed between $\mathbf{OP}\left(\mathbf{D}\left(O^i_{tst}, \mathcal{M}^i\right)\right)$ and each DS-OP, in order to find its K-nearest neighbors. Third, the SVMs that correctly classify the K corresponding DS vectors are selected and used to classify $\mathbf{D}\left(O^i_{tst}, \mathcal{M}^i\right)$. Finally, the SVMs decisions are fused through majority voting. The two following variants of KNORA are proposed to manage output profiles:

1) **OP-ELIMINATE.** Given the test vector $\mathbf{D}\left(O^i_{tst}, \mathcal{M}^i\right)$, the objective of this first variant is to find an ensemble of up to K SVMs that simultaneously classify its K-nearest neighbors in the DS database correctly. After obtaining $\mathbf{OP}\left(\mathbf{D}\left(O^i_{tst}, \mathcal{M}^i\right)\right)$, its K-nearest DS-OPs, $\mathbf{OP}\left(\mathbf{D}\left(O^i_{ds(k)}, \mathcal{M}^i\right)\right)$, $1 \le k \le K$, are found via Euclidean distance. For each SVM c_r, $r = 1, 2, ..., \mathcal{R}$, in the pool $\mathcal{C}$, the OP-ELIMINATE algorithm verifies if c_r has previously classified all corresponding DS vectors $\mathbf{D}\left(O^i_{ds(k)}, \mathcal{M}^i\right)$, $1 \le k \le K$, correctly. If so, c_r is added to the ensemble E; otherwise, the next SVM in the pool is verified. In the case where no classifier ensemble can correctly classify all K DS vectors, the value of K is decreased until at least one SVM can correctly classify a DS

---

[1] During dynamic selection, $\mathbf{D}\left(O^i_{ds(j)}, \mathcal{M}^i\right)$ is refered as a DS vector.

vector. Finally, each SVM in the ensemble E submites a vote on the test vector $\mathbf{D}\left(O_{tst}^i, \mathcal{M}^i\right)$, where the final classification label $\mathcal{L}$ is obtained by using the majority vote rule.

2) OP-UNION. Given the test vector $\mathbf{D}\left(O_{tst}^i, \mathcal{M}^i\right)$ and its K-nearest neighbors in the DS database, the objective of this variant is to find for each neighbor k, $1 \le k \le K$, an ensemble of up to K SVMs that correctly classify it. First, the test output profile, $\mathbf{OP}\left(\mathbf{D}\left(O_{tst}^i, \mathcal{M}^i\right)\right)$, and its K-nearest DS-OPs, $\mathbf{OP}\left(\mathbf{D}\left(O_{ds(k)}^i, \mathcal{M}^i\right)\right)$, $1 \le k \le K$, are obtained, such as performed for OP-ELIMINATE. For each neighbor k, and for each SVM c_r, $r = 1, 2, ..., \mathcal{R}$, in the pool $\mathcal{C}$, the OP-UNION algorithm then verifies if c_r has previously classified the DS vector $\mathbf{D}\left(O_{ds(k)}^i, \mathcal{M}^i\right)$ correctly. If so, c_r is added to the ensemble E_k; otherwise, the next SVM in the pool is verified. After applying this procedure to all K-NNs, the SVMs in each ensemble E_k are combined in order to classify the test vector $\mathbf{D}\left(O_{tst}^i, \mathcal{M}^i\right)$. Finally, the final classification label $\mathcal{L}$ is obtained by using the majority vote rule. Note that a same SVM can give more than one vote if it correctly classifies more than one DS vectors.

4 Experimental Methodology

Brazilian SV database. It contains 7920 samples of signatures that were digitized as 8-bit greyscale images over 400X1000 pixels, at resolution of 300 dpi. The signatures were provided by 168 writers and are organized in two sets: the development database (DB_{dev}) and the exploitation database (DB_{exp}). DB_{exp}contains signature samples from writers enrolled to the system, and is used to model the genuine class. It is composed of 3600 signatures supplied by 60 writers. Each writer i has 40 genuine samples, 10 simple forgeries and 10 skilled forgeries. 20 genuine samples are available for training ($\mathcal{T}_{exp(20)}^i$) and 10 for validation ($\mathcal{V}_{exp(10)}^i$). As test set, each writer i has 10 genuine samples, 10 random forgeries, 10 simple forgeries and 10 skilled forgeries; where the random forgeries are genuine samples randomly selected from other writers in DB_{exp}.

DB_{dev}contains signature samples from writers not enrolled to the system, and is used as *prior* knowledge to design the codebooks and the impostor class. It is composed of 4320 genuine samples supplied by 108 writers. Each writer j has 40 genuine samples, where 20 are available for training ($\mathcal{T}_{dev(20)}^j$) and 10 for validation ($\mathcal{V}_{dev(10)}^j$). The remaining 10 samples, available for test, are not employed in this work. Given a writer i enrolled to the system, $DB_{dev}$and $DB_{exp}$are used to compose different datasets employed in different phases of the system design (see Table 1).

After binarization, the signature images are divided in 62 columns of cells, where each cell is a rectangle composed of 40X16 pixels. To absorb the horizontal variability of the signatures, the images are aligned to the left and the blank cells in the end of the images are discarded. Then, feature extraction is performed.

Table 1. Datasets for a specific writer i, using the Brazilian SV database

Dataset Name	Task	Genuine Samples	Random Forgery Samples
DB^i_{hmm}	HMM Training	$T^i_{exp(20)} + V^i_{exp(10)}$	$T^{j=1:108}_{dev(20)} + V^{j=1:108}_{dev(10)}$
DB^i_{svm}	SVM Training	$T^i_{exp(20)}$	20 from $T^{j=1:180}_{dev(20)}$
DB^i_{grid}	SVM Grid Search		10 from $V^{j=1:180}_{dev(10)}$
DB^i_{roc}	ROC Curve	$V^i_{exp(10)}$	$V^{j=1:180}_{dev(10)}$
DB^i_{ds}	Dynamic Selection		

HMM Training. 29 different codebooks q $(1 \leq q \leq 29)$ are generated by varying the number of clusters from 10 to 150, in steps of 5; using all training and validation signatures of DB_{dev}. Given a writer i and a codebook q, DB^i_{hmm} is employed to train a set of discrete *left-to-right* HMMs with different number of states, using the Baum-Welch algorithm. As the number of states varies from 2 to the lenght of the smallest sequence used for training (L_{min}), the genuine class is composed of a variable number HMMs (i.e., 29x(L_{min}-1)) that depends on the writer's signature size. On the other hand, to compose the impostor class, there are thousands of HMMs taken from the writers in DB_{dev}.

SVM Training. By using the RSM, 100 subspaces composed of 30 dimensions each (i.e., 15 likelihoods randomly selected from each class) are used to train 100 different SVMs (RBF kernel) per writer. For a same writer i, the training set, DB^i_{svm}, remains the same for all 100 SVMs.

Comparison of Techniques. In this paper, the simulation results obtained with OP-UNION/ELIMINATE are compared with KNORA-UNION/ELIMINA-TE [6], with the standard combination of all classifiers, and with Decision Templates (DT) [7] – a well-known DS method in the multi-classifier system (MCS) community. With OP-UNION/ELIMINATE, the search for the K-nearest neighbors is performed using the output labels provided by all 100 SVMs; while with KNORA-UNION/ELIMINATE, only the SVM input subspace providing the lowest error rates on DB^i_{ds} is used during the search. The value of K is defined as being half of the number of genuine samples in DB^i_{ds}, that is, 5. Comparisons are performed as well with two reference systems proposed in our previous research, that is, (i) a traditional generative system based on HMMs [2] (refered in this paper as baseline system), and (ii) a hybrid system based on the static selection of generative-discriminative ensembles [3]. In both systems, HMMs are trained by using $T^i_{exp(20)} + V^i_{exp(10)}$, from the Brazilian SV database.

Performance Evaluation. The overall system performance is measured through different operating points of an averaged ROC curve. To obtain this curve using DB^i_{roc}, the operating points $\{TPR_i, FPR_i\}$ related to different users, $i = 1, .., N$, are averaged if they have a same true negative rate (TNR, or

$1 - FPR$) – refered in this paper as γ. The average error rate (AER) indicates the total error of the system for a specific γ, where the false negative rate (FNR) and the false positive rate (FPR) are averaged taking into account the *a priori* probabilities. In the experiments, the FPR is calculated with respect to three forgery types: random, simple and skilled.

5   Simulation Results

Figure 3 presents the $AERs$ curves obtained on test data (DB_{tst}^i) as function of operating points (γ). These results indicate that the proposed DS strategies, i.e., OP-UNION and OP-ELIMINATE, provided the lowest $AERs$, demonstrating the advantage of using a DS approach based on output profiles – as opposed to KNORA, where the input feature space is used to find the K-nearest DS samples. It is also beneficial to employ EoCs composed of a small set of base classifiers – in contrast to DTs and to the standard technique of combination of classifiers, where all base classifiers in the pool are part of the ensemble.

OP-UNION and OP-ELIMINATE strategies also achieved $AERs$ that are lower than those obtained with SS. This represents a situation where DS is superior to SS, i.e., in a problem where a significant level of uncertainty emerges due a considerable amount of intra-class variability. Finally, the lowest performance of the baseline system is obtained because a pure generative approach is adopted for system design, i.e., only the genuine class is modeled, and a single HMM is employed per writer.

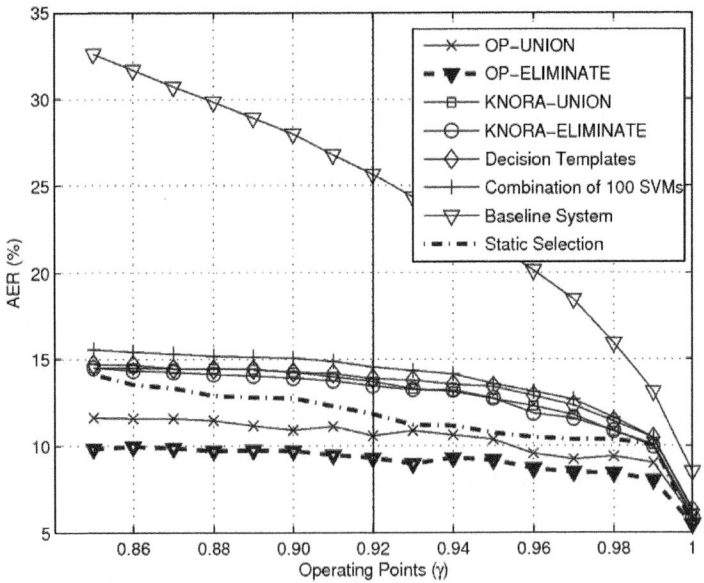

Fig. 3. *AERs versus* operating points (γ) obtained on Brazilian test data using off-line SV systems that employ different ensemble selection techniques

Table 2. Overall error rates (%) obtained on test data for $\gamma=0.92$

Method	FNR	FPR_{random}	FPR_{simple}	$FPR_{skilled}$	AER
OP-UNION	**4.17**	**1.50**	**3.50**	**33.17**	**10.58**
OP-ELIMINATE	**5.17**	**1.00**	**2.83**	**28.17**	**9.29**
KNORA-UNION	2.17	2.50	6.83	43.33	13.71
KNORA-ELIMINATE	2.33	2.67	6.33	42.50	13.46
Decision Templates	2.67	2.17	6.33	44.33	13.88
Combination of 100 SVMs	2.33	2.67	7.33	45.83	14.54
Static Selection [3]	2.17	5.67	5.00	34.50	11.83
Baseline [2]	0.33	9.83	18.17	74.33	25.67

Table 3. Overall error rates (%) compared to reference SV systems from literature

Method	FNR	FPR_{random}	FPR_{simple}	$FPR_{skilled}$	AER
Batista et al. [2]	9.83	0.00	1.00	20.33	7.79
Bertolini et al. [4]	11.32	4.32	3.00	6.48	6.28
Justino et al. [5]	2.17	1.23	3.17	36.57	7.87
Santos et al. [8]	10.33	4.41	1.67	15.67	8.02
OP-UNION $(\gamma = 1.0)$	**8.17**	**0.67**	**0.67**	**14.00**	**5.88**
OP-ELIMINATE $(\gamma = 1.0)$	**7.50**	**0.33**	**0.50**	**13.50**	**5.46**

Table 2 presents the overall results for $\gamma=0.92$. Note that the proposed DS strategies provide lower $FPRs$ at the expense of higher $FNRs$. In practice, the trade-off between FPR and FNR can be adjusted according to the risk linked to an input sample. In banking applications, for example, the decision to use a specific operating point may be associated with the amount of the check. In the simplest case, for a user that rarely signs high value checks, big amounts would require operating points related to low $FPRs$, such as would be provided by a γclose to 1; while lower amounts would require operating points related to low $FNRs$, since the user would not feel comfortable with frequent rejections.

Finally, Table 3 presents a comparison with other systems designed with the Brazilian SV database. By assuming that the objective of these systems is to minimize the AER, the comparison is performed with $\gamma= 1$.

6   Conclusions

In this paper, two new DS strategies based on KNORA, namely OP-UNION and OP-ELIMINATE, are proposed to improve performance of off-line SV systems. These strategies employ the classifier outputs (i.e., the output profile), instead of the input feature space, to find the most accurate EoC for a given input sample. To compare ensemble selection strategies, a hybrid off-line SV system is considered. In this system, HMMs are employed as feature extractors followed by SVMs as two-class classifiers.

Experiments performed using real world signature data indicate that OP-UNION/ELIMINATE can achieve a higher level of accuracy in off-line SV than other reference DS and SS strategies. This is especially true in problems where a significant level of uncertainty emerges due a considerable amount of intra-class variability. Future work consists of investigating the adaptive capabilities of the proposed strategies for incremental learning of new reference samples.

References

1. Bajaj, R., Chaudhury, S.: Signature verification using multiple neural classifiers. Pattern Recognition 30, 1–7 (1997)
2. Batista, L., Granger, E., Sabourin, R.: Improving performance of hmm-based off-line signature verification systems through a multi-hypothesis approach. International Journal on Document Analysis and Recognition 13, 33–47 (2010)
3. Batista, L., Granger, E., Sabourin, R.: A multi-classifier system for off-line signature verification based on dissimilarity representation. In: Ninth International Workshop on Multiple Clasifier System, pp. 264–273 (2010)
4. Bertolini, D., Oliveira, L., Justino, E., Sabourin, R.: Reducing forgeries in writer-independent off-line signature verification through ensemble of classifiers. Pattern Recognition 43, 387–396 (2010)
5. Justino, E., Bortolozzi, F., Sabourin, R.: Off-line signature verification using hmm for random, simple and skilled forgeries. In: International Conference on Document Analysis and Recognition, pp. 105–110 (2001)
6. Ko, A., Sabourin, R., Britto, A.: From dynamic classifier selection to dynamic ensemble selection. Pattern Recognition 41(5), 1718–1731 (2008)
7. Kuncheva, L., Bezdek, J., Duin, R.: Decision templates for multiple classifier fusion: an experimental comparison. Pattern Recognition 34, 299–314 (2001)
8. Santos, C., Justino, E., Bortolozzi, F., Sabourin, R.: An off-line signature verification method based on the questioned document expert's approach and a neural network classifier. In: International Workshop on Frontiers in Handwriting Recognition, pp. 498–502 (2004)
9. Tumer, K., Ghosh, J.: Analysis of decision boundaries in linearly combined neural classifiers. Pattern Recognition 29(2), 341–348 (1996)

Classifier Selection Approaches
for Multi-label Problems

Ignazio Pillai, Giorgio Fumera, and Fabio Roli

Deparment of Electrical and Electronic Engineering, Univ. of Cagliari
Piazza d'Armi, 09123 Cagliari, Italy
{pillai,fumera,roli}@diee.unica.it

Abstract. While it is known that multiple classifier systems can be effective also in multi-label problems, only the classifier fusion approach has been considered so far. In this paper we focus on the classifier selection approach instead. We propose an implementation of this approach specific to multi-label classifiers, based on selecting the outputs of a possibly *different* subset of multi-label classifiers for each class. We then derive static selection criteria for the macro- and micro-averaged F measure, which is widely used in multi-label problems. Preliminary experimental results show that the considered selection strategy can exploit the complementarity of an ensemble of multi-label classifiers more effectively than selection approaches analogous to the ones used in single-label problems, which select the outputs of the *same* classifier subset for all classes. Our results also show that the derived selection criteria can provide a better trade-off between the macro- and micro-averaged F measure, despite it is known that an increase in either of them is usually attained at the expense of the other one.

Keywords: Multi-label classification, Multiple classifier systems, Classifier selection.

1 Introduction

In multi-label classification problems each sample can belong to more than one class, contrary to traditional, single-label problems. Multi-label problems occur in several applications related to retrieval tasks, like text categorisation, image annotation, protein function classification and music classification, and are receiving an increasing interest from the pattern recognition and machine learning literature. So far, several works have shown that multiple classifier systems (MCSs) can be effectively exploited to improve classification performance also in multi-label problems [7,6,9,10]. In [9] it was also claimed that MCSs can be useful to deal with imbalanced class distribution, which often occurs in multi-label problems.

To our knowledge, all previous works considered the classifier *fusion* approach, which consists in the combination of the outputs of *all* the available classifiers. In this work we focus on the classifier *selection* approach instead. We first present

C. Sansone, J. Kittler, and F. Roli (Eds.): MCS 2011, LNCS 6713, pp. 167–176, 2011.

in Sect. 2 a selection approach specific to multi-label classifiers. It is based on the selection of the output of a possibly *different* subset of multi-label classifiers for each class. This potentially allows one to better exploit the complementarity between the available multi-label classifiers on the different classes. We then focus in Sect. 4 on static selection, and develop selection criteria based on the macro- and micro-averaged Van Rijsbergen's F measure, which is a widely used performance measure in multi-label problems [2,8,9,11]. The F measure is described in detail in Sect. 3. From the analysis of the proposed selection criteria, we argue that our method may also provide a good trade-off between the macro- and micro-averaged F measure. This is an interesting result, as it is known that an improvement in the macro-averaged F measure can be usually be attained only at the expense of the micro-averaged one, and vice versa [13]. In Sect. 5 we give an empirical evaluation of the proposed static classifier selection approach on three multi-label data sets related to text categorisation and gene functionality classification tasks. Conclusions are finally drawn in Sect. 6.

2   A Classifier Selection Approach for Multi-label Problems

In the following we denote with $\Omega = \{\omega_1, \ldots, \omega_N\}$ the set of classes of a given problem, and the feature vector of a sample as $\mathbf{x} \in X \subseteq \mathbb{R}^n$, where n is the size of the feature space X. In a single-label problem each sample belongs to exactly one class, and a classifier implements a decision function $f : X \to \Omega$.

Given an ensemble of single-label classifiers $f^1, \ldots, f^L$, the rationale of selection methods is that different classifiers can be more accurate than others in different regions of the feature space [12,4]. Accordingly, given a testing sample $\mathbf{x}$, these methods aim to select one of the classifiers that correctly classify $\mathbf{x}$ (if any). This can be done "statically", by defining at the design phase the so-called region of competence of each classifier in the feature space. Each testing sample is then labelled by the classifier associated to the region in which it falls. Classifier selection can also be done "dynamically". In this case, the classifier which exhibits the highest accuracy in a neighbourhood of the testing sample is selected. Such "local" accuracy is estimated online, at the classification phase. To the scope of this work, a somewhat related approach is the selection of a *subset* of classifiers to be fused, out of a larger ensemble. It is based on a similar rationale as the one above: a different subset of classifiers can be more effective than the whole ensemble, on different testing samples [14]. This approach is usually implemented statically.

As in multi-label problems each sample can belong to more than one class, a multi-label classifier implements a decision function $f : X \to \{+1, -1\}^N$, where the value $+1$ (-1) in the k-th element of the vector $f(\mathbf{x})$ means that the sample $\mathbf{x}$ is labelled as (not) belonging to ω_k. The classifier (subset) selection approaches described above can be used with multi-label classifiers as well. In addition, for multi-label classifiers it is possible to implement a different selection approach. Consider an ensemble of multi-label classifiers $f^1, \ldots, f^L$, and let us denote with $f_k^i(\mathbf{x})$ the output of the i-th classifier for the k-th class. Instead of selecting the same subset of L' classifiers ($1 \leq L' < L$) to label a testing sample $\mathbf{x}$, as happens in single-label problems, one can also select a *different* subset of L' classifiers for each class. In other words, the decision whether $\mathbf{x}$ belongs or not to ω_k is taken by combining the outputs $f_k^{i_1(k)}, f_k^{i_2(k)}, \ldots, f_k^{i_{L'}(k)}$, where a different subset $\{i_1(k), i_2(k), \ldots, i_{L'}(k)\} \subset \{1, \ldots, L\}$ can be chosen for each ω_k. We will denote this approach as "Hybrid Selection of Multi-label classifiers" (HSM), where "hybrid" refers to the selection a possibly different classifier subset for each class.

To implement the HSM approach, a seemingly reasonable criterion is to select for each class the subset of L' classifiers which exhibit the highest classification performance *on that class* (either a local performance measure around a given testing sample, in the case of dynamic selection, or the performance in pre-defined regions of the feature space, in the case of static selection). However, as we will show in the next sections, it turns out that this criterion is not always suitable to maximise the *overall* classification performance of a multi-label classifier evaluated on all classes. This is due to the particular performance measures usually used in multi-label problems, which are based on precision and recall. In particular, the above criterion is suitable for macro-averaged performance measures, but not for micro-averaged ones. This implies that, to maximise a micro-averaged performance measure, an exhaustive search over all the possible $\left(\frac{L}{L'}\right)^N$ choices of L' out of L classifiers for each of the N classes is required, which is clearly impractical. To address this issue, we first describe in detail performance measures based on precision and recall in Sect. 3, and then derive a suboptimal selection criterion for micro-averaged measures in Sect. 4.

3   Performance Measures for Multi-label Classifiers

The performance of multi-label classifiers is measured in terms of precision and recall, as multi-label problems usually occur in retrieval tasks. In the field of information retrieval, precision and recall are respectively defined as the probability that a retrieved document is relevant to a given query or topic, and as the probability that a relevant document is retrieved. In a multi-label classification problem, each class corresponds to a distinct topic. Accordingly, precision and recall for the k-th class are defined respectively as $p_k = \mathrm{P}(\mathbf{x} \in \omega_k \mid f_k(\mathbf{x}) = 1)$, and $r_k = \mathrm{P}(f_k(\mathbf{x}) = 1 \mid \mathbf{x} \in \omega_k)$. Ideally, both measures should equal 1. However, in practice a higher precision can be attained only at the expense of a lower recall, and vice versa. In practice, they can be estimated from a multi-label data set as:

$$\hat{p}_k = \frac{TP_k}{TP_k + FP_k}, \hat{r}_k = \frac{TP_k}{TP_k + FN_k}, \tag{1}$$

where TP_k (true positive) is the number of samples that are correctly labelled as belonging to ω_k, while FP_k (false positive) and FN_k (false negative) are defined analogously.

To obtain a scalar performance measure, the Van Rijsbergen's F measure is often used. For a class ω_k it is defined as:

$$\hat{F}_{\beta,k} = \frac{1 + \beta^2}{1/\hat{p}_k + \beta^2/\hat{r}_k}, \tag{2}$$

where $\beta \in [0, +\infty]$ allows to weigh the relative importance of precision and recall. In particular, for $\beta = 1$ the F measure equals their harmonic mean.

The overall precision and recall on all categories can be computed either by macro- or micro-averaging the class-related values, depending on application requirements [8]. We will denote macro- and micro-averaged values respectively with the superscripts 'M' and 'm'. Macro-averaging simply consists in averaging the category-related values. The corresponding F measure is:

$$\hat{F}_{\beta}^{M} = \frac{1}{N} \sum_{k=1}^{N} \hat{F}_{\beta,k} = \frac{1}{N} \sum_{k=1}^{N} (1 + \beta^2) / \left((1 + \beta^2) + \frac{FP_k + \beta^2 FN_k}{TP_k} \right). \tag{3}$$

The micro-averaged precision and recall are instead defined as:

$$\hat{p}^{m} = \frac{\sum_{k=1}^{N} TP_k}{\sum_{k=1}^{N} (TP_k + FP_k)}, \quad \hat{r}^{m} = \frac{\sum_{k=1}^{N} TP_k}{\sum_{k=1}^{N} (TP_k + FN_k)}, \tag{4}$$

and the corresponding F measure is defined as $\hat{F}_{\beta}^{m} = \frac{1+\beta^2}{1/\hat{p}^m + \beta^2/\hat{r}^m}$ [13], which after some algebra leads to:

$$\hat{F}_{\beta}^{m} = (1 + \beta^2) / \left((1 + \beta^2) + \frac{\sum_{k=1}^{N} (FP_k + \beta^2 FN_k)}{\sum_{k=1}^{N} TP_k} \right). \tag{5}$$

It is known that macro-averaged measures are dominated by the performance on rare classes, while the opposite happens in the case of micro-averaging [8,13]. Furthermore, an improvement on one of them can be usually attained only at the expense of the other one, especially when there are rare categories [2]. In the rest of this paper we will consider only the F measure, as it is widely used in multi-label tasks, is easier to handle being a scalar measure, and can be used to find a trade-off between precision and recall [13].

4   Criteria Based on the F Measure for Static Multi-label Classifier Selection

In this section we discuss selection criteria for the static implementation of the HSM method, when classification accuracy is evaluated in terms of the F measure. We consider first the case of $L' = 1$. In this case, for each class ω_k we want to select a single, possibly different classifier $f^{i(k)}$, $i(k) \in \{1, \ldots L\}$. What we obtain can be seen as a new, single multi-label classifier whose outputs for the N classes are given by $f_1^{i(1)}(\mathbf{x}), f_2^{i(2)}(\mathbf{x}), \ldots, f_N^{i(N)}(\mathbf{x})$. The final goal is to maximise the overall F measure (either macro- or micro-averaged).

From Eq. (2) it can be seen that maximising the macro-averaged F measure $\hat{F}_\beta^M$ amounts to independently maximise the $\hat{F}_{\beta,k}$ measure of each class. This implies that the classifier for each class can be chosen independently on the other classes. It is also easy to see that to maximise $\hat{F}_{\beta,k}$ one should choose the classifier $f^{i(k)}$ such that:

$$i(k) = \arg \min_{i \in \{1,...,L\}} \frac{FP_k^i + \beta^2 FN_k^i}{TP_k^i}, \qquad (6)$$

where FP_k^i denotes the number of samples erroneously labelled as belonging to ω_k by f^i, and similarly for FN_k^i and TP_k^i. Obviously, these terms have to be estimated from validation samples. We name the above selection criterion HSM^M (the superscript 'M' stands for "macro-averaging").

In the case of the micro-averaged $\hat{F}_\beta^m$ measure instead, Eq. (5) shows that it can not be maximised by independently considering the contribution of the different classes. This can be seen by noting that maximising $\hat{F}_\beta^m$ amounts to minimise the ratio between the two summands in Eq. (5). Let us rewrite this term by separating the contribution of any class ω_k:

$$\frac{(FP_k + \beta^2 FN_k) + \sum_{j \neq k}(FP_j + \beta^2 FN_j)}{TP_k + \sum_{j \neq k} TP_j}. \qquad (7)$$

It is clear that the contribution of the terms related to ω_k, $(FP_k + \beta^2 FN_k)$ and TP_k, to the overall value of expression (7) is not independent on the contribution of the remaining terms, related to all the other classes. It follows that, to select the classifiers $f_1^{i(1)}, f_2^{i(2)}, \ldots, f_N^{i(N)}$ which maximise $\hat{F}_\beta^m$, an exhaustive search over all the possible L^N choices is required, which is clearly impractical.

Nevertheless, the analysis of expression (7) reveals that, under some conditions on the values of the four terms $(FP_k + \beta^2 FN_k)$, $\sum_{j \neq k}(FP_j + \beta^2 FN_j)$, TP_k and $\sum_{j \neq k} TP_j$, the contribution of the terms related to ω_k is independent on that of the remaining terms. Under such conditions, it turns out that the classifier $f_k^{i(k)}$ that maximises $\hat{F}_\beta^m$ can be chosen independently on the other ones, using the following criterion:

$$i(k) = \arg \min_{i \in \{1,...,L\}} \frac{(FP_k^i + \beta^2 FN_k^i) + A_k}{TP_k^i + B_k}, \qquad (8)$$

where A_k and B_k are two arbitrary, positive constants. It also turns out that, under slightly stricter conditions, both A_k and B_k can be zero. Due to the lack of space, the proof is not reported in this paper.[1]

According to the above result, when the micro-averaged F measure is used we propose to use the criterion (8) to select a classifier for each class. We will call this criterion HSM^m (the superscript stands for "micro-averaging"). Clearly, this is a suboptimal choice, as the conditions under which (8) is the optimal criterion may not hold for all classes simultaneously, and anyway in practice one can not

---

[1] The proof can be found at http://prag.diee.unica.it/pra/bib/pillai_mcs2011

know whether they hold or not, for any class. The choice of the constants A_k and B_k can be made in such a way to limit the consequences of the non-optimality of HSM^m. To this aim, we propose to set A_k and B_k to a value that approximates the corresponding terms $\sum_{j \neq k}(FP_j + \beta^2 FN_j)$ and $\sum_{j \neq k} TP_j$ in Eq. (7), which can be estimated from validation data together with the terms FP_k^i, FN_k^i and TP_k^i of (8).

Consider now the case of $L' > 1$, namely when two or more classifiers have to be selected for each class. To maximise $\hat{F}_\beta^M$, the best subset of L' classifiers can be chosen independently for each class, for the same reason explained above. The corresponding criterion is the same as HSM^M of (6), where the FP, FN and TP values now refer to the combination of L' classifiers instead of a single classifier. However, this requires to evaluate all possible $\binom{L}{L'}$ combinations of classifiers, which may be impractical.

Similar considerations apply to the case of the $\hat{F}_\beta^M$ measure. Even in the conditions under which the criterion HSM^m of (8) is optimal for all classes (where the FP, FN and TP values now refer to an ensemble of L' classifiers as above), all possible $\binom{L}{L'}$ combinations of classifiers must be evaluated for each class. In principle, in the worst case when such conditions do not hold for any class, the number of classifier ensembles to evaluate becomes $\binom{L}{L'}^N$.

To keep computational complexity low when $L' > 1$, in this paper we will consider the simplest sub-optimal criterion for both $\hat{F}_\beta^M$ and $\hat{F}_\beta^m$. It consists in selecting the top L' classifiers for each class, ranked in terms of the corresponding HSM^M or HSM^m criterion of (6) and (8).

We finally discuss an interesting by-product of the above results. We mentioned above that under the conditions when HSM^m is optimal, the positive constants A_k and B_k of (8) can be arbitrarily small. Accordingly, as A_k and B_k approach zero, HSM^m tends to the HSM^M criterion of (6). This leads us to argue that HSM^M may also provide a good micro-averaged F measure. On the other hand, HSM^m requires to maximise for each class a quantity that does not depend on the other classes, analogously to HSM^M. It may thus provide in turn also a good macro-averaged F measure. In other words, we argue that both HSM^m and HSM^M may provide a good trade-off between the macro- and micro-averaged F measure, with respect to the performance of the individual multi-label classifiers.

5   Experimental Evaluation

The experiments presented in this section are aimed at investigating whether, given a set of multi-label classifiers, the HSM method can outperform a "standard" selection method which selects the same classifier subset for each class, as well as the fusion of all the available classifiers.

The experiments have been carried out on three widely used benchmark data sets: the "ModApte" version of "Reuters 21578";[2] the Heart Disease subset of the

---

[2] http://www.daviddlewis.com/resources/testcollections/reuters21578/

Table 1. Characteristics of the three data sets used in the experiments

Data set	Reuters	Ohsumed	Yeast
N. of training samples	7769	12775	1500
N. of testing samples	3019	3750	917
Feature set size	18157	17341	104
N. of classes	90	99	14
Distinct sets of classes	365	1392	164
N. of labels per sample (avg./max.)	1.234 / 15	1.492 / 11	4.228 / 11
N. of samples per class (min./max.)	1.3E-4 / 0.37	2.4E-4 / 0.25	0.07 / 0.75

Ohsumed data set [5]; and the Yeast data set.[3] Reuters and Ohsumed are text categorisation tasks, while Yeast is a gene functionality classification problem. Their main characteristics are reported in Table 1.

For Reuters and Ohsumed we used the *bag-of-words* feature model with the term frequency-inverse document frequency (tf-idf) kind of feature [8]. A feature selection step has also been carried out through a four-fold cross-validation on training samples, by applying stemming, stop-word removal and the information gain criterion. A feature set of 15,000 features was obtained for both data sets.

We implemented multi-label classifiers using the well known *binary relevance* (BR) approach. It consists in independently training N two-class classifiers (one per class) using the one-vs-all strategy: each classifiers independently decides whether labelling an input sample as belonging or not to the corresponding class [6,8,11]. We used as base two-class classifier a support vector machine (SVM) with a linear kernel for Reuters and Ohsumed (as it is considered the state of the art classifier for text categorisation tasks) and a SVM with a radial-basis function (RBF) kernel for Yeast. We used the libsvm software to implement SVMs [1]. The C parameter of the SVM learning algorithm was set to the libsvm default value of 1. The σ parameter of the RBF kernel, defined as $K(\mathbf{x}, \mathbf{y}) = \exp\left(-\|\mathbf{x} - \mathbf{y}\|^2/2\sigma\right)$, was set to 1 according to a four-fold cross-validation on training samples.

Since the output of a SVM is a real number, a threshold has to be set to decide whether labelling or not an input sample as belonging to the corresponding class. The threshold values can be set to optimise the considered performance measure. To maximise the macro-averaged F measure, it is known that the threshold can be set by independently maximising the F measure of each class [13]. No optimal criterion exists for maximising the micro-averaged F measure instead. To this aim we used a sub-optimal algorithm proposed in [2]. In both cases we estimated the thresholds through a five-fold cross-validation on training samples.

All the quantities involved in the selection criteria HSM^M and HSM^m were estimated through a five-fold cross-validation on training samples. Classification performance was evaluated using the F_1 measure, namely $\beta = 1$.

To generate an ensemble of multi-label classifiers for each data set, we used the random subspace method of [3]: each individual multi-label classifier was

---

[3] http://www.csie.ntu.edu.tw/~cjlin/libsvmtools/datasets/multilabel.html

trained on the whole set of training samples, by using a different, randomly chosen feature subset. We used a fraction of 3/4 of the original feature set, and set the ensemble size to 10. We used as combining rules majority voting and simple averaging. For the latter, we also estimated the decision thresholds on the outputs of the fused classifiers, using again a five-fold cross-validation on training samples.

Five runs of the experiments were carried out. The average F_1 values and the standard deviation are shown in Table 2, related to the majority voting rule. The different columns contain the F_1 measure attained by all the considered methods. The two left-most and the two right-most columns correspond to the case when the criteria for classifier selection and for threshold optimisation were based respectively on the micro- and on the macro-averaged F_1. In both cases we report both the resulting macro- and micro-averaged F_1. For each data set we show the performance of the "standard" selection method when one, three and five two-class classifiers are selected for each class, the performance of the HSM method for the same number of selected classifiers, and the performance attained by the fusion of all the multi-label classifiers of the ensemble. The comparison between the "standard" selection method and HSM has to be done being equal the number of selected classifiers, and within the same column.

Table 2 shows that, being equal the number of selected classifiers, HSM almost always performed better or at least as good as the "standard" selection method, both in terms of the macro- and the micro-averaged F_1 measure. This shows that it can be capable to better exploit the complementarity between the multi-label classifiers on the different classes.

HSM also attained a better or at least a very similar performance as the fusion of all the available multi-label classifiers. This shows that, besides being more efficient at the classification phase, HSM can also attain a higher classification performance. On the contrary, the "standard" selection method was almost always outperformed by the fusion of all the available classifiers, with some exceptions on the Yeast data sets only.

Finally, HSM attained a higher macro-averaged F_1 measure than the "standard" selection method, even when the selection criterion based on the micro-averaged F_1 was used (see the second column of each table), and vice versa (third column). Moreover, we observed that, regardless on the selection criterion (either HSM^M or HSM^m), the resulting macro-averaged F_1 attained by HSM was very similar. The same result was observed in the case of the micro-averaged F_1 (these results are not reported, due to lack of space). The micro-averaged F_1 attained by the "standard" selection method was instead significantly higher than the macro-averaged F_1, if the selection criterion was based on the former measure, and vice-versa. This result provides evidence that, as argued in Sect. 4, both the HSM^M and the HSM^m criteria can be capable to attain a good trade-off between the macro- and micro-averaged F measure.

Similar results have been obtained using the simple average combining rule, as well as two different base classifiers, k-nearest neighbours and Naive Bayes (these results are not reported here due to lack of space).

Table 2. Average F_1 measure and standard deviation attained on the three data sets by the "standard" selection method and by HSM (see text for the details)

Selection method and ensemble size	Selection based on F_1^m		Selection based on F_1^M	
	F_1^m	F_1^M	F_1^m	F_1^M
Best single	0.863 ± 0.005	0.520 ± 0.020	0.261 ± 0.037	0.564 ± 0.018
HSM 1	0.878 ± 0.002	0.586 ± 0.011	0.427 ± 0.028	0.609 ± 0.010
Best 3	0.874 ± 0.001	0.534 ± 0.013	0.267 ± 0.030	0.586 ± 0.019
HSM 3	0.881 ± 0.002	0.588 ± 0.009	0.333 ± 0.030	0.613 ± 0.004
Best 5	0.878 ± 0.002	0.557 ± 0.011	0.270 ± 0.038	0.587 ± 0.015
HSM 5	0.882 ± 0.001	0.582 ± 0.005	0.308 ± 0.029	0.617 ± 0.005
All	0.880 ± 0.002	0.560 ± 0.006	0.274 ± 0.037	0.603 ± 0.010
Best single	0.667 ± 0.005	0.536 ± 0.021	0.653 ± 0.003	0.573 ± 0.012
HSM 1	0.683 ± 0.001	0.591 ± 0.010	0.672 ± 0.002	0.612 ± 0.010
Best 3	0.681 ± 0.003	0.560 ± 0.016	0.672 ± 0.005	0.607 ± 0.003
HSM 3	0.688 ± 0.001	0.595 ± 0.007	0.682 ± 0.002	0.624 ± 0.010
Best 5	0.685 ± 0.003	0.576 ± 0.006	0.677 ± 0.001	0.609 ± 0.005
HSM 5	0.690 ± 0.001	0.596 ± 0.006	0.684 ± 0.002	0.623 ± 0.008
All	0.684 ± 0.004	0.573 ± 0.005	0.681 ± 0.003	0.617 ± 0.005
Best single	0.675 ± 0.004	0.439 ± 0.009	0.618 ± 0.008	0.494 ± 0.002
HSM 1	0.676 ± 0.002	0.449 ± 0.004	0.642 ± 0.005	0.497 ± 0.004
Best 3	0.679 ± 0.003	0.443 ± 0.006	0.623 ± 0.005	0.496 ± 0.004
HSM 3	0.680 ± 0.002	0.449 ± 0.005	0.641 ± 0.003	0.501 ± 0.004
Best 5	0.680 ± 0.002	0.444 ± 0.003	0.624 ± 0.006	0.498 ± 0.002
HSM 5	0.680 ± 0.002	0.445 ± 0.003	0.636 ± 0.003	0.499 ± 0.001
All	0.680 ± 0.002	0.438 ± 0.003	0.631 ± 0.006	0.498 ± 0.002

The three data set groups are labelled (vertically) Reuters, Ohsumed, and Yeast respectively.

6   Conclusions

In this work we proposed a classifier selection approach specific to ensembles of multi-label classifiers, which is based on selecting a possibly different subset of classifiers for each class. This allows in principle to better exploit the complementarity between the multi-label classifiers, on the different classes. Moreover, we developed two static classifier selection criteria based on the macro- and the micro-averaged F measure, which is widely used in multi-label tasks.

Our experimental results provided evidence that the proposed selection approach can be more effective than a "standard" approach based on selecting the same classifier subset for each class, as well as than fusing all the available multi-label classifiers. An interesting by-product is that both the proposed selection can also attain a good trade-off between the macro- and micro-averaged F measure, despite it is known that an increase in either of them is usually attained at the expense of the other one.

In light of these results, it becomes interesting to further investigate the following issues: investigating the characteristics of an ensemble of multi-label classifiers that make the proposed selection approach more effective; evaluating dynamic selection methods based on this approach; analysing the behaviour of these static and dynamic selection methods as a function of the training set size, as well as the effect of class imbalance, which is a typical problem in multi-label tasks involving a high number of classes.

Acknowledgements. This work was partly supported by a grant from Regione Autonoma della Sardegna awarded to Ignazio Pillai, PO Sardegna FSE 2007-2013, L.R.7/2007 "Promotion of the scientific research and technological innovation in Sardinia".

References

1. Chang, C.C., Lin, C.J.: LIBSVM: a library for support vector machines (2001), http://www.csie.ntu.edu.tw/~cjlin/libsvm
2. Fan, R.E., Lin, C.J.: A study on threshold selection for multi-label. Tech. Rep., National Taiwan University (2007)
3. Ho, T.K.: The random subspace method for constructing decision forests. IEEE Trans. Pattern Anal. Mach. Intell. 20, 832–844 (1998)
4. Kuncheva, L.I.: Combining Pattern Classifiers: Methods and Algorithms. Wiley Interscience, Hoboken (2004)
5. Lewis, D.D., Schapire, R.E., Callan, J.P., Papka, R.: Training algorithms for linear text classifiers. In: SIGIR, pp. 298–306 (1996)
6. Read, J., Pfahringer, B., Holmes, G., Frank, E.: Classifier chains for multi-label classification. In: Buntine, W., Grobelnik, M., Mladenić, D., Shawe-Taylor, J. (eds.) ECML PKDD 2009. LNCS, vol. 5782, pp. 254–269. Springer, Heidelberg (2009)
7. Schapire, R.E., Singer, Y.: Boostexter: A boosting-based system for text categorization. Machine Learning 39(2/3), 135–168 (2000)
8. Sebastiani, F.: Machine learning in automated text categorization. ACM Computing Surveys 34(1), 1–47 (2002)
9. Tahir, M., Kittler, J., Mikolajczyk, K., Yan, F.: Improving Multilabel Classification Performance by Using Ensemble of Multi-label Classifiers. In: Proc. of Multiple Classifier Systems (2010)
10. Tsoumakas, G., Katakis, I.: Multi label classification: An overview. Int. Journal of Data Warehousing and Mining 3(3), 1–13 (2007)
11. Tsoumakas, G., Katakis, I., Vlahavas, I.: Mining multi-label data. In: Data Mining and Knowledge Discovery Handbook, pp. 667–685 (2010)
12. Woods, K., Kegelmeyer, W.P., Bowyer, K.W.: Combination of multiple classifiers using local accuracy estimates. IEEE Trans. Pattern Anal. Mach. Intell. (1997)
13. Yang, Y.: A study of thresholding strategies for text categorization. In: Int. Conf. on Research and development in information retrieval, New York, USA, (2001)
14. Zhou, Z.-H., Jianxin, Z., Wei, W.: Ensembling neural networks: many could be better than all. Artificial Intelligence 137(1/2), 239–263 (2002)

Selection Strategies for pAUC-Based Combination of Dichotomizers

Maria Teresa Ricamato, Mario Molinara, and Francesco Tortorella

DAEIMI - Università degli Studi di Cassino
via G. Di Biasio 43, 03043 Cassino, Italy
{mt.ricamato,m.molinara,tortorella}@unicas.it

Abstract. In recent years, classifier combination has been of great interest for the pattern recognition community as a method to improve classification performance. Several combination rules have been proposed based on maximizing the accuracy and the Area under the ROC curve (AUC). Taking into account that there are several applications which focus only on a part of the ROC curve, i.e. the one most relevant for the problem, we recently proposed a new algorithm aimed at finding the linear combination of dichotomizers which maximizes only the interesting part of the AUC. Since the algorithm uses a greedy approach, in this paper we define and evaluate some possible strategies which select the dichotomizers to combine at each step of the greedy approach. An experimental comparison is drawn on a multibiometric database.

Keywords: Classifiers combination, ROC curve, partial AUC.

1 Introduction

Classifier combination has become an established technique for building proficient classification systems. Among the various combination methods proposed up to now, linear classifier combination has been used mainly for its simplicity and effectiveness. In particular, some methods have been designed to increase the Area under the ROC curve (AUC), a more suitable performance measure than the classification accuracy [1], specially for those applications characterized by imprecise environment or imbalanced class priors [2]. AUC resumes in a single quantitative index the performance exhibited by the classifier over all the false positive rate (FPR) values.

However, there are many applications that are interested only to a particular range of FPRs. For example, in a biometric authentication system used to identify people, or to verify the claimed identity of registered users when entering in a protected area, a false positive is considered the most serious error, since it gives unauthorized users access to the systems that expressly are trying to keep them out. In such case, the FPR values considered are the ones that correspond to lower values, and the partial AUC [3] is the most indicate index to use, since it allows us to focus on particular regions of the ROC space. In [4] we have proposed a new method aimed at calculating the weight vector in a

C. Sansone, J. Kittler, and F. Roli (Eds.): MCS 2011, LNCS 6713, pp. 177–186, 2011.

linear combination of $K \geq 2$ dichotomizers, such that the pAUC is maximized. In particular, we have provided an algorithm for finding the optimal weight in a combination of two dichotomizers and then have extended to the combination of $K > 2$ dichotomizers by means of a greedy approach which divides the whole K-combination problem into a series of pairwise combination problems.

In such a case, making the right local choice at each stage is of fundamental importance since it affects the performance of the whole algorithm. For this reason, in this paper we define and evaluate some possible strategies which select the dichotomizers to combine at each step of the greedy approach. The strategies considered are based both on the evaluation of the best single dichotomizer and of the best pair of dichotomizers. Such strategies are then experimentally compared on a biometric database.

The paper is organized as follow. The next section presents the pAUC index and its main properties while section 3 analyzes the combination of two dichotomizers. The combination of $K > 2$ dichotomizers is presented in section 4; in the same section the proposed selection strategies are described and analyzed. The performed experiments and obtained results are shown in section 5, while section 6 concludes the paper.

2   ROC Analysis and Partial Area Under the ROC Curve

Receiver Operating Characteristics (ROC) graphs are useful for visualizing, organizing and selecting classifiers based on their performance. Given a two-class classification model, the ROC curve describes the trade-off between the fraction of correctly classified actually-positive cases (True Positive Rate, TPR) and the fraction of wrongly classified actually-negative cases (False Positive Rate, FPR), giving a description of the performance of the decision rule at different operating points.

In some cases, it is preferable to use the Area under the ROC Curve (AUC) [5] [6], a single metric able to summarize the performance of the classifiers system:

$$AUC = \int_0^1 ROC(\tau)d\tau \qquad (1)$$

As said before, some applications do not use all the range of false positive rates: in particular, the most part of biometric and medical applications [7] work on false positive rate close to the zero value. In such cases it is worth to consider a different summary index measuring the area under the part of the ROC curve with FPRs between 0 and a maximal acceptable value t. This index is called *partial AUC* (pAUC) and defined as:

$$pAUC = \int_0^t ROC(\tau)d\tau \qquad (2)$$

where the interval $(0, t)$ denotes the false positive rates of interest. Its choice depends on the particular application, and it is related to the involved cost of a false positive diagnosis.

Moreover, the pAUC can be also defined as the probability that a classifier will rank a randomly chosen positive instance higher than a randomly chosen negative one, such that this latter is higher than the $1 - t$ quantile[1] q_y^t:

$$pAUC = P\left\{x_i > y_j, y_j > q_y^t\right\} \tag{3}$$

where $x_i = f(\mathbf{p}_i)$ and $y_j = f(\mathbf{n}_j)$ are the outcomes of the dichotomizer f on a positive sample $\mathbf{p}_i \in P$ and a negative sample $\mathbf{n}_j \in N$.

In order to evaluate the pAUC of a dichotomizer avoiding to perform a numerical integration on the ROC curve, we use the non-parametric estimator [3], which is defined as:

$$pAUC = \frac{1}{m_P \cdot m_N} \sum_i^{m_P} \sum_j^{m_N} V_{ij}^{q_y^t} \tag{4}$$

where m_P and m_N are the cardinalities of the positive and negative subsets, respectively, and

$$V_{ij}^{q_y^t} = I\{x_i > y_j, y_j > q_y^t\} = \begin{cases} 1 & \text{if } x_i > y_j \bigwedge y_j > q_y^t; \\ 0.5 & \text{if } x_i = y_j \bigwedge y_j > q_y^t; \\ 0 & \text{if } x_i < y_j \bigwedge y_j > q_y^t. \end{cases} \tag{5}$$

3   Combination of Two Dichotomizers

As a first step, let us consider a set $T = P \cup N$ of samples, and define the outputs of two generic dichotomizers f_h and f_k on two positive and negative samples $\mathbf{p}_i$ and $\mathbf{n}_j$:

$$x_i^h = f_h(\mathbf{p}_i), \quad x_i^k = f_k(\mathbf{p}_i), \quad y_j^h = f_h(\mathbf{n}_j), \quad y_j^k = f_k(\mathbf{n}_j).$$

The pAUCs for the two dichotomizers, considering the FPR interval $(0, t)$, are:

$$pAUC_h = \frac{\sum_{i=1}^{m_P} \sum_{j=1}^{m_N} I\left(x_i^h > y_j^h, y_j^h > q_{y^h}^t\right)}{m_P \cdot m_N}, \quad pAUC_k = \frac{\sum_{i=1}^{m_P} \sum_{j=1}^{m_N} I\left(x_i^k > y_j^k, y_j^k > q_{y^k}^t\right)}{m_P \cdot m_N} \tag{6}$$

It is worth noting that the linear combination of two generic dichotomizers $f_{lc} = \alpha_h f_h + \alpha_k f_k$ can be put as $f_{lc} = f_h + \alpha f_k$ without loss of generalization, with $\alpha = \frac{\alpha_k}{\alpha_h} \in (-\infty, +\infty)$. Therefore, considering the linear combination, the outcomes on $\mathbf{p}_i$ and $\mathbf{n}_j$ are:

$$\xi_i = f_{lc}(\mathbf{p}_i) = x_i^h + \alpha x_i^k, \qquad \eta_j = f_{lc}(\mathbf{n}_j) = y_j^h + \alpha y_j^k. \tag{7}$$

and the pAUC is:

$$pAUC_{lc} = \frac{1}{m_P \cdot m_N} \left(\sum_{i=1}^{m_P} \sum_{j=1}^{m_N} I\left(\xi_i > \eta_j, \left(\eta_j > q_\eta^t(\alpha)\right)\right) \right) \tag{8}$$

---

[1] The quantile function returns the value below which random draws from the negative population would fall, $(1 - t) \times 100$ percent of the time.

To have an insight into the formulation of the $pAUC_{lc}$, let us analyze the term $I(\xi_i > \eta_j)$ without considering the constraint on the quantile. In particular, let us consider how it depends on the values of $I(x_i^h, y_j^h)$ and $I(x_i^k, y_j^k)$:

– $I(x_i^h, y_j^h) = 1$ and $I(x_i^k, y_j^k) = 1$. In this case the two samples are correctly ranked by the two dichotomizers, and $I(\xi_i > \eta_j) = 1$.
– $I(x_i^h, y_j^h) = 0$ and $I(x_i^k, y_j^k) = 0$. In this case neither dichotomizer ranks correctly the samples and thus the contribution for the $pAUC$ is 0.
– $I(x_i^h, y_j^h)$ xor $I(x_i^k, y_j^k) = 1$. Only one dichotomizer ranks correctly the samples while the other one is wrong. In this case the value of $I(\xi_i > \eta_j)$ depends on the weight α.

The subset T can be divided into four subsets: T_{hk}, $T_{h\bar{k}}$, $T_{\bar{h}k}$ and $T_{\bar{h}\bar{k}}$ defined as:

$$T_{hk} = \{(\mathbf{p}_i, \mathbf{n}_j) | I(x_i^h, y_j^h) = 1 \text{ and } I(x_i^k, y_j^k) = 1\},$$
$$T_{\bar{h}k} = \{(\mathbf{p}_i, \mathbf{n}_j) | I(x_i^h, y_j^h) = 0 \text{ and } I(x_i^k, y_j^k) = 1\},$$
$$T_{h\bar{k}} = \{(\mathbf{p}_i, \mathbf{n}_j) | I(x_i^h, y_j^h) = 1 \text{ and } I(x_i^k, y_j^k) = 0\},$$
$$T_{\bar{h}\bar{k}} = \{(\mathbf{p}_i, \mathbf{n}_j) | I(x_i^h, y_j^h) = 0 \text{ and } I(x_i^k, y_j^k) = 0\}$$

Now, let us consider the constraint on the negative samples related to the quantile, and define the following set:

$$\Gamma_\alpha = \{(\mathbf{p}_i, \mathbf{n}_j) \in P \times N | y_j^h + \alpha y_j^k > q_\eta^t\} \tag{9}$$

where q_η^t is the $1 - t$ quantile of η, which depends on the weight α. If we define the sets $T'_{hk}, T'_{\bar{h}k}, T'_{h\bar{k}}, T'_{\bar{h}\bar{k}}$ as:

$$T'_{hk} = T_{hk} \cap \Gamma_\alpha, \quad T'_{\bar{h}k} = T_{\bar{h}k} \cap \Gamma_\alpha,$$
$$T'_{h\bar{k}} = T_{h\bar{k}} \cap \Gamma_\alpha, \quad T'_{\bar{h}\bar{k}} = T_{\bar{h}\bar{k}} \cap \Gamma_\alpha,$$

the expression for $pAUC_{lc}$ in equation 8 can be written as:

$$pAUC_{lc} = \frac{1}{m_P \cdot m_N} \left(\sum_{(\mathbf{p}_i, \mathbf{n}_j) \in T'_{\bar{h}\bar{k}}} I(\xi_i > \eta_j) + \sum_{(\mathbf{p}_i, \mathbf{n}_j) \in T'_{hk}} I(\xi_i > \eta_j) + \sum_{(\mathbf{p}_i, \mathbf{n}_j) \in T'_{h\bar{k}} \cup T'_{\bar{h}k}} I(\xi_i > \eta_j) \right).$$

Starting from equation above, the value of α which maximizes $pAUC_{lc}$ can be found by means of a linear search; the details are described in [4].

4   Combination of $K > 2$ Dichotomizers

Let us now consider the linear combination of $K > 2$ dichotomizers which is defined as:

$$f_{lc}(x) = \alpha_1 f_1(x) + \alpha_2 f_2(x) + ... + \alpha_K f_K(x) = \sum_{i=1}^{K} \alpha_i f_i(x) \tag{10}$$

In order to find the weight vector $\alpha_{opt} = (\alpha_1, ..., \alpha_K)$ that maximizes the pAUC associated to $f_{lc}(x)$, we can consider a greedy approach which divides the whole K-combination problem into a series of pairwise combination problems. Even though suboptimal, such approach provides a computationally feasible algorithm for the problem of combining of K dichotomizers which would be intractable if tackled directly. In particular, the first step of the algorithm chooses the two "most promising" dichotomizers and combines them so that the number of dichotomizers decreases from K to $K-1$. This procedure is repeated until all the dichotomizers have been combined.

The choice of the dichotomizers that should be combined in each iteration plays an important role since it affects the performance of the algorithm. To this aim, we have considered various selection strategies that differ in the way the greedy approach is accomplished: in the *Single Classifier Selection* the best candidate dichotomizer is chosen, while with the *Pair Selection* the best candidate pair of dichotomizers is taken.

4.1   Single Classifier Selection

This is the most immediate approach, that selects the dichotomizer with the best performance index. The related implementation first sorts the K dichotomizers into decreasing order of an individual performance measure, and the first two classifiers are combined. The remaining dichotomizers are then singularly added to the group of the combined dichotomizers.

pAUC based selection. A first way to implement this strategy is to use the pAUC of the dichotomizers as single performance measure. In this way, one looks at the behavior of the single classifier in the range of interest of the false positives assuming that it could be a sufficiently good estimate of how the classifier contributes to the combination. In other words, we are assuming that it is sufficient to take into account the performance of the dichotomizer, say f_h, on the set of the negative samples $N_h = \{\mathbf{n}_j \in N | y_j^h > q_h^t\}$ even though this set likely does not coincide with the set of negative samples contributing to the value of $pAUC_{lc}$.

AUC based selection. A second way is to employ the whole AUC of the dichotomizer. In this case we don't assume that the samples in N_h are sufficient to predict the performance of the combination and thus consider the behavior of the dichotomizer in the whole FPR range.

4.2   Pair Selection

This approach is based on the estimation of the joint characteristics of a pair of dichotomizers, so as to predict how proficient is their combination. In particular, we rely on the idea that combining dichotomizers with different characteristics should provide good performance. Therefore, a pairwise measure is evaluated which estimates the diversity of the dichotomizers; at each step, the pair of classifiers exhibiting the best index is combined. The procedure is repeated $K-1$ times until a single classifier is obtained. Two different measures are considered which estimate the diversity in the ranking capabilities between two dichotomizers.

Kendall Rank Coefficient. The first index we consider is the *Kendall rank correlation coefficient* [8] that evaluates the degree of agreement between two sets of ranks with respect to the relative ordering of all possible pairs of objects. Given K the sum of concordant pairs and l the number of considered items, the Kendall rank correlation coefficient is defined as:

$$\frac{2K}{\frac{1}{2}l(l-1)} - 1 \tag{11}$$

where $\frac{1}{2}l(l-1) = \binom{l}{2}$ is the total amount of pairs.

In our case the subsets are the ones defined in the previous section and thus the correlation coefficient can be redefined as:

$$\tau' = \frac{2\left(|T'_{hk}| + |T'_{\bar{h}\bar{k}}|\right)}{t \cdot m_P \cdot m_N} - 1 \tag{12}$$

However, to evaluate τ' we should previously know the optimal value of the coefficient α chosen for the combination, but this would be computationally heavy. Therefore we use a surrogate index τ defined as

$$\tau = \frac{2\left(|T_{hk}| + |T_{\bar{h}\bar{k}}|\right)}{m_P \cdot m_N} - 1 \tag{13}$$

which is an upper bound for τ' since $|T_{hk}| \geq |T'_{hk}|$ and $|T_{\bar{h}\bar{k}}| \geq |T'_{\bar{h}\bar{k}}|$. At each step the pair with the lowest τ is chosen.

Ranking Double Fault. The second index comes from an analysis of the expression of the $pAUC_{lc}$ given in eq. 10. The maximum allowable pAUC of a linear combination depends on the cardinality of the subsets T'_{hk}, $T'_{h\bar{k}}$ and $T'_{\bar{h}k}$. Since the number of pairs in each subset depends on the value of the quantile, it is not possible to compute a priori the value of $pAUC_{lc}^{max}$. However, it is feasible to obtain a lower bound for it. To this aim, let us consider the relation between the number of pairs of positive and negative samples obtained without using the quantile, and its reduction after using the quantile. The $(1-t)$ quantile is the value which divides a set of samples such that there is the given proportion $(1-t)$ of observations below it. Therefore, when the quantile is applied, the number of the considered negative samples decreases, with the consequent change of the number of the total pairs:

$$\begin{aligned} m'_{tot} &= m_{tot} - |\{(\mathbf{p}_i, \mathbf{n}_j)|\eta_j < q_\eta^t(\alpha)\}| \\ &= m_P \cdot m_N - m_P[(1-t)m_N] = t \cdot m_P \cdot m_N = t \cdot m_{tot} \end{aligned} \tag{14}$$

where $m_{tot} = m_P \cdot m_N$ is the number of pairs considered without the constraint of the quantile.

Therefore, the $pAUC_{lc}^{max}$ can be rewritten as:

$$pAUC_{lc}^{max} = \frac{1}{m_P \cdot m_N} \left(m'_{tot} - |T'_{\bar{h}\bar{k}}| \right) = \frac{1}{m_P \cdot m_N} \left(t \cdot m_P \cdot m_N - |T'_{\bar{h}\bar{k}}| \right) \quad (15)$$

It is obvious that :

$$|T'_{\bar{h}\bar{k}}| \leq |T_{\bar{h}\bar{k}}| \quad (16)$$

since the number of the considered negative samples decreases.

Therefore:

$$pAUC_{lc}^{max} \geq \frac{1}{m_P \cdot m_N} \left(t \cdot m_P \cdot m_N - |T_{\bar{h}\bar{k}}| \right) \quad (17)$$

In particular, the lower bound for $pAUC_{lc}^{max}$ is high when we have a low number of pairs that have been misranked by both the classifiers. This quantity can be interestingly related to the *double fault* measure [9] that is used to evaluate the diversity between classifiers. For this reason we define *Ranking Double Fault* the index:

$$DF = \frac{|T_{\bar{h}\bar{k}}|}{m_P \cdot m_N} \quad (18)$$

and adopt it as second diversity index. Also in this case, the pair with the lowest DF is chosen at each step.

5   Experimental Results

In order to compare the selection strategies proposed in the previous section, some experiments have been performed on the public-domain biometric dataset XM2VTS [10], characterized by 8 matchers. We used the partition of the scores into training and test set proposed in [10] and shown in table 1. The XM2VTS is a multimodal database containing video sequences and speech data of 295 subjects recorded in four sessions in a period of 1 month. In order to assess its performance the Lausanne protocol has been used to randomly divide all the subjects into positive and negative classes: 200 positive, 25 evaluation negatives and 70 test negatives. All the details about the procedure used to obtain the final dichotomizers are described in [10].

For each considered strategy the vector of coefficients for the linear combination is evaluated on the validation set, and then applied to the test set. The results are analyzed in term of partial AUC, considering the false positive ranges: $FPR_{0.1} = (0, 0.1)$, $FPR_{0.05} = (0, 0.05)$ and $FPR_{0.01} = (0, 0.01)$. For the sake

Table 1. XM2VTS database properties

	# Sample	# Positive	# Negative
Validation Set	40600	600	40000
Test Set	112200	400	111800

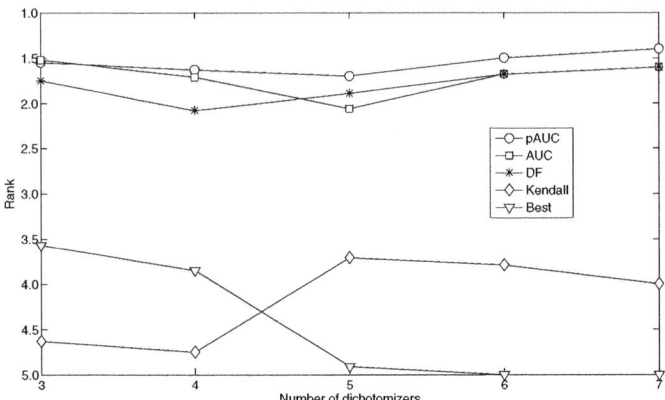

Fig. 1. Mean rank of the selection strategies for t=0.1

of comparison, we consider, besides the combinations obtained with the four selection strategies described in the previous section, also the best single classifier chosen by looking at the highest pAUC value on the validation set.

The number of combined dichotomizers varies from 2 to 7. For each of those experiments we obtain different number of possible combinations that are independent from each other. Therefore, we use an approach based on giving a rank to each method compared to the others, for each independent experiment. Let us consider the pAUC values $\{pAUC_{ij}\}_{M \times L}$, for $i = 1, \ldots, M$ with M the number of combinations, and for $j = 1, \ldots, L$ with L number of the strategies compared. For each row we assign a rank value r_j^i from 1 to L to each column depending on the pAUC values: the highest pAUC gets rank 1, the second highest the rank 2, and so on until L (in our case $L = 5$). If there are tied pAUCs, the average of the ranks involved is assigned to all pAUCs tied for a given rank. Only in this case it is appropriate to average the obtained ranks on the number of combinations:

$$\bar{r}_j = \frac{1}{M} \sum_{i=1}^{M} r_j^i \qquad (19)$$

Figures 1-3 show the results obtained varying the FPR ranges. The higher the curve, i.e. the lower the average rank value, the better the related method.

The results show a clear predominance of the strategies based on single classifier selection. In particular, only the pair choice selection based on the DF index is comparable with pAUC and AUC based selection, specially when the number of the dichotomizers to be combined grows. A probable reason for such outcome is that in the pair selection strategies we use upper bound surrogates of the actual indices and this could sensibly affect the effectiveness of the selection strategy.

A comparison between the two single classifier selection strategies reveals how, even though pAUC is almost always better than AUC, the difference is

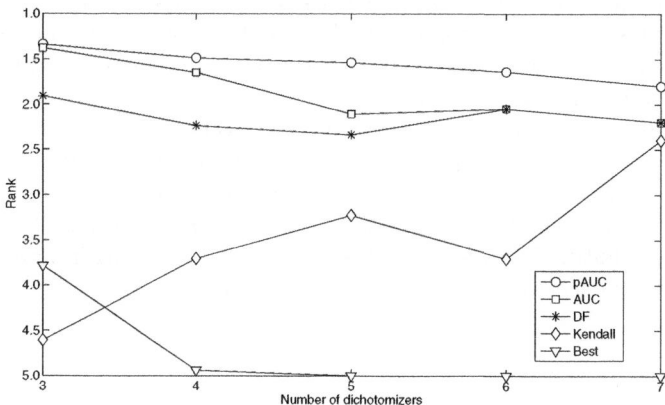

Fig. 2. Mean rank of the selection strategies for t=0.05

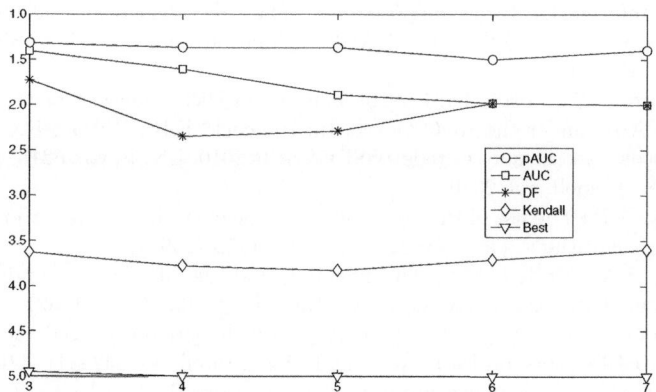

Fig. 3. Mean rank of the selection strategies for t=0.01

not so high for $t = 0.1$. As the FPR range becomes smaller and smaller, the pAUC based strategy clearly outperforms the AUC counterpart and this trend becomes more evident when the number of dichotomizers grows. This suggests that the behavior of the single classifier in the FPR range of interest provides a sufficiently good estimate of how the classifier contributes to the combination.

In summary, a selection strategy which chooses the single dichotomizer with the highest pAUC at each step of the greedy approach described in Sect. 4 seems to ensure the best results on a large extent of situations.

6   Conclusions

In this paper, we have defined and evaluated some strategies which select the dichotomizers to combine at each step of a pAUC combination method based on a greedy approach. The strategies considered are based both on the evaluation of the best single dichotomizer and of the best pair of dichotomizers. Such strategies have been experimentally compared on a biometric database, i.e. an application for which the use of the pAUC is particularly important. The results obtained have shown clearly that the single classifier selection strategies seem the most proficient ones, in particular the selection based on the pAUC of the single dichotomizer. However, it should be taken into account that the indices used by the pair strategies are actually replaced by computationally feasible approximations. Future investigations will be aimed at verifying if more tight (and hopefully more effective) approximations are attainable.

References

1. Huang, J., Ling, C.X.: Using AUC and accuracy in evaluating learning algorithms. IEEE Trans. on Knowledge and Data Engineering 17, 299–310 (2005)
2. Cortes, C., Mohri, M.: AUC optimization vs. error rate minimization advances. In: Neural Information Processing Systems. MIT Press, Cambridge (2003)
3. Dodd, L.E., Pepe, M.S.: Partial AUC estimation and regression. Biometrics 59, 614–623 (2003)
4. Ricamato, M.T., Tortorella, F.: Combination of Dichotomizers for Maximizing the Partial Area under the ROC Curve. In: Hancock, E.R., Wilson, R.C., Windeatt, T., Ulusoy, I., Escolano, F. (eds.) SSPR&SPR 2010. LNCS, vol. 6218, pp. 660–669. Springer, Heidelberg (2010)
5. Bradley, A.P.: The use of the area under the roc curve in the evaluation of machine learning algorithms. Patt. Recogn. 30, 1145–1159 (1997)
6. Hanley, J.A., McNeil, B.J.: The meaning and use of the area under a receiver operating characteristic (ROC) curve. Radiology 143, 29–36 (1982)
7. Nandakumar, K., Dass, S.C., Jain, A.K.: Likelihood ratio-based biometric score fusion. IEEE Trans. on Patt. Anal. and Mach. Intell. 30, 342–347 (2008)
8. Kendall, M.G.: A new measure of rank correlation. Biometrika 30, 81–93 (1938)
9. Giacinto, G., Roli, F.: Design of effective neural network ensembles for image classification processes. Image and Vision Computing 19, 699–707 (2001)
10. Poh, N., Bengio, S.: Database, protocol and tools for evaluating score-level fusion algorithms in biometric authentication. Patt. Recogn. 39, 223–233 (2006)

Sequential Classifier Combination for Pattern Recognition in Wireless Sensor Networks

Janos Csirik[1], Peter Bertholet[2], and Horst Bunke[2]

[1] Institute of Informatics, University of Szeged, Hungary
[2] Institute of Informatics and Applied Mathematics, University of Bern, Switzerland
csirik@inf.u-szeged.hu,
peter_bertholet@students.unibe.ch,
bunke@iam.unibe.ch

Abstract. In the current paper we consider the task of object classification in wireless sensor networks. Due to restricted battery capacity, minimizing the energy consumption is a main concern in wireless sensor networks. Assuming that each feature needed for classification is acquired by a sensor, a sequential classifier combination approach is proposed that aims at minimizing the number of features used for classification while maintaining a given correct classification rate. In experiments with data from the UCI repository, the feasibility of this approach is demonstrated.

Keywords: Sequential classifier combination, wireless sensor networks, feature ranking, feature selection, system lifetime.

1 Introduction

Multiple classifier systems have become an intensive area of research in the last decade [1]. Usually, the parallel architecture is adopted for the combination of a number of individual classifiers into a single system. This means that n experts process a given input pattern in parallel and their decisions are combined. A large number of combination rules have been proposed in the literature. In [2] an in-depth analysis of some of these rules is provided.

In addition to parallel classifier combination, there also exists a serial, or sequential, approach [3]. It is characterized by activating a number of classifiers in sequential order, where the later classifiers use, in some way, the results derived by the classifiers that were applied earlier in the processing chain. Various applications of sequential classifier combination have been reported in the literature, including text categorization [4], handwritten phrase and digit recognition [5], human face recognition [6], and fingerprint classification [7].

As discussed in detail in [3], there are two principal methods of sequential classifier combination. The first category, called *Class Set Reduction*, is characterized by aiming at a successive reduction of the number of classes under consideration until, in the ideal case, only a single class remains in the end. By contrast, combination strategies of the second type, called *Reevaluation* methods, analyze the results of each classifier C_i in the processing chain and activate

C. Sansone, J. Kittler, and F. Roli (Eds.): MCS 2011, LNCS 6713, pp. 187–196, 2011.
© Springer-Verlag Berlin Heidelberg 2011

the next classifier C_{i+1} if the confidence of C_i is below a given threshold. The overall aim of multiple classifier combination is either to increase the recognition accuracy of a system or to reduce the computational effort to reach a certain classification rate [8].

In the current paper we propose a sequential classifier combination approach for pattern recognition in wireless sensor networks [9]. Pattern recognition tasks, such as vehicle or target classification, localization, and tracking [10,11,12,13,14], as well as surveillance [15], are common in wireless sensor networks. Typically, a wireless sensor network consists of a base station and a (large) number of sensors. While the base station usually has sufficient computational power and energy supply, the individual sensors are characterised by limited computational resources and very restricted battery capacity. Therefore, to keep the number of sensor measurements as low as possible is a major concern in the design of a wireless sensor network.

In this paper we address the problem of object classification in wireless sensor networks. It is assumed that the classification algorithm runs on the base station of the network, where also classifier training and parameter optimization are conducted. The individual sensors are used only for the acquisition of features, where each sensor corresponds to exactly one feature. Each sensor acquires a feature only upon request from the base station. Once a request is received, the value of the corresponding feature is measured and transmitted back to the base station. In such a scenario, keeping the energy consumption of the sensors low is equivalent to minimizing the number of features used by the classifier. In order to achieve classification accuracy as high as possible and minimize the number of sensors, i.e. features, at the same time, we propose a sequential multiple classifier combination approach following the reevaluation strategy. In a preprocessing step we perform a ranking of all available features according to their ability to discriminate between different classes. Then, in the operational phase of the system, we sequentially apply classifiers $C_1, C_2, ..., C_n$ where classifier C_i uses all features of classifier C_{i-1} plus the best ranked individual feature not yet used, i.e. not used by C_{i-1}. The first classifier works with only a single feature, which is the best ranked element out of all features. In our model, it is possible that individual features are not available at a certain point in time, due to battery exhaustion or some other sensor fault.

Minimizing the energy consumption of sensors and maximizing the lifetime of a wireless sensor network has been addressed in a number of papers before. For a survey see [16]. To the knowledge of the authors, the current paper is the first one where a sequential multiple classifier system is proposed in this context.

2   General Approach

We assume that a pattern $\mathbf{x}$ is represented by an N-dimensional feature vector, i.e. $\mathbf{x} = (x_1, \ldots, x_N)$, where x_i is the value of the i-th feature; $i = 1, \ldots, N$. Let $S = \{s_1, \ldots, s_N\}$ be the set of available sensors, where each sensor s_i measures exactly one particular feature $f(s_i) = x_i$ to be used by the classifier. Hence, the maximal set of features possibly available to the classifier is $\{x_1, \ldots, x_N\}$. Furthermore, let $\varphi : S \to \mathbb{R}$ be a function that assigns a utility value $\varphi(x_i)$ to each feature x_i. Concrete examples of utility functions will be discussed below. For the moment, let us assume that the utility of a feature x_i is proportional to its ability to discriminate between the different classes an unknown object may belong to.

The basic structure of the algorithm for object classification proposed in this paper is given in Fig. 1. The system uses a base classifier. This base classifier can be a classifier of any type, in principle. For the purpose of simplicity, however, we use a k-nearest neighbor (k-NN) classifier in this paper.

1: **begin**
2: rank sensors $s_1, \ldots, s_N$ according to the utility of the their features such that
$\varphi(x_1) \geq \varphi(x_2) \geq \ldots \geq \varphi(x_N)$
3: $F = \emptyset$
4: **for** $i = 1$ to N **do**
5:    **if** sensor s_i is available **then**
6:       read feature $f(s_i) = x_i$
7:       $F = F \cup \{x_i\}$
8:       $classify(F)$
9:       **if** $confidence(classify(F)) \geq \theta$ **then**
10:          output result of $classify(F)$ and **terminate**
11:       **end if**
12:    **end if**
13: **end for**
14: output result of $classify(F)$
15: **end**

Fig. 1. Basic algorithm for sequential classifier combination

Having a base classifier at its disposition, the algorithm starts with ranking the sensors in line 2. After this step, the sensors $s_1, \ldots, s_N$ are ordered according to the utility of their features $x_1, \ldots, x_N$, such that $\varphi(x_1) \geq \varphi(x_2) \geq \ldots \geq \varphi(x_N)$. That is, the first sensor yields the most discriminating feature, the second sensor the second most, and so on. Then the algorithm initializes the set F of features to be used by the classifier to the empty set (line 3). Next it iteratively activates one sensor after the other, reads in each sensor's measurement, and adds it to feature set F (lines 4 to 7). Once a new feature has been obtained, statement $classify(F)$ is executed, which means that the base classifier is applied, using feature set F (line 8). Note that a k-NN classifier is particularly suitable for such an incremental mode of operation where new features are iteratively added, because the distance computations can be performed in an incremental fashion, processing one feature after the other and accumulating the individual features' distances. In line 9, it is checked whether the confidence of the classification result is equal to or larger than a threshold θ. If this is the case the classification result is considered final. It is output and the algorithm terminates (line 10).

Otherwise, if the confidence is below the given threshold θ, the next sensor is activated.

Obviously, in order to classify an unknown object, the base classifier uses nested subsets of features $\{x_1\}, \{x_1, x_2\}, \ldots, \{x_1, x_2, \ldots, x_i\}$ until its confidence in a decision becomes equal to or larger than threshold θ. While running through the for-loop from line 4 to 13, it may happen that a sensor s_i becomes unavailable due to battery exhaustion or some other cause. In this case, sensor s_i will be simply skipped and the algorithm continues with sensor s_{i+1}. In case none of the considered feature subsets leads to a classification result with enough confidence, the classifier outputs, in line 14, the result obtained with the set F of features considered in the last iteration through the for-loop, i.e. for $i = N$.

An important issue in the algorithm of Fig. 1 is how one determines the confidence of the classifier. Many solutions to this problem can be found in the literature [17,18,19]. In the current paper, our base classifier is of the k-NN type. This means that it determines, for an unknown object represented by a subset of features $\{x_1, \ldots, x_i\}$, the k nearest neighbors in the training set. Then it assigns the unknown object to that class that is represented most often among the k nearest neighbors. In case of a tie, a random decision is made. Let $k' \leq k$ be the number of training samples that are among the k nearest neighbors and belong to the majority class. Then we can say that the larger k', the more confident is the classifier in its decision. Consequently, we can define $confidence(classify(F)) = k'$. That is, if there are $k' > \theta$ nearest neighbors from the majority class, then the classification result is output and the system terminates. Otherwise, if the number of nearest neighbors belonging to the majority class is less than or equal to θ, the next sensor is activated. If M denotes the number of classes to be distinguished, then the range of feasible thresholds is the set of integers from the interval $[\lceil k/M \rceil, \ldots, k - 1]$.

In order to rank the features in line 2 of the algorithm, three well-known methods have been used. The first method is Relief [20], which directly yields a ranking of the given features. Secondly, a wrapper approach (WA) in conjunction with k-NN classifiers is applied. The k-NN classifiers use only a single feature each. The features are finally ordered according to their performance on an independent validation set. Thirdly, sequential forward search [21] in conjunction with a k-NN wrapper is applied (WA-SFS). Here nested subsets of features $\{x_{i_1}\}$, $\{x_{i_1}, x_{i_2}\},\ldots,\{x_{i_1}, \ldots, x_{i_N}\}$ are generated and the ranking is given by the order $x_{i_1}, \ldots, x_{i_N}$ in which the features are added. For more details of features ranking, we refer to [22].

Table 1. Datasets and some of their characteristic properties

Data Sets	# Instances	# Features	# Classes	Training Set	Test Set
Isolet	7797	617	26	6237	1560
Multiple Features	2000	649	10	1500	500

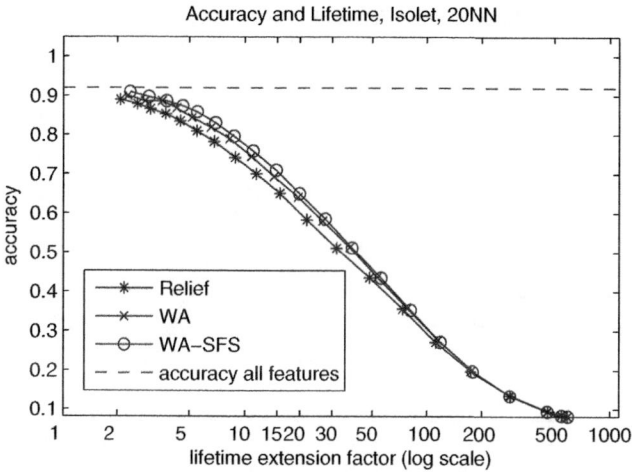

Fig. 2. Accuracy and lifetime as a function of the threshold on Isolet

3   Experiments

The algorithm described in Section 2 was implemented and experimentally evaluated. In the field of wireless sensor networks, there are not many data sets publicly available, especially not for pattern recognition problems. In [23], a dataset for activity recognition of humans equipped with body worn sensors as well as environmental sensors is described, and in [24] the PlaceLab datasets are introduced, which were acquired in the context of research on ubiquitous computing in a home setting. However, the authors of these papers do not mention any use of the data sets for pattern classification problems. Moreover, no classification benchmarks have been defined for any of these data sets. For this reason, it was decided to use datasets from the UCI Machine Learning Repository [25]. The sensors were simulated by assuming that each feature in any of these datasets is delivered by a sensor. The experiments reported in this paper were conducted on dataset Isolet and Multiple Features (see Table 1). These dataset pose classification problems with only numerical and no missing feature values. These conditions are necessary for the k-NN classifier being applicable in a straightforward way. Moreover, these datasets have a rather large number of features which makes them suitable to test the approach proposed in this paper. Experiments on other datasets from the UCI repository with similar characteristics gave similar results but are not reported here because of lack of space.

3.1   First Experiment

The purpose of the first experiment was to study how the value of threshold θ (see line 9 of the algorithm in Fig. 1) influences the classification accuracy and

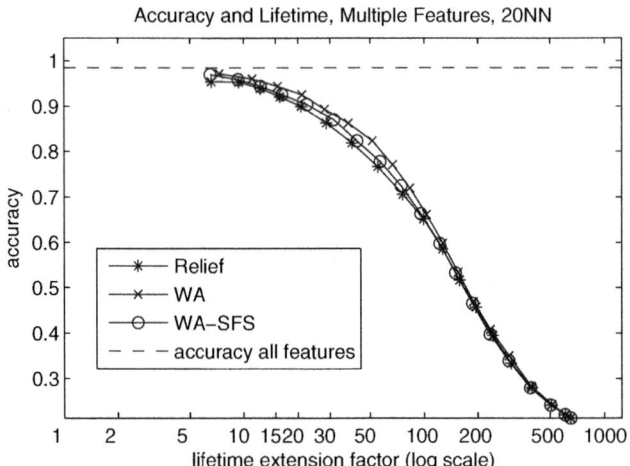

Fig. 3. Accuracy and lifetime as a function of the threshold on Multiple Features

the total number of sensor measurements (i.e. features) used for classification. Instead of computing the total number of sensor readings we measure the lifetime of the considered system, i.e. the number of classifications a system is able to perform before the sensors become unavailable, because of battery exhaustion. Hence the aim of the experiment is to measure the lifetime of the system and analyze the trade-off between lifetime and accuracy depending on threshold θ.

We assume that the test set consists of M patterns and each feature x_i can be used exactly M times before the battery of its sensor is exhausted. This means that with a conventional pattern recognition system, which uses the full set of features for each pattern to be classified, the test set can be classified exactly once before all sensors become unavailable. By contrast, with the system proposed in this paper, not all features will be used in each classification step, which allows one to classify the test set multiple times.

In this experiment, we classify the test set multiple times until all sensors become unavailable. Let $M' \geq M$ be the number of pattern instances actually classified, where we count an element of the test set as often as it has been classified. Now we define $lifetime\_\ extension\_\ factor\ =\ M'/M$. Clearly, the $lifetime\_\ extension\_\ factor$ is bounded by 1 from below. According to our assumption that each feature can be used exactly M times before it becomes unavailable, the case $lifetime\_\ extension\_\ factor = 1$ occurs if the underlying system always uses all features in each classification step. However, if less than N features are used, the value of the $lifetime\_\ extension\_\ factor$ will be greater than 1.

We measure the *accuracy* and the $lifetime\_\ extension\_\ factor$ both as a function of threshold θ. A representation of the results appears in Figs. 2 and 3. In these figures, the value of θ is decreased as we go from left to right. Obviously, we observe a trade-off between *accuracy* and $lifetime\_\ extension\_\ factor$.

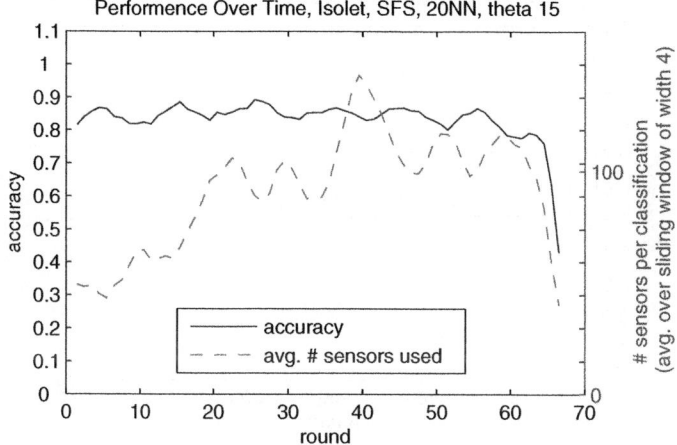

Fig. 4. Performance of SFS-WA over Time on Isolet

Clearly, in neither of the two datasets the proposed system reaches the accuracy obtained with the full set of features, but for large values of θ (at the left end of the curve) it gets quite close. Without loosing much recognition accuracy, the lifetime of the system can be extended by a factor of about 2 on Isolet and 5 on Multiple Features. For lower values of θ a much higher lifetime extension factor can be achieved, though at a price of a more pronounced loss of recognition accuracy. Comparing the different ranking strategies we note on Isolet that WA-SFS performs best and Relief worst. One may also conclude that the choice of a proper feature ranking strategy is not a critical issue.

3.2   Second Experiment

From the first experiment one can conclude that the lifetime of a system can be increased at the cost of decreased *accuracy*. However, no quantitative statement can be made about how the decrease in *accuracy* takes place over time. In the second experiment we proceed similarly to Experiment 1 and classify the test set several times. Yet we do not report the accuracy in the global sense, i.e. in one number for all runs together, but want to see how it changes as the system evolves over time and more sensors become unavailable.

In Experiment 2 the test set was divided into smaller portions of size one tenth of the original test set size. Then the algorithm of Fig. 1 was applied until all sensors became unavailable. For each portion of the test data the recognition rate and the number of sensors used were recorded. For the sake of brevity, we show only results for threshold $\theta = 15$ and the feature ranking strategy WA-SFS.

In Figs. 4 and 5, the results of the second experiment are shown. The x-axis depicts the number of rounds through the partitions of the test set, while on the left and right y-axis the *accuracy* and the number of sensors actually used is

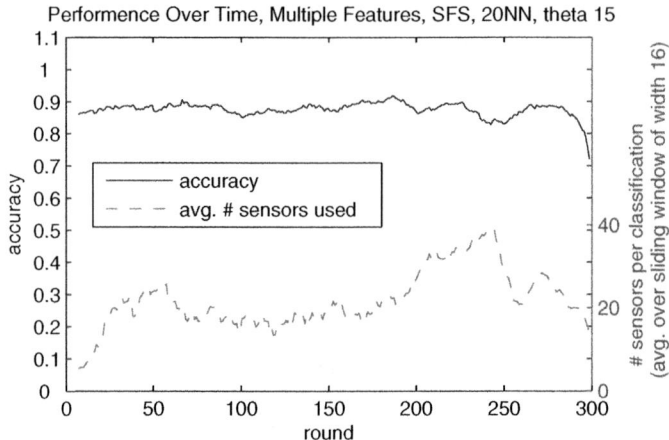

Fig. 5. Performance of SFS-WA over Time on Multiple Features

given, respectively. On both datasets we observe a similar behavior. On data set Isolet, the accuracy does not decrease much until about round 65. Afterwards it decays very rapidly. The number of features fluctuates remarkably, but shows an upward trend until round 60. Then it quickly declines. The second phase of this decline, when only very few sensors are left, is paralleled by a steep decline of the *accuracy*. From the qualitative point of view, a similar behavior can be observed on data set Multiple Features.

4   Summary, Discussion and Conclusions

In this paper, a sequential multiple classifier system is proposed for reducing the number of features used by a classifier. It is motivated by applications in wireless sensor networks. The procedure can be applied in conjunction with any known method for feature ranking. In the current paper three well known methods, viz. Relief, a wrapper approach based on evaluating each feature individually with a k-nearest neighbor classifier, and a wrapper approach in conjunction with sequential forward search are applied. The underlying base classifier is a k-nearest neighbor classifier.

The proposed procedure was implemented and experimentally tested. As test data, two datasets from the UCI Machine Learning repository were used. A wireless sensor network scenario was simulated by assuming that the individual features are delivered by independent sensors. The results of the experiments revealed that the system behaves very well. Its lifetime can be noticeably increased without loosing much recognition accuracy. During most of its lifetime, the system behaves quite stable. That is, the recognition rate only slightly decreases over the system's lifetime, and a drastic drop happens only towards the very end when almost all sensors are no longer available.

The proposed system can be applied to pattern classification tasks in real wireless sensor networks provided that the objects or events to be classified behave in some stationary way.[1] Because features are acquired in a sequential fashion and the decision of the classifier about the class label of an unknown object or event is only available after the first i features $(1 \leq i \leq N)$ have been processed, it is required that the object or event to be recognized does not change until sensor s_i has delivered feature $f(s_i) = x_i$. This may be a problem when quickly moving objects or rapidly changing events are to be classified. However, there are many potential applications of wireless sensor networks where this stationary assumption is typically satisfied. Examples include environment monitoring and surveillance.

There are many ways in which the work described in this paper can be extended. First of all one can think of investigating classifiers other than the k-nearest neighbor classifier.[2] Similarly, in addition to the three feature ranking strategies considered in this paper, there are many alternative methods known from the literature [26]. It would be certainly worthwhile to extend the experiments to these methods and compare them to the ones applied in this paper. Moreover, an extension of the experiments to more datasets would be desirable, in particular datasets obtained from real wireless sensor networks.

Acknowledgments. This research has been supported by the T'AMOP-4.2.2/ 08/1/2008-0008 program of the Hungarian National Development Agency.

References

1. Kuncheva, L.I.: Combining Pattern Classifiers: Methods and Algorithms. John Wiley & Sons, Inc., Hoboken (2004)
2. Kittler, J., Hatef, M., Duin, R.P.W., Matas, J.: On combining classifiers. IEEE Transactions on Pattern Analysis and Machine Intelligence 20(3), 226–239 (1998)
3. Rahman, A.F.R., Fairhurst, M.C.: Serial Combination of Multiple Experts: A Unified Evaluation. Pattern Anal. Appl. 2(4), 292–311 (1999)
4. Zhang, Z., Zhou, S., Zhou, A.: Sequential Classifiers Combination for Text Categorization: An Experimental Study. In: Li, Q., Wang, G., Feng, L. (eds.) WAIM 2004. LNCS, vol. 3129, pp. 509–518. Springer, Heidelberg (2004)
5. Chellapilla, K., Shilman, M., Simard, P.: Combining Multiple Classifiers for Faster Optical Character Recognition. In: Bunke, H., Spitz, A.L. (eds.) DAS 2006. LNCS, vol. 3872, pp. 358–367. Springer, Heidelberg (2006)
6. Viola, P., Jones, M.: Robust Real-Time Face Detection. Int. Journal of Computer Vision 57(2), 137–154 (2004)
7. Capelli, R., Maio, D., Maltoni, D., Nanni, L.: A Two-Stage Fingerprint Classification System. In: Proc. of the ACM SIGMM Workshop on Biometrics Methods and Applications, pp. 95–99. ACM, New York (2003)

---

[1] *Stationary* means that an object to be classified does not change (or disappear) before all necessary features have been measured.

[2] Clearly, such an extension is not obvious for most classifiers.

8. Last, M., Bunke, H., Kandel, A.: A Feature-Based Serial Approach to Classifier Combination. Pattern Analysis and Applications 5, 385–398 (2002)
9. Xiao, Y., Chen, H., Li, F.: Handbook on Sensor Networks, World Scientific Publishing Co., Singapore (2010)
10. Duarte, M.F., Hu, Y.H.: Vehicle Classification in Distributed Sensor Networks. Journal of Parallel and Distributed Computing 64(7), 826–838 (2004)
11. Brooks, R.R., Ramanathan, P., Sayeed, A.M.: Distributed Target Classification and Tracking in Sensor Networks. Proceedings of the IEEE 91(8), 1163–1171 (2003)
12. Meesookho, C., Narayanan, S., Raghavendra, C.S.: Collaborative Classification Applications in Sensor Networks. In: IEEE Sensor Array and Multichannel Signal Processing Workshop Proceedings, pp. 370–374 (2002)
13. Wang, X., Bi, D., Ding, L., Wang, S.: Agent Collaborative Target Localization and Classification in Wireless Sensor Networks. Sensors 7, 1359–1386 (2007)
14. Sun, Y., Qi, H.: Dynamic Target Classification in Wireless Sensor Networks. In:19th International Conference Pattern Recognition, ICPR 2008 (2008)
15. Gu, L., et al.: Lightweight Detection and Classification for Wireless Sensor Networks in Realistic Environments. In: Proc. of the Third Int. Conference on Embedded Networked Sensor Systems, ACM, New York (2005)
16. Wang, L., Xiao, Y.: A Survey of Energy-Efficient Scheduling Mechanisms in Sensor Networks. Mobile Networks and Applications 11, 723–740 (2006)
17. Chow, C.K.: On Optimum Recognition Error and Reject Trade-Off. IEEE Transactions on Information Theory, IT 16(1), 41–46 (1970)
18. Fumera, G., Roli, F., Giacinto, G.: Reject Option with Multiple Thresholds. Pattern Recognition 33(12), 2099–2101 (2000)
19. Hanczar, B., Dougherty, E.R.: Classification with Reject Option in Gene Expression Data. Bioinformatics 24(17), 1889–1895 (2008)
20. Kira, K., Rendell, L.A.: The Feature Selection Problem Traditional Methods and a new Algorithm. In: Proceedings of the tenth National Conference on Artificial Intelligence, pp. 129–134. AAAI Press, Menlo Park (1992)
21. Pudil, P., Novovicova, J., Kittler, J.: Floating Search Methods in Feature Selection. Pattern Recognition Letters 15(11), 1119–1125 (1994)
22. Csirik, J., Bunke, H.: Feature Selection and Ranking for Pattern Classification in Wireless Sensor Networks. In: Csirik, J., Bunke, H. (eds.) Pattern Recognition, Machine Intelligence and Biometrics Expanding Frontiers, Springer, Heidelberg (2011)
23. Zappi, P., Lombriser, C., Farelle, E., Roggen, D., Benini, L., Troester, G.: Experiences with Experiments in Ambient Intelligence Environments. In: IADIS Int. Conference Wireless Applications and Computing (2009)
24. Intille, S., Larson, K., Tapia, E.M., Beaudin, J.S., Kaushik, P., Nawyn, J., Rockinson, R.: Using a Live-In Laboratory for Ubiquitous Computing Research. In: Fishkin, K.P., Schiele, B., Nixon, P., Quigley, A. (eds.) PERVASIVE 2006. LNCS, vol. 3968, pp. 349–365. Springer, Heidelberg (2006)
25. Frank, A., Asuncion, A.U.: Machine Learning Repository. University of California, School of Information and Computer Science, Irvine CA (2010), http://archive.ics.uci.edu/ml
26. Gyon, I., Gunn, S., Nikravesh, M., Zadeh, L. (eds.): Feature Extraction, Foundations and Applications. Springer, Heidelberg (2006)

Multi-class Multi-scale Stacked Sequential Learning

Eloi Puertas[1,2], Sergio Escalera[1,2], and Oriol Pujol[1,2]

[1] Dept. Matemàtica Aplicada i Anàlisi, Universitat de Barcelona,
Gran Via 585, 08007, Barcelona, Spain
[2] Computer Vision Center, Campus UAB, Edifici O, 08193, Bellaterra, Spain
eloi@maia.ub.es, sergio@maia.ub.es, oriol@maia.ub.es

Abstract. One assumption in supervised learning is that data is independent and identically distributed. However, this assumption does not hold true in many real cases. Sequential learning is that discipline of machine learning that deals with dependent data.

In this paper, we revise the Multi-Scale Sequential Learning approach (MSSL) for applying it in the multi-class case (MMSSL). We have introduced the ECOC framework in the MSSL base classifiers and a formulation for calculating confidence maps from the margins of the base classifiers. Another important contribution of this papers is the MMSSL compression approach for reducing the number of features in the extended data set. The proposed methods are tested on 5-class and 9-class image databases.

1 Introduction

Sequential learning [3] assumes that samples are not independently drawn from a joint distribution of the data samples X and their labels Y. In sequential learning, training data actually consists of sequences of pairs (x, y), so that neighboring examples on a support lattice display some correlation. Usually sequential learning applications consider one-dimensional relationship support, but this kind of relationships appear very frequently in other domains, such as images, or video. Consider the case of object recognition in image understanding. It is clear that if one pixel belongs to a certain object category, it is very likely that neighboring pixels also belong to the same object (with the exception of its borders).

In literature, sequential learning has been addressed from different perspectives: from the point of view of graphical models, using Hidden Markov Models or Conditional Random Fields (CRF) [10,7,4,15] for inferring the joint or conditional probability of the sequence. From the point of view of meta-learning, sequential learning has been addressed by means of sliding window techniques, recurrent sliding windows [3] or stacked sequential learning (SSL) [6]. In SSL, a first base classifier is used to produce predictions. A sliding window among the predictions is applied and it is concatenated with the original data, building an extended dataset. Finally, a second base classifier predicts the final output from the extended dataset. In our previous work [11], we identified that the main

C. Sansone, J. Kittler, and F. Roli (Eds.): MCS 2011, LNCS 6713, pp. 197–206, 2011.

step of the relationship modeling proposed in [6] is precisely how this extended
dataset is created. In consequence, we formalized a general framework for the
SSL called Multi-scale Stacked Sequential Learning (MSSL), where a multi-scale
decomposition is used in the relationship modeling step.

Previous approaches addres bi-class problems. Few of them have been extended to the multi-class case. However in many applications, like image segmentation, problems are inherently multi-class. In this work, our contribution
is an efficient extension of MSSL to the multi-class case. We revise the general
stacked sequential learning scheme for applying to multi-class as well as bi-class
problems. We introduce the ECOC framework [2] in the base classifiers to extend them to the multi-class case, defining the Multi-class Multi-scale Stacked
Sequential Learning (MMSSL). An important issue considered in this work is
how the number of features in the extended data set increases with the number of classes. We propose a feature compression approach for mitigating this
problem.

The paper is organized as follows: first, we review the original MSSL for the bi-
class case. In the next section, we formulate the MSSL for the multi-class case
(MMSSL), introducing the ECOC framework. Each step in MSSL is revised
for the multi-class case and a compression approach for the extended dataset
is explained. Then experiments and results of our methodology are shown in
Section 4. Finally, Section 5 concludes the paper.

2   Related Work: Multi-scale Stacked Sequential Learning

SSL [6] is a meta-learning framework [9] consisting in two steps. In the first
step, a base classifier is trained and tested with the original data. Then, an
extended data set is created which joins the original training data features with
the predicted labels produced by the base classifier considering a window around
the example. At the second step, another classifier is trained with this new feature
set. In [11] SSL is generalized by emphasizing the key role of neighborhood
relationship modeling. The framework presented includes a new block in the
pipeline of the basic SSL. Figure 1 shows the Generalized Stacked Sequential
Learning process. A classifier $h_1(x)$ is trained with the input data set $(\mathbf{x}, y)$ and
the set of predicted labels y' is obtained. The next block defines the policy for
creating the neighborhood model of the predicted labels. It is represented by
$z = J(y', \rho, \theta) : \mathcal{R} \rightarrow \mathcal{R}^w$, a function that captures the data interaction with a
model parameterized by θ in a neighborhood ρ. The result of this function is a

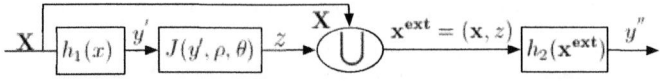

Fig. 1. Generalized stacked sequential learning

w-dimensional value, where w is the number of elements in the support lattice of the neighborhood ρ. In the case of defining the neighborhood by means of a window, w is the number of elements in the window. Then, the output of $J(y', \rho, \theta)$ is joined with the original training data creating the extended training set $(\mathbf{x}^{\mathbf{ext}}, y) = ((\mathbf{x}, z), y)$. This new set is used to train a second classifier $h_2(\mathbf{x}^{\mathbf{ext}})$ with the goal of producing the final prediction y''.

The definition of $J(y', \rho, \theta)$ proposed in [11] consists of two steps: first the multi-scale decomposition that answers how to model the relationship between neighboring locations, and second, the sampling that answers how to define the support lattice to produce the final set z.

3   Multi-class Multi-scale Stacked Sequential Learning

To extend the generalized stacked sequential learning scheme to the multi-class case, it is necessary that base classifiers $h_1(x)$ and $h_2(x^{ext})$ can deal with data belonging to several classes instead of just two. This can be achieved by using as base classifier a built-in multi-class classifier. However, it is also possible to use binary base classifiers by applying an ensemble of classifiers. One of the most successful ensemble strategies is the Error-Correct Output Codes framework [2,14]. In essence, ECOC is a methodology used for reducing a multi-class problem into a sequence of two-class problems. It is composed by two phases: a coding phase, where a codeword is assigned to each class, and a decoding phase, where, given a test sample, it looks for the most similar class codeword. The codeword is a sequence of bits, where each bit identifies the membership of the class for a given binary classifier. The most used coding strategy is the *one-versus-all* [5], where each class is discriminated against the rest, obtaining a codeword of length equal to the number of classes. Another standard coding strategy is the *one-versus-one* [1] which considers all possible pairs of classes, with a codeword length of $\frac{N(N-1)}{2}$.

The decoding phase of the ECOC framework is based on error-correcting principles under the assumption that the learning task can be modeled as a communication problem. Decoding strategies based on distances measurements between the output code and the target codeword are the most frequently applied. Among these, Hamming measure and Euclidean measure are the most used [14].

Apart from the extension of the base classifiers, the neighborhood function J has to be also modified. Figure 2 shows the multi-class multi-scale stacked sequential learning scheme presented in this work. Now, from an input sample, the first classifier produces not only a prediction, but a measure of confidence for belonging to each class. These confidences maps are the input of the neighborhood function. This function performs a multi-class decomposition over the confidence maps. Over this decomposition, a sampling z around each input example is returned. The extended data set is built up using the original samples as well as the set of features selected in z. Additionally, in order to reduce the number of features in the extended data set, we propose a compression approach for encoding the resulting multi-class decomposition of the confidence maps. Finally, having the extended data set $\mathbf{x}^{\mathbf{ext}}$ as input and using again the ECOC framework, the second classifier will predict to which class belongs the input sample $\mathbf{x}$. In the next subsections we explain in detail how the generalized stacked sequential learning can be extended to the multi-class case.

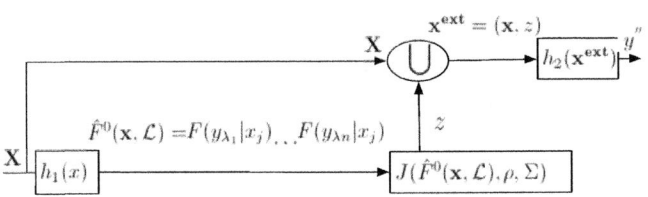

Fig. 2. Multi-class multi-scale stacked sequential learning

3.1 Extending the Base Classifiers

For the multi-class case, we need base classifiers that can handle with multiple classes. For this purpose, we use the ECOC framework explained before. Given a set of N classes, n different bipartitions (groups of classes) are formed, and n binary problems over the partitions are trained. As a result, a codeword of length n is obtained for each class, where each position (bit) of the code corresponds to a possible prediction value of a given binary classifier (generally $\{+1, -1\}$). We define a coding matrix M where each row is a codeword, where $M \in \{-1, +1\}^{N \times n}$ in the binary code case. Nonetheless, binary as well as ternary codes can be used in the coding strategy [14]. How to construct this matrix is an open problem, but in our experiments we will use the *one-versus-one* strategy cited above.

During the decoding process, applying the n binary classifiers, a code x is obtained for each data sample in the test set. In the SSL model, this code is obtained using the mere predicted label. However, it would be desirable that the base classifiers can provide not only a prediction but a measure of confidence for each class. The use of confidences gives a more precise information about the decisions of the first classifier than just its prediction. Given a set of possible labels $\mathcal{L} = \{\lambda_1, \ldots, \lambda_n\}$, we have n membership confidences,

$$\hat{F}^0(\mathbf{x}, \mathcal{L}) = \{F(y = \lambda_1 | \mathbf{x}), \ldots, F(y = \lambda_n | \mathbf{x})\} \tag{1}$$

In order to obtain these confidence maps we need base classifiers to generate not only a class prediction, but also its confidence. Unfortunately, not all kind of classifiers can give a confidence for its predictions. However, classifiers that work with margins such as Adaboost or SVM can be used [8]. In these cases, it is necessary to convert the margins used by these classifiers to a measure of confidence. In the Adaboost case, we apply a sigmoid function that normalizes

Adaboost margins from the interval $[-\infty, \infty]$ to $[-1, 1]$ (the same as codeword interval) by means of the following equation,

$$f(x) = \frac{1 - e^{-\beta m_x}}{1 + e^{-\beta m_x}} \tag{2}$$

where m_x is the margin of the predicted label given by Adaboost algorithm for the example x, and a constant that governs the transition: $\beta = \frac{-\ln(0.5\epsilon)}{0.25t}$, which depends on the number of iterations t that Adaboost performs, and an arbitrary small constant ϵ.

Once we have the normalized code by applying Equation 2, we use a soft distance d for decoding. The obtained code is compared to the codewords $C_i, i \in [1, \ldots, N]$ of each class defined in the matrix M, as follows:

$$F(y = \lambda_1 | x_j) = e^{-\alpha d(C_1, f(x_j))}, \ldots, F(y = \lambda_n | x_j) = e^{-\alpha d(C_N, f(x_j))}$$

where d may be any soft distance, as the Euclidean distance for example, $\alpha = -\ln(\epsilon)/2$, and ϵ is arbitrarily small positive quantity. By applying this to the all data samples $x_j \in \mathbf{x}$ we have the confidence maps for each class as expressed in Equation 1 that will be used in the next step.

3.2 Extending the Neighborhood Function J

We define the neighborhood function J in two stages: 1) a multi-scale decomposition over the confidence maps and 2) a sampling performed over the multi-scale representation. This function is extended in order to deal with multiple classes. Now it is formulated as: $z = J(\hat{F}^0(\mathbf{x}, \mathcal{L}), \rho, \Sigma)$.

Starting from the confidence maps $\hat{F}^0(\mathbf{x}, \mathcal{L})$, we apply a multi-scale decomposition upon them, resulting in as many decomposition sequences as labels. For the decomposition we use a multi-resolution gaussian approach. Each level of the decomposition (scale) is generated by the convolution of the confidence field by a gaussian mask of variable σ, where σ defines the grade of the decomposition. This means that the bigger the sigma is, the longer interactions are considered. Thus, at each level of decomposition all the points have information from the rest accordingly to the sigma parameter. Given a set of $\Sigma = \{\sigma_0, \ldots, \sigma_n\} \in \mathbb{R}^+$ and all the predicted confidence maps $\hat{F}^0(\mathbf{x}, \mathcal{L})$, each level of the decomposition is computed as follows,

$$\hat{F}^{s_i}(\mathbf{x}, \mathcal{L}) = g^{\sigma_i}(\mathbf{x}) * \hat{F}^0(\mathbf{x}, \mathcal{L})$$

where $g^{\sigma_i}(\mathbf{x})$ is defined as a multidimensional isotropic gaussian filter with zero mean,

$$g^{\sigma_i}(\mathbf{x}) = \frac{1}{(2\pi)^{d/2}\sigma_i^{1/2}} e^{-\frac{1}{2}\mathbf{x}^T \sigma_i^{-1} \mathbf{x}}$$

Once we have the multi-scale decomposition, we define the support lattice z. This is, the sampling performed over the multi-scale representation in order to obtain the extended data. Our choice is to use a scale-space sliding window over each label multi-scale decomposition. The selected window has a fixed radius with length defined by ρ in each dimension and with origin in the current prediction example. Thus, the elements covered by the window is $w = (2\rho + 1)^d$ around the origin. For the sake of simplicity, we use a fixed radius of length $\rho = 1$. Then, for each scale i considered in the previous decomposition (σ_i $i = 1 \ldots n$), the window is stretched in each direction using a displacement proportional to the scale we are analyzing. We use a displacement $\delta_i = 3\sigma_i + 0.5$. This displacement at each scale forces that each point considered around the current prediction has very small influence from previous neighbor points. In this way, the number of features of z, that are appended to the extended data set, is equal to $(2\rho + 1)^d \times |\Sigma| \times |\mathcal{L}|$, where $|\mathcal{L}|$ is the number of classes. According to this, we can see that the extended data set increases with the number of classes the problem has. This can be a problem if the number of classes is large, since the second classifier have to deal with many features, and therefore, the learning time is increased. In the next subsection, a grouping approach for the extended features is explained.

3.3   Extended Data Set Grouping: A Compression Approach

The goal of grouping the extended data set is to compress its number of features. Using the above approach, we can see that a confidence map is obtained for each class.Then, for each map, a multi-scale decomposition is computed, and finally, a sampling around each input example is performed. To reduce the number of confidence maps, we add a compression process between the multi-scale decomposition and the sampling process. This compression is done following information theory by means of partitions. Let be $\{P^1, P^2\}$ a partition of groups of classes and $\mathcal{L} = \{\lambda_i\}$ the set of all the classes, such that $P^1 \subseteq \mathcal{L}$ and $P^2 \subseteq \mathcal{L} \mid P^1 \cup P^2 = \mathcal{L}$ and $P^1 \cap P^2 = \emptyset$, the confidence maps are encoded as

$$\mathcal{F}^{s_j}(\{P^1, P^2\}) = \sum_i \hat{F}^{s_j}(\mathbf{x}, \lambda_i \in P^1) - \sum_i \hat{F}^{s_j}(\mathbf{x}, \lambda_i \in P^2)$$

for all the scales $s_j \in \Sigma$. Then, using a coding strategy, a minimum set of partitions $\mathcal{P} = \{P^1, P^2\}_1, \cdots, \{P^1, P^2\}_\alpha$ is defined, where $\alpha = \lceil \log_2 |\mathcal{L}| \rceil$ is the minimum number of partitions for compressing all the classes.

Following this compression approach, now the support lattice z is defined over $\mathcal{F}^\Sigma(\mathcal{P})$, this is, over all the scales in Σ and all the partitions in $\mathcal{P}$. Therefore, the number of features in z is reduced to $(2\rho + 1)^d \times |\Sigma| \times \lceil \log_2 |\mathcal{L}| \rceil$.

4   Experiments and Results

Before presenting the results, we discuss the data, methods and validation protocol of the experiment.

- **Data:** We test our multi-class methodology on the e-trims database [12]. The e-trims database is comprised of two datasets, 4-class with four annotated object classes and 8-class with eight annotated object classes. There are 60 annotated images in each of the dataset. The object classes considered in 4-class dataset are: (1) building, (2) pavement/road, (3) sky and (4) vegetation. In 8-class dataset the object classes considered are: (1) building, (2) car, (3) door, (4) pavement, (5) road, (6) sky, (7) vegetation and (8) window. Additionally, for each database we have a background class (0) for any other object. Examples of this database and ground-truth for 4-class and 8-class are shown in Figure 5.

- **Methods:** We test our method using all the confidences maps, named *MMSSL Standard* and using the compression approach, named *MMSSL Compressed*. The settings for both experiments are the same, the only difference is the number of generated features in the extended dataset. We have used as base classifier a Real Adaboost ensemble of 100 decision stumps. For each image in the training set we perform a stratified sampling of 3000 pixels per image. For each example (pixel) the feature vector is only composed of RGB attributes. This data is used for training the first classifier using leave-one-image-out to produce confidence maps. The coding strategy for the ECOC framework in each classifier is *one versus one* and the decoding measure is euclidean distance. The neighborhood function performs a gaussian multi-resolution decomposition in 4 scales, using $\Sigma = \{1, 2, 4, 8\}$. Observe that in order to compute the neighbors of each pixel the whole left-out image is classified. Finally, both classifiers are trained using the same feature samples without and with the extended set, respectively.

  Furthermore, we have performed an experiment using multi-label optimization via α-expansion [13]. Using the confidence maps for each class obtained from the first classifier, we have applied the α-expansion optimization. For the neighborhood term, we have took into account the intensity of the 4-connected pixels.

- **Validation:** We have performed each experiment using six disjoint folds, where 50 images are used as train set and 10 images used as test.

Tables 3 and 4 show accuracy, overlapping, sensitivity and specifity averaged for all the classes. The accuracy of both MMSSL approaches in both databases are very similar. This fact reinforce that our idea of grouping features without losing performance is correct. The main advantage for using the compression approach is that reducing the number of features in the extended dataset, the time of the learning phase for the second classifier is reduced as well. In this way if the number of classes increases the approach of MMSSL is still feasible.

Figures 3 and 4 show the overlapping for each class in 4-class and 8-class database respectively. Large classes as building or sky have more overlapping than the rest. Observe that class road/pavement in 4-class or road in 8-class have a significantly larger overlapping than the Adaboost approach, this means that using only the RGB features, this class is difficult to distinguish, but the

Table 1. Method results for 4-class database

Method	Accuracy	Overlapping	Sensivity	Specifity
MMSSL Standard	0.8380	0, 6425	0, 7398	0, 795
MMSSL Compression	0.8358	0, 6390	0, 7342	0, 8001
Optimization α-expansion	0.5503	0, 3940	0, 6121	0, 5471
Adaboost	0.6200	0, 4366	0, 5634	0, 6278

Table 2. Method results for 8-class database

Method	Accuracy	Overlapping	Sensivity	Specifity
MMSSL Standard	0.7015	0, 4248	0, 5390	0, 5941
MMSSL Compression	0.7070	0, 4528	0, 5685	0, 6443
Optimization α-expansion	0.5138	0, 2570	0, 3927	0, 3529
Adaboost	0.6200	0, 2915	0, 4031	0, 4347

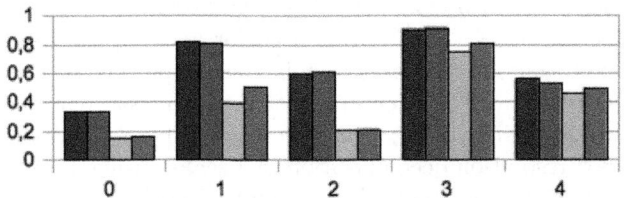

Fig. 3. Method Overlapping for 4-class

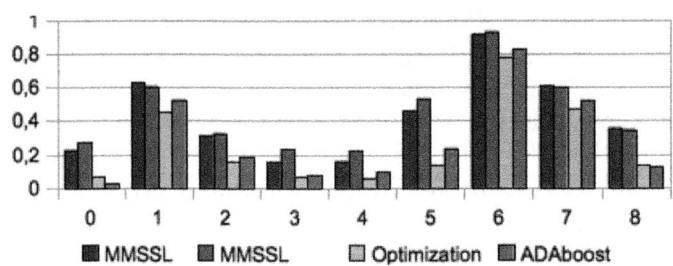

Fig. 4. Method Overlapping for 8-class

relationship among the rest classes makes it easier (road/pavement class is at bottom of the building).

Figure 5 shows results for some images in the 4-class and 8-class dataset. Observe that MMSSL methods grasp the relationship between adjacent classes and ADAboost can not. In addition, we have observed that using the optimization α-expansion, the results are poor. This means that the classification performed by the first classifier is not good enough to generalize from the surrounding pixels. On contrary, our MMSSL method can extract longer interrelations between samples and, thus, obtain better results.

Original   GroundThrough   MMSSL   MMSSL compression   Optimization   ADAboost

Fig. 5. Etrims 4-class (four top rows) and 8-class (four bottom rows) results

5 Conclusions

In this paper we adapted multi-scale sequential learning (MSSL) to the multi-class case (MMSSL). First, we have introduced the ECOC Framework in the base classifiers. Next, we show how to compute the confidence maps using the normalized margins obtained from the base classifiers. Another important contribution is the compression approach we have used for reducing the number of features in the extended data set. Taking into account that the experiments was done using only RGB features, the effect of sequential features from MMSSL is highly remarkable in our results. Moreover, they also shows that in terms of accuracy the loss of information during de compression process is negligible, but the amount of reduced features is considerable.

References

1. Allwein, E., Schapire, R., Singer, Y.: Reducing Multiclass to Binary: A Unifying Approach for Margin Classifiers. J. Machine Learning Research 1, 113–141 (2002)
2. Dietterich, T.G., Bakiri, G.: Solving Multiclass Learning Problems via Error-Correcting Output Codes. J. Artificial Intelligence Research 2, 263–286 (1995)
3. Dietterich, T.G.: Machine Learning for Sequential Data: A Review. In: Caelli, T.M., Amin, A., Duin, R.P.W., Kamel, M.S., de Ridder, D. (eds.) SPR 2002 and SSPR 2002. LNCS, vol. 2396, pp. 15–30. Springer, Heidelberg (2002)
4. Dietterich, T.G., Ashenfelter, A., Bulatov, Y.: Training conditional random fields via gradient tree boosting. In: Proc. of the 21th ICML (2004)
5. Nilsson, N.J.: Learning Machines. McGraw-Hill, New York (1965)
6. Cohen, W.W., de Carvalho, V.R.: Stacked sequential learning. In: IJCAI 2005, pp. 671–676 (2005)
7. McCallum, A., Freitag, D., Pereira, F.: Maximum entropy markov models for information extraction and segmentation. In: Proc. of ICML 2000, pp. 591–598 (2000)
8. Friedman, J., Hastie, T., Tibshirani, R.: Additive logistic regression: a statistical view of boosting. Annals of Statistics 28 (2000)
9. Wolpert, D.H.: Stacked generalization. Neural Networks 5(2), 241–259 (1992)
10. Lafferty, J.D., McCallum, A., Pereira, F.: Conditional random fields: Probabilistic models for segmenting and labeling sequence data. In: Proc. of ICML 2001, pp. 282–289 (2001)
11. Pujol, O., Puertas, E., Gatta, C.: Multi-scale stacked sequential learning. In: Benediktsson, J.A., Kittler, J., Roli, F. (eds.) MCS 2009. LNCS, vol. 5519, pp. 262–271. Springer, Heidelberg (2009)
12. Korč, F., Förstner, W.: eTRIMS Image Database for Interpreting Images of Man-Made Scenes, TR-IGG-P-2009-01, University of Bonn (2009)
13. Boykov, Y., Funka-Lea, G.: Graph Cuts and Efficient N-D Image Segmentation. I. Journal of Computer Vision 70(2), 109–131 (2006)
14. Escalera, S., Tax, D., Pujol, O., Radeva, P., Duin, R.: Subclass Problem-dependent Design of Error-Correcting Output Codes. IEEE T. in Pattern Analysis and Machine Intelligence 30(6), 1041–1054 (2008)
15. Mottl, V., Dvoenko, S., Kopylov, A.: Pattern recognition in interrelated data: The problem, fundamental assumptions, recognition algorithms. In: Proc. of the 17th ICPR, Cambridge, UK vol. 1, pp. 188–191 (2004)

A Comparison of Random Forest with
ECOC-Based Classifiers

R.S. Smith[1], M. Bober[2], and T. Windeatt[1]

[1] Centre for Vision, Speech and Signal Processing, University of Surrey, Guildford,
Surrey GU2 7XH, UK
[2] Mitsubishi Electric R&D Centre Europe B.V, 20 Frederick Sanger Road,
The Surrey Research Park, Guildford, Surrey GU2 7YD, UK
{Raymond.Smith,T.Windeatt}@surrey.ac.uk

Abstract. We compare experimentally the performance of three approaches to ensemble-based classification on general multi-class datasets. These are the methods of random forest, error-correcting output codes (ECOC) and ECOC enhanced by the use of bootstrapping and class-separability weighting (ECOC-BW). These experiments suggest that ECOC-BW yields better generalisation performance than either random forest or unmodified ECOC. A bias-variance analysis indicates that ECOC benefits from reduced bias, when compared to random forest, and that ECOC-BW benefits additionally from reduced variance. One disadvantage of ECOC-based algorithms, however, when compared with random forest, is that they impose a greater computational demand leading to longer training times.

1 Introduction

Two of the most popular approaches to constructing multiple classifier systems (MCS) to solve multi-class classification problems are random forest [1] and error-correcting output codes (ECOC) [2,3]. In this paper we present the result of an experimental comparison of these two methods when applied to a selection of real-world datasets taken from the UCI repository [4]. We also consider an enhanced version of ECOC, referred to as ECOC-BW, in which bootstrapping[1] is applied when constructing base-classifier training sets and weighting is applied to base-classifier decisions. Previous work has shown these enhancements to be beneficial [5,6].

The random forest algorithm was introduced by Breiman in 2001 [1]. A number of variants of random forest have been proposed but here we focus on the method that is often cited as a reference in the literature, known as Forest-RI. This consists of building an ensemble of unpruned decision tree classifiers whose

---

[1] Bootstrapping is a technique whereby new training sets are constructed from a given training set by repeated sampling with replacement. Each new training set (referred to as a bootstrap replicate) has, on average, 63% of the patterns in the original set but with some patterns repeated so as to form a set of the same size.

C. Sansone, J. Kittler, and F. Roli (Eds.): MCS 2011, LNCS 6713, pp. 207–216, 2011.
© Springer-Verlag Berlin Heidelberg 2011

classification decisions are combined by a voting procedure. Each decision tree is randomised in two ways: firstly, the training set is modified by constructing a bootstrap replicate and secondly, at each node of the tree, the search for the best split is limited to a subset of features that is randomly selected (without replacement) from the full set of features. Forest-RI thus aims to achieve good classification performance by combining the principles of bagging and random feature selection.

In the ECOC approach, first described by Dietterich in 1991 [7], a multi-class problem is decomposed into a series of 2-class problems, or dichotomies, and a separate base classifier trained to solve each one. These 2-class problems are constructed by repeatedly partitioning the set of target classes into pairs of super-classes so that, given a large enough number of such partitions, each target class can be uniquely represented as the intersection of the super-classes to which it belongs. The classification of a previously unseen pattern is then performed by applying each of the base classifiers so as to make decisions about the super-class membership of the pattern. Redundancy can be introduced into the scheme by using more than the minimum number of base classifiers and this allows errors made by some of the classifiers to be corrected by the ensemble as a whole.

The operation of the ECOC algorithm can be broken down into two distinct stages - the coding stage and the decoding stage. The coding stage consists of applying the base classifiers to the input pattern $\mathbf{x}$ so as to construct a vector of base classifier outputs $\mathbf{s}(\mathbf{x})$ and the decoding stage consists of applying some decoding rule to this vector so as to make an estimate of the class label that should be assigned to the input pattern. A commonly used decoding method is to base the classification decision on the minimum distance between $\mathbf{s}(\mathbf{x})$ and the vector of target outputs for each of the classes, using a distance metric such as Hamming or L^1. This, however, treats all base classifiers as equal, and takes no account of variations in their reliability. In the ECOC-BW variant of ECOC we assign different weights to each base classifier and target class combination so as to obtain improved ensemble accuracy. The weighting algorithm is referred to as class-separability weighting (CSEP) because the weights are computed in such a way that they measure the ability of a base classifier to distinguish between examples belonging to and not belonging to a given class [8].

Although, unlike random forest, bootstrapping is not a standard feature of the ECOC algorithm, we have shown [5,6] that it can be beneficial, particularly when combined with the CSEP weighting scheme. For this reason, in ECOC-BW we apply bootstrapping to the training set when each base classifier is trained. The effect of bootstrapping is to increase the desirable property of diversity [9] among the base classifiers in the ensemble. By this is meant that the errors made by component classifiers should, as far as possible, be uncorrelated so that the error correcting properties of the ensemble can have maximum effect. A further potential benefit of bootstrapping is that each base classifier is trained on only a subset of the available training data and this leaves the remaining data, known as the out-of-bootstrap (OOB) set, to be used for other purposes such as parameter tuning. Note, however, that the OOB set is unique to each base classifier.

When considering the errors made by statistical pattern classifiers it is useful to group them under three headings. Firstly there is the unavoidable error, known as *Bayes error*, which is caused by noise in the process that generates the patterns. A second source of error is *variance*; this is caused by the sensitivity of a learning algorithm to the chance details of a particular training set and causes slightly different training sets to produce classifiers that give different predictions for some patterns. Thirdly there are errors caused by *bias* in a learning algorithm[2]; here the problem is that the classifier is unable, for whatever reason, to adequately model the class decision boundaries in the pattern feature space.

In this paper we use the concepts of bias and variance to investigate the reasons for the differences in the accuracy achieved by different classification methods.

The ideas of bias, variance and noise originally emerged from regression theory. In this context they can be defined in such a way that the squared loss can be expressed as the sum of noise, bias (squared) and variance. The goal of generalising these concepts to classification problems, using a 0-1 or other loss function, has proved elusive and several alternative definitions have been proposed (see [10] for a summary). In fact it is shown in [10] that, for a general loss function, these concepts cannot be defined in such a way as to possess all desirable properties simultaneously. For example the different sources of error may not be additive, or it may be possible for variance to take negative values. In this study we adopt the Kohavi-Wolpert definitions [11]. These have the advantage that bias and variance are non-negative and additive. A disadvantage, however, is that no explicit allowance is made for Bayes error and it is, in effect, incorporated into the bias term.

The remainder of this paper is structured as follows. The technique of CSEP weighting is described in detail in section 2. Here we also derive a novel probabilistic interpretation of the method. Section 3 then describes the other main novelty of this paper which is an experimental comparison of each of the three classification methods: random forest, ECOC and ECOC-BW; this comparison is made in terms of classifier accuracy and is also broken down into bias and variance components. Finally, section 4 summarises the conclusions to be drawn from this work.

2   ECOC Weighted Decoding

The ECOC method consists of repeatedly partitioning the full set of N classes Ω into L super-class pairs. The choice of partitions is represented by an $N \times L$ binary *coding matrix* $\mathbf{Z}$. The rows $\mathbf{Z}_i$ are unique *codewords* that are associated with the individual target classes ω_i and the columns $\mathbf{Z}^j$ represent the different super-class partitions. Denoting the jth super-class pair by S^j and $\overline{S^j}$, element Z_{ij} of the coding matrix is set to 1 or 0 depending on whether class ω_i has been

---

[2] Bias is actually measured as the quantity bias^2 .

put into S^j or its complement[3]. A separate base classifier is trained to solve each of these 2-class problems.

Given an input pattern vector $\mathbf{x}$ whose true class $y(\mathbf{x}) \in \Omega$ is unknown, let the soft output from the jth base classifier be $s_j(\mathbf{x}) \in [0,1]$. The set of outputs from all the classifiers can be assembled into a vector $\mathbf{s}(\mathbf{x}) = [s_1(\mathbf{x}), \ldots, s_L(\mathbf{x})]^{\mathrm{T}} \in [0,1]^L$ called the *output code* for $\mathbf{x}$. Instead of working with the soft base classifier outputs, we may also first harden them, by rounding to 0 or 1, to obtain the binary vector $\mathbf{h}(\mathbf{x}) = [h_1(\mathbf{x}), \ldots, h_L(\mathbf{x})]^{\mathrm{T}} \in \{0,1\}^L$. The principle of the ECOC technique is to obtain an estimate $\hat{y}(\mathbf{x}) \in \Omega$ of the class label for $\mathbf{x}$ from a knowledge of the output code $\mathbf{s}(\mathbf{x})$ or $\mathbf{h}(\mathbf{x})$.

In its general form, a *weighted* decoding procedure makes use of an $N \times L$ weights matrix $\mathbf{W}$ that assigns a different weight to each target class and base classifier combination. The class decision, based on the L^1 metric, is made as follows:

$$\hat{y}(\mathbf{x}) = \arg\min_{\omega_i} \sum_{j=1}^{L} \mathbf{W}_{ij} \left| s_j(\mathbf{x}) - \mathbf{Z}_{ij} \right|, \tag{1}$$

where it is assumed that the rows of $\mathbf{W}$ are normalised so that $\sum_{j=1}^{L} \mathbf{W}_{ij} = 1$ for $i = 1 \ldots N$. If the base classifier outputs $s_j(\mathbf{x})$ in Eqn. 1 are replaced by hardened values $h_j(\mathbf{x})$ then this describes the weighted Hamming decoding procedure.

The values of $\mathbf{W}$ may be chosen in different ways. For example, if $\mathbf{W}_{ij} = \frac{1}{L}$ for all i, j then the decoding procedure of Eqn. 1 is equivalent to the standard unweighted L^1 or Hamming decoding scheme. In this paper we make use of the class separability measure [8,5] to obtain weight values that express the ability of each base classifier to distinguish members of a given class from those of any other class.

In order to describe the class-separability weighting scheme, the concept of a correctness function must first be introduced: given a pattern $\mathbf{x}$ which is known to belong to class ω_i, the correctness function for the j'th base classifier takes the value 1 if the base classifier makes a correct prediction for $\mathbf{x}$ and 0 otherwise:

$$C_j(\mathbf{x}) = \begin{cases} 1 & \text{if } h_j(\mathbf{x}) = \mathbf{Z}_{ij} \\ 0 & \text{if } h_j(\mathbf{x}) \neq \mathbf{Z}_{ij} \end{cases}. \tag{2}$$

We also consider the complement of the correctness function $\overline{C}_j(\mathbf{x}) = 1 - C_j(\mathbf{x})$ which takes the value 1 for an incorrect prediction and 0 otherwise.

For a given class index i and base classifier index j, the class-separability weight measures the difference between the positive and negative correlations of base classifier predictions, ignoring any base classifiers for which this difference is negative:

---

[3] Alternatively, the values +1 and -1 are often used.

$$\mathbf{W}_{ij} = \max \left\{ 0, \frac{1}{K_i} \left[\sum_{\substack{\mathbf{p} \in \omega_i \\ \mathbf{q} \notin \omega_i}} C_j(\mathbf{p}) C_j(\mathbf{q}) - \sum_{\substack{\mathbf{p} \in \omega_i \\ \mathbf{q} \notin \omega_i}} \overline{C}_j(\mathbf{p}) \overline{C}_j(\mathbf{q}) \right] \right\}, \quad (3)$$

where patterns $\mathbf{p}$ and $\mathbf{q}$ are taken from a fixed training set T and K_i is a normalisation constant that ensures that the i'th row of $\mathbf{W}$ sums to 1. The algorithm for computing $\mathbf{W}$ is summarised in fig. 1.

Inputs: matrix of training patterns $\mathbf{T} \in \mathbb{R}^{P \times M}$, binary coding matrix $\mathbf{Z} \in \{0,1\}^{N \times L}$, trained ECOC coding function $E : \mathbb{R}^M \mapsto [0,1]^L$.
Outputs: weight matrix $\mathbf{W} \in [0,1]^{N \times L}$ where $\sum_{j=1}^L \mathbf{W}_{ij} = 1$, for $i = 1 \ldots N$.
Apply E to each row of $\mathbf{T}$ and round to give prediction matrix $\mathbf{H} \in \{0,1\}^{P \times L}$.
Initialise $\mathbf{W}$ to $\mathbf{0}$.
for $c = 1$ to N
   for $i =$ indices of training patterns belonging to class c
      for $j =$ indices of training patterns *not* belonging to class c
        let d be the true class of the pattern $\mathbf{T}_j$.
        for k $= 1$ to L
           if $\mathbf{H}_{ik} = \mathbf{Z}_{ck}$ and $\mathbf{H}_{jk} = \mathbf{Z}_{dk}$, add 1 to $\mathbf{W}_{ck}$
               as the predictions for both patterns $\mathbf{T}_i$ and $\mathbf{T}_j$ are correct.
           if $\mathbf{H}_{ik} \neq \mathbf{Z}_{ck}$ and $\mathbf{H}_{jk} \neq \mathbf{Z}_{dk}$, subtract 1 from $\mathbf{W}_{ck}$
               as the predictions for both patterns $\mathbf{T}_i$ and $\mathbf{T}_j$ are incorrect.
        end
      end
   end
end
Reset all negative entries in $\mathbf{W}$ to 0.
Normalise $\mathbf{W}$ so that each row sums to 1.

Fig. 1. Pseudo-code for computing the class-separability weight matrix for ECOC.

The weights matrix $\mathbf{W}_{ij}$ of Eqn. 3 was derived from a consideration of the spectral properties of the Boolean functions that map base classifier outputs to the ensemble decisions. In this interpretation base classifiers are weighted by their ability to distinguish the members of a given class from patterns which do not belong to that class. An alternative interpretation may also be given in terms of base classifier accuracy probabilities. Let

$$P_{ij} = \frac{1}{M_i} \sum_{\mathbf{x} \in \omega_i} c_j(\mathbf{x}), \ Q_{ij} = \frac{1}{(M - M_i)} \sum_{\mathbf{x}' \notin \omega_i} c_j(\mathbf{x}') \quad (4)$$

where M is the total number of training patterns and M_i is the number of belonging to class ω_i. Then P_{ij} and Q_{ij} respectively represent estimates of the probability that the jth base classifier makes correct decisions for patterns belonging to and not belonging to class ω_i. By substituting $M_i P_{ij}$ and $(M - M_i) Q_{ij}$ for $\sum_{\mathbf{x} \in \omega_i} c_j(\mathbf{x})$ and $\sum_{\mathbf{x}' \notin \omega_i} c_j(\mathbf{x}')$ in Eqn. 3 and making use of the fact that $\bar{c}_j(\mathbf{x}) = 1 - c_j(\mathbf{x})$, it can be easily shown that an alternative definition of the CSEP weights is given by:

$$\mathbf{W}_{ij} = \max \left\{ 0, \frac{1}{K'_i} [P_{ij} + Q_{ij} - 1] \right\}, \tag{5}$$

where $K'_i = K_i / M_i (M - M_i)$ is a modified normalisation constant.

From Eqn. 5 it can be seen that CSEP weighting rewards those base classifiers that have a high true detection rate and a low false detection rate for class ω_i. Any base classifier that cannot outperform random guessing, where $P_{ij} = Q_{ij} = 0.5$, will be zero weighted under this algorithm.

3   Experiments

In this section we present the results of performing classification experiments on 11 multi-class datasets obtained from the publicly available UCI repository [4]. The characteristics of these datasets in terms of size, number of classes and number of features are given in table 1 All experiments were based on a 20/80 training/test set split and each run used a different randomly chosen stratified training set. These training sets were first normalised to have zero mean and unit variance.

Table 1. Experimental datasets showing the number of patterns, classes, continuous and categorical features

Dataset	Num. Patterns	Num. Classes	Cont. Features	Cat. Features
dermatology	366	6	1	33
ecoli	336	8	5	2
glass	214	6	9	0
iris	150	3	4	0
segment	2310	7	19	0
soybean	683	19	0	35
thyroid	7200	3	6	15
vehicle	846	4	18	0
vowel	990	11	10	1
waveform	5000	3	40	0
yeast	1484	10	7	1

For each dataset, ECOC ensembles of 200 base classifiers were constructed. Each base classifier consisted of a multi-layer perceptron (MLP) neural network with one hidden layer. The Levenberg-Marquardt algorithm was used for base classifier training as this has been shown to converge more rapidly than back-propagation. Base classifier complexity was adjusted by varying the hidden node counts and training epochs. Each such combination was repeated 10 times and the lowest mean test error was obtained. Random variations between each run were introduced by generating a different random code matrix and by randomly setting the initial MLP weights. The code matrices were constructed in such a way as to place an approximately equal number of target classes in the super-classes S^j and $\overline{S^j}$.

The ECOC experiments were repeated using ECOC-BW (i.e. with CSEP weighting and bootstrapping being applied). Each base classifier was trained on a separate bootstrap replicate drawn from the full training set for that run. The CSEP weight matrix was computed from the full training set each time so its value was determined in part by patterns (the OOB set) that were not used in the training of the base classifier.

The random forest experiments were conducted in a similar way to ECOC except that it was found necessary to repeat each experiment 100 times in order to obtain stable results. The number of decision trees in each forest was varied up to 400 and the optimal number required to minimise test error was obtained. The number of random features selected at each node was chosen using Breiman's heuristic $\log_2 F + 1$ where F is the total number of features available[4]. This has been shown to yield near optimal results [12].

The outcome of these experiments on individual datasets is shown graphically in Fig. 2. This shows a bar chart of the lowest test error attained by the three classification methods. Also shown is the average test error taken over all datasets. Table 2 summarises these results and shows the number of experiments that were found to be statistically significant at the 5% level using a t-test.

Table 2. A comparison of classifier types showing the number of favourable and non-favourable experiments in support of the proposition that the first classifier is more accurate than the second. Figures in brackets show how many experiments were significant at the 5% level.

First Classifier	Second Classifier	Favourable (out of 11)	Non-Favourable (out of 11)
ECOC-BW	RF	9 (8)	2 (1)
ECOC	RF	7 (6)	4 (4)
ECOC-BW	ECOC	9 (3)	2 (0)

Inspection of Fig. 2 and table 2 shows that no single classification algorithm gave the best results on all datasets. Indeed, random forest gave the lowest generalisation error on *glass* and *soybean*, ECOC gave the lowest error on *segment*

---

[4] Another commonly used heuristic is $\sqrt{F}$. For these datasets, however, both formulae selected a very similar, and in many cases identical, number of features.

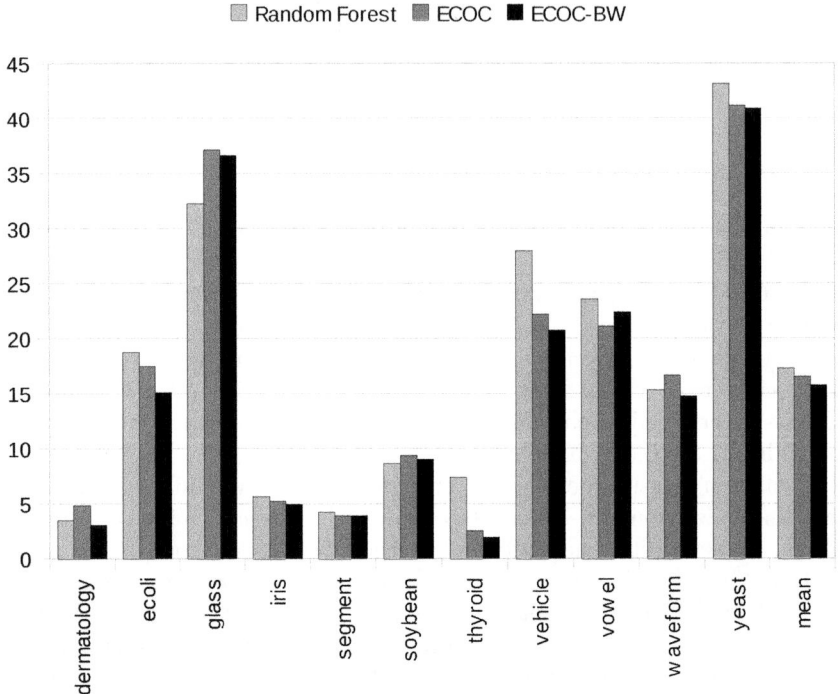

Fig. 2. The lowest percentage ensemble test error attained by different classifiers on 11 datasets

and *vowel* whilst ECOC-BW was optimal on *dermatology, ecoli, iris, thyroid, vehicle, waveform* and *yeast*. Comparing the different algorithms, it can be seen that ECOC gave a lower error than random forest on 7/11 datasets and also had a lower average error taken over all datasets. ECOC-BW yielded lower error than random forest on 9/11 datasets and also beat standard ECOC on 9/11 datasets. ECOC-BW also had the lowest mean error over all datasets. The evidence from these experiments is then that the ECOC-BW algorithm tends to give the best generalisation performance out of the three methods. There is also evidence that standard ECOC tends to perform a little better random forest but the advantage is not so consistent.

One further consideration that is worth taking into account when comparing these classification methods is that of computational overheads. The decision tree base classsifiers used by random forest are of a more lightweight nature that the MLP base classifiers that were used in ECOC classification and this was reflected in the elapsed times of the experiments which were, typically, about 15 times greater for the ECOC-based methods than for random forest.

It is interesting to look at the performance of these classifiers in terms of a bias-variance decomposition of the error. Table 3 shows a breakdown of the error incurred by each ensemble type when averaged over all datasets.

Inspection of table 3 suggests that when ECOC is compared with random forest, the variances of the two algorithms are the same and the slightly greater accuracy of ECOC may be attributed to a lower level of bias. It seems likely that lower bias of ECOC is due to the fact that the MLP base classifiers themselves will tend to have lower bias due to the fact that they are able to model non-linear decision boundaries between classes. By contrast, random forest uses decision trees as base classifiers are thus constrained to model decision boundaries as segments of hyperplanes that run parallel to the feature-space axes.

When ECOC-BW is used, variance is also reduced, leading to a further reduction in classification error. This is consistent with previous work [6] which has demonstrated that the use of CSEP weighting and bootstrapping tends to make the ECOC ensemble less prone to over-fitting the data so that classifier decisions become less sensitive to variations in the training set.

Table 3. A comparison of percentage bias, variance and total error incurred by different classifiers. The values are averaged over 11 datasets.

	Bias^2	Variance	Total Error
RF	9.8	7.5	17.4
ECOC	9.1	7.5	16.5
ECOC-BW	9.1	6.7	15.8

4   Discussion and Conclusions

In this paper we have compared experimentally the generalisation performance of three types of ensemble classifier on general multi-class datasets. The classifier types were random forest, ECOC and ECOC-BW in which ECOC is enhanced by the application of class-separability weighting and bootstrapping. The evidence from this set of experiments is that, although each classifier type can be optimal on some datasets, in general ECOC-BW tends to yield better accuracy than either random forest or ECOC. There is evidence that accuracy of ECOC is slightly better than that of random forest but the advantage cannot be so consistently observed as for ECOC-BW.

A breakdown of the error into bias and variance components reveals that standard ECOC has similar variance properties to random forest but benefits from slightly lower bias. It is suggested that this is due to the lower bias of the MLP base classifiers that were used with ECOC, when compared to the decision tree base classifiers of the random forest algorithm. For ECOC-BW the variance is also lower than for random forest and this leads to a further reduction in overall classification error.

Although ECOC-BW tends to yield greater classification accuracy, it is worth noting that random forest has an advantage over the ECOC-based algorithms in the sense that it has substantially reduced computational requirements.

Acknowledgements

This work was supported by EPSRC grant E061664/1. The authors would also like to thank the providers of the PRTools [13] and Weka [14] software.

References

1. Breiman, L.: Random Forests. Journal of Machine Learning 45(1), 5–32 Springer, Heidelberg (2001)
2. Dietterich, T.G., Bakiri, G.: Solving Multiclass Learning Problems via Error-Correcting Output Codes. Journal of Artificial Intelligence Research 2, 263–286 (1995)
3. James, G.: Majority Vote Classifiers: Theory and Applications. PhD Dissertation, Stanford University (1998)
4. Merz, C.J., Murphy, P.M.: UCI Repository of Machine Learning Databases (1998), http://www.ics.uci.edu/~mlearn/MLRepository.html
5. Smith, R.S., Windeatt, T.: Class-Separability Weighting and Bootstrapping in Error Correcting Output Code Ensembles. In: El Gayar, N., Kittler, J., Roli, F. (eds.) MCS 2010. LNCS, vol. 5997, pp. 185–194. Springer, Heidelberg (2010)
6. Smith, R.S., Windeatt, T.: A Bias-Variance Analysis of Bootstrapped Class-Separability Weighting for Error-Correcting Output Code Ensembles. In: Proc. 20th Int. Conf. on Pattern Recognition (ICPR), pp. 61–64 (2010)
7. Dietterich, T.G., Bakiri, G.: Error-correcting output codes: A general method for improving multiclass inductive learning programs. In: Proceedings of the Ninth National Conference on Artificial Intelligence (AAAI-1991), pp. 572–577. AAAI Press, Anaheim (1991)
8. Windeatt, T.: Accuracy/ Diversity and Ensemble Classifier Design. IEEE Trans. Neural Networks 17(4) (2006)
9. Brown, G., Wyatt, J., Harris, R., Yao, X.: Diversity Creation Methods: A Survey and Categorisation. Journal of Information Fusion 6(1) (2005)
10. James, G.: Variance and Bias for General Loss Functions. Machine Learning 51(2), 115–135 (2003)
11. Kohavi, R., Wolpert, D.: Bias plus variance decomposition for zero-one loss functions. In: Proc. 13th International Conference on Machine Learning, pp. 275–283 (1996)
12. Bernard, S., Heutte, L., Adam, S.: Influence of Hyperparameters on Random Forest Accuracy. In: Benediktsson, J.A., Kittler, J., Roli, F. (eds.) MCS 2009. LNCS, vol. 5519, pp. 171–180. Springer, Heidelberg (2009)
13. Duin, R.P.W., Juszczak, P., Paclik, P., Pekalska, E., de Ridder, D., Tax, D.M.J., Verzakov, S.: PRTools 4.1, A Matlab Toolbox for Pattern Recognition. Delft University of Technology (2007)
14. Witten, I.H., Frank, E.: Data Mining: Practical machine learning tools and techniques, 2nd edn. Morgan Kaufmann, San Francisco (2005)

Two Stage Reject Rule
for ECOC Classification Systems

Paolo Simeone, Claudio Marrocco, and Francesco Tortorella

DAEIMI, Università degli Studi di Cassino,
Via G. Di Biasio 43, 03043 Cassino, Italy
{paolo.simeone,c.marrocco,tortorella}@unicas.it

Abstract. The original task of a multiclass classification problem can be decomposed using Error Correcting Output Coding in several two-class problems which can be solved with dichotomizers. A reject rule can be set on the classification system to improve the reliability of decision through an external threshold on the decoding outcomes before the decision is taken. If a loss-based decoding rule is used, more can be done to make such external scheme works better introducing a further reject stage in the system. This internal approach is meant to single out unreliable decisions for each classifier in order to proficiently exploit the properties of loss decoding techniques for ECOC as proved by experimental results on popular benchmarks.

Keywords: ECOC, reject option, multiple classifiers systems.

1 Introduction

Error Correcting Output Coding (ECOC) is a successful technique to face a multiclass problem by decomposing it in several two-class problems. The main idea is to create a certain number of different binary tasks which aggregate the original classes in only two classes according to a coding matrix. Each row of this matrix associates a binary string to each class of the original problem while each column defines a binary problem for a dichotomizer. When a new sample is classified by the dichotomizers, a new binary string is obtained, which has to be matched with the existing class codewords using a suitable decoding technique. The motivation for such method founds on the error correcting capabilities of the codes used to group classes and on the stronger theoretical roots characterizing two-class classifiers. Moreover, it has also been proved that ECOC provides a reliable probability estimation and a concurrent reduction of both bias and variance [1] which motivate its good generalization capabilities. For such reasons it has been successfully applied to a wide range of real applications such as text and digit classification [2,3], face recognition and verification [4,5] or fault detection [6].

When used in real applications, ECOC systems can produce errors which could have serious consequences, typically expressed in terms of an error cost. A well known technique to reduce the error rate is to abstain from decision when

C. Sansone, J. Kittler, and F. Roli (Eds.): MCS 2011, LNCS 6713, pp. 217–226, 2011.
© Springer-Verlag Berlin Heidelberg 2011

the reliability of the classification is estimated to be not sufficient (reject option) [7]. Thus, a dichotomizer, in addition to the possible outputs, can be in the state of reject which possibly avoids a wrong decision. The reject option is actually employed in many applications since it can alleviate or remove the problem of a misclassification rate too high for the requirements of the application at hand.

This technique can be usefully applied to ECOC systems and a preliminary analysis of this problem has been presented in [8]. In that paper we modified the decoding stage of the system in order to introduce a reject threshold uniquely based on the observation of the output of the ensemble of classifiers. Here we carry out a more in depth analysis that takes explicitly into account the structure of the ECOC system and estimates the reliability of the output of each dichotomizer. In this way, the binary decisions are rejected if not sufficiently reliable and the rejected elements in the output vector are decoded in an appropriate way based on the loss distance [9]. The effectiveness of the proposed scheme has been verified on several benchmark data sets.

The paper is organized as follows: in sect. 2 we analyze ECOC and the loss decoding rule. Sect. 3 describes the different proposed approaches. The experimental results are reported in sect. 4 while sect. 5 concludes the paper and draws some possible future developments.

2   The ECOC Approach

Error Correcting Output Coding is a technique meant to solve multi class classification problems with a decomposition of the original task in binary problems. A bit string of length L (*codeword*) is associated to each class ω_i with $i = 1, \ldots, n$, where n is the number of the original classes. The set of codewords is arranged in a $n \times L$ coding matrix $\mathbf{C} = \{c_{hk}\}$ where $c_{hk} \in \{-1, +1\}$. Each column defines a binary problems which requires a specific dichotomizer (An example of $\mathbf{C}$ is in table 1).

Table 1. Example of a coding matrix for a 5 classes problem

classes	codewords														
ω_1	+1	+1	+1	+1	+1	+1	+1	+1	+1	+1	+1	+1	+1	+1	+1
ω_2	-1	-1	-1	-1	-1	-1	-1	+1	+1	+1	+1	+1	+1	+1	+1
ω_3	-1	-1	-1	-1	+1	+1	+1	+1	-1	-1	-1	-1	+1	+1	+1
ω_4	-1	-1	+1	+1	-1	-1	+1	+1	-1	-1	+1	+1	-1	-1	+1
ω_5	-1	+1	-1	+1	-1	+1	-1	+1	-1	+1	-1	+1	-1	+1	-1

Each sample $\mathbf{x}$ is fed to the group of dichotomizers and the outcomes are collected into the *output vector*. Such vector is then compared with the original collection of words from the coding matrix using a decoding procedure. Different decoding techniques are available in literature, but the loss based decoding proved to outperform the others because it considers the reliability of the decision by evaluating a specific loss function on the margin of each classifier [9].

In the case of ECOC, the margins associated to the choice of a particular codeword $\mathbf{c_j}$ are given by $c_{ih}f_h(\mathbf{x})$ with $h = 1, \ldots, L$; if we know the loss function $\mathcal{L}(\cdot)$ used for training the dichotomizers, we can evaluate the global loss associated to such codeword:

$$D_{\mathcal{L}}(\mathbf{c}_i, \mathbf{f}) = \sum_{h=1}^{L} \mathcal{L}(c_{ih}f_h(\mathbf{x})). \tag{1}$$

The combination of margin values into this *loss-based distance* gives the level of confidence on the output word and the following rule can be used to predict the label for the k-th class:

$$\omega_k = \arg\min_i D_{\mathcal{L}}(\mathbf{c}_i, \mathbf{f}). \tag{2}$$

Eq. (2) can be also considered when the loss function of the dichotomizers is not known and the L_1 or L_2 norm distance is used. In fact, it is easy to see that the loss-based distance in (1) reduces to L_1 or L_2 distance provided that the loss $\mathcal{L}(z) = |1 - z|$ or $\mathcal{L}(z) = (1 - z)^2$ is used, respectively.

3   From an External to an Internal Reject Approach

A popular approach in many classification tasks is to reduce the costs by turning as many errors as possible into rejects. A simple technique to reject on ECOC systems can be accomplished by evaluating the reliability of the decision. Such rule is typically based on the final decision criterion which can be the minimization of the distance between the output word and the original codewords of the coding matrix. A threshold can be externally set on the output value of the decision block of the system.

Such external schemes do not require any assumption neither on the dichotomizers nor on the coding matrix, because the application of the rule does not imply any modification on the internal structure of the ECOC system.

In a previous work[8] we have analyzed the reject option applied on a decoding technique based on the loss distance and found that it works sensibly better than more traditional decoding techniques, such as those based on Hamming distance. The reason is that the loss distance based decoding is able to more effectively estimate the reliability of each dichotomizer by considering the margin it provides on the sample to be classified. Assuming a loss value normalized in the range $[0, 1]$, such a criterion, that we indicate as *Loss Decoding*, can be formalized as:

$$r(\mathbf{f}, t_l) = \begin{cases} \omega_k & \text{if } D_{\mathcal{L}}(\mathbf{c}_k, \mathbf{f}) < t_l, \\ reject & \text{if } D_{\mathcal{L}}(\mathbf{c}_k, \mathbf{f}) \geq t_l. \end{cases} \tag{3}$$

where ω_k is the class chosen according to eq. (2) and $t_l \in [0, 1]$.

More can be done for an ECOC system by introducing a previous stage of reject. If each binary classifier outputs a real value in the range $[-1, +1]$, we can provide them with a reject option by fixing a threshold on their outcomes. In this situation, a new scheme (*internal* scheme) can be introduced which evaluates the reliability of the dichotomizers output and rejects the unreliable samples. The rationale is to process the binary decisions before arriving at the decoding stage.

Some considerations must be done on the design of the reject option for the ensemble of dichotomizers. Assuming that each one $f_h(\mathbf{x})$ outputs a real value in the range $[-1, +1]$, then such value is compared with a threshold τ_h in order to assign them to a class $+1$ if $f_h(\mathbf{x}) \geq \tau_h$ and to a class -1 otherwise.

Whichever the value of the decision threshold τ_h the majority of unreliable decisions corresponds to the closer outcomes for the boundary, i.e. where the distributions of the two classes overlap. Samples with an output falling in this region are characterized by some ambiguity in the allocation, since their corresponding outcomes are very similar and thus quite difficult to distinguish. Therefore, reliable results are obtained adopting a decision rule with two thresholds τ_{h1} and τ_{h2} with $\tau_{h1} \leq \tau_{h2}$ such that:

$$r(f_h, \tau_{h1}, \tau_{h2}) = \begin{cases} +1 & \text{if } f_h(\mathbf{x}) > \tau_{h2}, \\ -1 & \text{if } f_h(\mathbf{x}) < \tau_{h1}, \\ reject & \text{if } f_h(\mathbf{x}) \in [\tau_{h1}, \tau_{h2}]. \end{cases} \tag{4}$$

Such rule has to consider the class distributions overlap region into the *reject interval* $[\tau_{h1}, \tau_{h2}]$ which is the principal cause of misclassification. Thresholds must be chosen to satisfy two contrasting requirements: enlarging the reject region to eliminate more errors and limiting the reject region to preserve as many correct classifications as possible. Since all the dichotomizers have different distributions a single threshold for every classifier could lead to abnormal results as shown in fig. 1).

For this reason, we do not impose the same reject threshold, but the same reject rate ρ and evaluate for each classifier the corresponding pair of thresholds, according to the method described by Pietraszek in [10]. Such method requires to estimate the ROC curve of each dichotomizer and calculate the pair of thresholds (τ_{h1}, τ_{h2}) such that f_h abstains for no more than ρ at the lowest possible error rate to make all the dichotomizers working almost at the same level of reliability.

The outcomes of the entire system are now different from the previous case. The output word is constituted by a mixture of zero values corresponding to a rejected sample by some dichotomizers and some real values for the same sample on some other dichotomizers. Such situation can thus be synthesized:

$$f_h^{(\rho)}(\mathbf{x}, \tau_{h1}, \tau_{h2}) = \begin{cases} 0 & \text{if } f_h(\mathbf{x}) \in [\tau_{h1}, \tau_{h2}] \\ f_h(\mathbf{x}) & \text{otherwise} \end{cases}. \tag{5}$$

Loss distance can be still computed even though the presence of the zero values in the output word $\mathbf{f}^{(\rho)}$ modifies the range of possible outcomes. It is given by:

$$D_{\mathcal{L}}(\mathbf{c}_i, \mathbf{f}^{(\rho)}) = \sum_{h \in I_{nz}} \mathcal{L}(c_{ih} f_h(\mathbf{x})) + |I_z| \cdot \mathcal{L}(0) \tag{6}$$

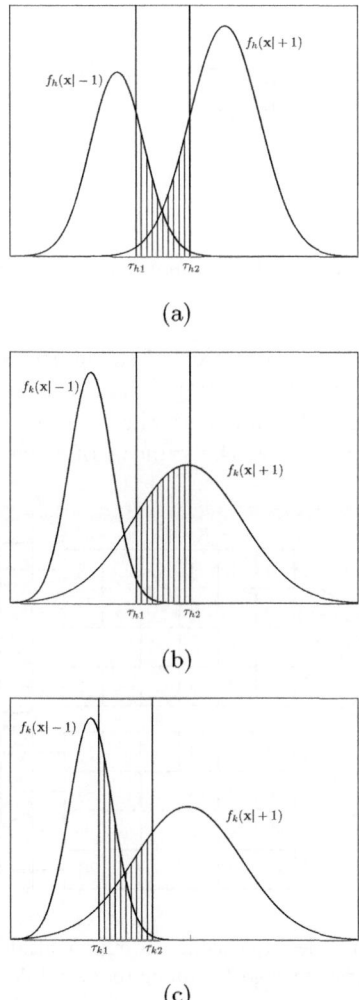

Fig. 1. (a) The output distributions of the dichotomizer f_h for the two classes and the pair of thresholds (τ_{h1}, τ_{h2}). (b) The same thresholds applied to a different dichotomizer f_k produce abnormal results. (c) The proper thresholds (τ_{k1}, τ_{k2}) evaluated for f_k.

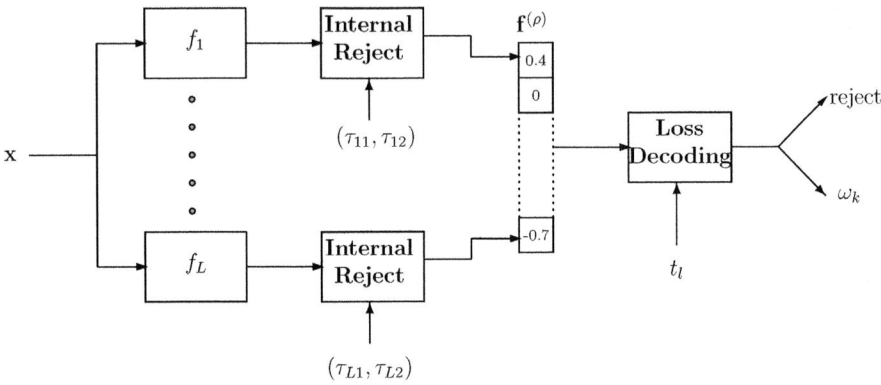

Fig. 2. The block diagram for the internal reject rule coupled with loss decoding

where I_{nz} and I_z are the sets of indexes of the nonzero values and zero values in the output word, respectively. In practice the loss is given by two contributions, where the second one is independent of the codeword that is compared to the output word. An example of how the values can range is shown in fig. 3.

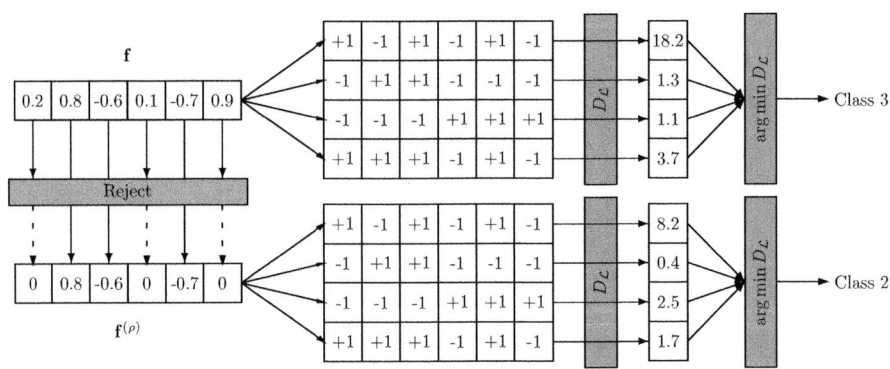

Fig. 3. Loss based distance decision using an exponential loss function: output word **f** is decoded as class 3 without reject (upper part), while is decoded as class 2 if an internal reject is applied (lower part)

If a threshold t_l is set the rejection rule, that we call *Trimmed Loss Decoding*, is defined as:

$$r(\mathbf{f}^{(\rho)}, t_l) = \begin{cases} \omega_k & \text{if } D_{\mathcal{L}}(\mathbf{c}_k, \mathbf{f}^{(\rho)}) > t_l, \\ reject & \text{if } D_{\mathcal{L}}(\mathbf{c}_k, \mathbf{f}^{(\rho)}) \leq t_l. \end{cases} \tag{7}$$

where ω_k is the class chosen according to eq. (2).

4   Experiments

Ten multiclass benchmark data sets publicly available at the UCI machine learning repository [11] have been used to compare the performance of the different reject rules. To avoid any bias in the comparison for each data set twelve runs of a multiple hold out procedure have been performed. In each run, the data set has been split into three subsets: a training set (containing the 70% of the samples of each class), a validation set and a test set (each containing the 15% of the samples of each class). The training set has been used for learning the base classifiers, the validation set for the normalization of the binary outputs into the range $[-1, 1]$ and the estimation of the thresholds (τ_{h1}, τ_{h2}) and the test set for the performance evaluation. A short description of the data sets is given in table 2. In the same table we also report the number of columns of the coding matrix chosen for each data set according to [12].

Three different classifiers have been used as base dichotomizer: Modest Adaboost (MA)[13] and Support Vector Machine (SVM) with both linear and RBF kernel. MA has been built using a decision tree with maximum depth equal to 10 as weak learner and 50 boosting steps. SVM was implemented through the SVM^{Light} [14] software library using a linear kernel and an RBF kernel with $\sigma = 1$. In both cases the C parameter has been set to the default value calculated by the learning procedure. The exponential loss $\mathcal{L}(z) = e^{-z}$ for the MA and the "hinge" loss $\mathcal{L}(z) = \max\{1 - z, 0\}$ for the SVM were adopted.

Experiments are meant to compare the two presented reject schemes: the Loss Decoding (LD) and Trimmed Loss Decoding (TLD). For the sake of comparison we also considered the basic reject rule [8] based on Hamming distance Decoding (HD). For HD the external threshold has been varied in the range $[0, L/2]$ with step 1, while for LD the loss output was normalized in the range $[0, 1]$ and consequently the thresholds were varied in this interval with the step 0.01. In the case of TLD we have varied the parameter ρ from 0 to 1 with step 0.05

Table 2. Data sets used in the experiments

Data Set	Classes	Features	Length (L)	Samples
Abalone	29	8	30	4177
Ecoli	8	7	62	341
Glass	6	9	31	214
Letter	26	16	63	5003
OptDigits	10	62	31	5620
Pendigits	10	12	31	10992
SatImage	6	36	31	6435
Segmentation	7	18	63	2310
Vowel	11	10	14	435
Yeast	10	8	31	1484

Table 3. Mean area under the error-reject curve and standard deviation for Modest Adaboost

Data Sets	HD	LD	TLD
Abalone	0.359(0.010)	0.348(0.009)	**0.342(0.010)**
Ecoli	0.032(0.016)	0.036(0.015)	**0.026(0.011)**
Glass	0.094(0.026)	0.090(0.019)	**0.069(0.017)**
Letter	0.061(0.005)	0.052(0.007)	**0.050(0.006)**
OptDigits	0.006(0.001)	0.004(0.001)	**0.004(0.001)**
Pendigits	0.003(0.001)	0.003(0.001)	**0.002(0.001)**
SatImage	0.025(0.005)	0.021(0.003)	**0.021(0.003)**
Segmentation	0.002(0.001)	0.001(0.001)	**0.001(0.001)**
Vowel	0.052(0.011)	0.037(0.010)	**0.032(0.010)**
Yeast	0.169(0.015)	0.160(0.015)	**0.151(0.012)**

and, as in the LD case, the loss output is normalized in the range $[0, 1]$ and the external threshold is varied with step 0.01 into the same range.

To make a complete comparison among the different strategies, we have considered the *Error-Reject (ER) curve* which gives a complete description of a classification system with the reject option by plotting the error rate $E(t)$ against the reject rate $R(t)$ when varying the threshold t on the reliability estimate.

In these experiments to consider a synthetic and clear performance figure for the classification systems, we have used the area under the error-reject curves which gives the average error rate when varying the reject rate. This is a single numerical value which provides a simple measure for the performance of a classification system with the reject option: the lower its value, the better the overall classification performance.

Tables 3, 4 and 5 report the results respectively for MA, linear SVM and RBF SVM classifier, obtained with the multiple hold out procedure before described. Each cell contains the values of the mean area under the error-reject curve and its standard deviation for each data set and reject rule. To verify if there is a statistical difference between the obtained results, the Wilcoxon signed rankstest [15]) has been performed with a significance level equal to 0.05. Since this statistical test is appropriate only for pairs of classifiers and we want to analyze the behavior of TLD we have performed it twice for the two pairs: HD versus TLD and LD versus TLD. The outcomes are shown in the reported tables where a bold value in TLD column indicates that this method has obtained a statistically lowest mean area in both tests. The Wilcoxon test assesses the superiority of TLD method in all the analyzed cases with respect to LD and HD since it allows us to control the individual errors of the base classifiers. For this reason, using an internal reject rule and thus the knowledge of the nature of the base classifier, the system has the possibility to face the uncertainty of wrong predictions in a more precise and effective way than external techniques.

Table 4. Mean area under the error-reject curve and standard deviation for linear SVM

Data Sets	HD	LD	TLD
Abalone	0.386(0.010)	0.362(0.010)	**0.353(0.006)**
Ecoli	0.029(0.017)	0.036(0.012)	**0.025(0.010)**
Glass	0.185(0.035)	0.189(0.027)	**0.143(0.024)**
Letter	0.174(0.011)	0.162(0.008)	**0.158(0.007)**
OptDigits	0.007(0.001)	0.005(0.001)	**0.005(0.001)**
Pendigits	0.021(0.003)	0.020(0.003)	**0.019(0.003)**
SatImage	0.041(0.003)	0.035(0.002)	**0.034(0.002)**
Segmentation	0.020(0.004)	0.018(0.004)	**0.016(0.004)**
Vowel	0.301(0.026)	0.238(0.026)	**0.220(0.022)**
Yeast	0.187(0.013)	0.192(0.012)	**0.177(0.011)**

Table 5. Mean area under the error-reject curve and standard deviation for Radial Basis Function SVM

Data Sets	HD	LD	TLD
Abalone	0.356(0.008)	0.344(0.006)	**0.338(0.006)**
Ecoli	0.035(0.022)	0.037(0.017)	**0.028(0.016)**
Glass	0.145(0.049)	0.122(0.032)	**0.102(0.032)**
Letter	0.031(0.005)	0.027(0.004)	**0.026(0.004)**
OptDigits	0.003(0.001)	0.001(0.000)	**0.001(0.000)**
Pendigits	0.002(0.001)	0.001(0.000)	**0.001(0.000)**
SatImage	0.037(0.003)	0.022(0.003)	**0.020(0.003)**
Segmentation	0.008(0.002)	0.006(0.002)	**0.005(0.002)**
Vowel	0.007(0.005)	0.003(0.003)	**0.002(0.002)**
Yeast	0.175(0.021)	0.179(0.019)	**0.169(0.018)**

5   Conclusions and Future Works

In this paper we have compared two techniques to provide the ECOC classification system with a reject option. The first and more immediate one consists in modifying the final decoding stage, where the multi class decision is accomplished and where it is possible to evaluate a reject threshold on the system output. The second solution, instead, is to work where base classifiers takes their binary decisions and to apply a reject rule to each one of them. The main difference is in the simplicity and ease of use of the external methods against the flexibility of internal methods. The experiments showed that the more complex technique attains higher reduction of the error rate confirming that an internal reject is useful when coupled with loss decoding.

Directions for future work include investigating the relation of the rejection rule with the characteristics of the coding, taking into special account new problem-dependent or data-dependent designs.

References

1. Kong, E.B., Dietterich, T.G.: Error-correcting output coding corrects bias and variance. In: ICML, pp. 313–321 (1995)
2. Ghani, R.: Combining labeled and unlabeled data for text classification with a large number of categories. In: Cercone, N., Lin, T.Y., Wu, X. (eds.) ICDM, pp. 597–598. IEEE Computer Society Press, Los Alamitos (2001)
3. Zhou, J., Suen, C.Y.: Unconstrained numeral pair recognition using enhanced error correcting output coding: A holistic approach. In: ICDAR, pp. 484–488. IEEE Computer Society Press, Los Alamitos (2005)
4. Windeatt, T., Ardeshir, G.: Boosted ECOC ensembles for face recognition. In: Proceedings of the International Conference on Visual Information Engineering, pp. 165–168 (2003)
5. Kittler, J., Ghaderi, R., Windeatt, T., Matas, J.: Face verification via error correcting output codes. Image Vision Comput. 21(13-14), 1163–1169 (2003)
6. Singh, S., Kodali, A., Choi, K., Krishna, R., Pattipati, Namburu, S.M., Sean, S.C., Prokhorov, D.V., Qiao, L.: Dynamic multiple fault diagnosis: Mathematical formulations and solution techniques. IEEE Transactions on Systems, Man, and Cybernetics, Part A 39(1), 160–176 (2009)
7. Chow, C.: On optimum recognition error and reject tradeoff. IEEE Transactions on Information Theory 16(1), 41–46 (1970)
8. Simeone, P., Marrocco, C., Tortorella, F.: Exploiting system knowledge to improve ecoc reject rules. International Conference on Pattern Recognition 4340–4343 (2010)
9. Allwein, E.L., Schapire, R.E., Singer, Y.: Reducing multiclass to binary: A unifying approach for margin classifiers. Journal of Machine Learning Research 1, 113–141 (2000)
10. Pietraszek, T.: On the use of ROC analysis for the optimization of abstaining classifiers. Machine Learning 68(2), 137–169 (2007)
11. Asuncion, A., Newman, D.J.: UCI machine learning repository (2007)
12. Dietterich, T.G., Bakiri, G.: Solving multiclass learning problems via error-correcting output codes. Journal of Artificial Intelligence Research 2, 263–286 (1995)
13. Vezhnevets, A., Vezhnevets, V.: Modest adaboost - teaching adaboost to generalize better. Graphicon 2005 (2005)
14. Joachims, T.: Making large-scale SVM learning practical. In: Schölkopf, B., Burges, C., Smola, A. (eds.) Advances in Kernel Methods - Support Vector Learning, ch. 11, MIT Press, Cambridge (1999)
15. Wilcoxon, F.: Individual comparisons by ranking methods. Biometrics Bulletin 1(6), 80–83 (1945)

Introducing the Separability Matrix for Error Correcting Output Codes Coding

Miguel Ángel Bautista[1,2], Oriol Pujol[1,2],
Xavier Baró[1,2,3], and Sergio Escalera[1,2]

[1] Applied Math and Analisis Dept, University of Barcelona, Gran Via de les Corts Catalanes. 585, 08007 Barcelona, Spain
[2] Computer Vision Center, Campus UAB, Edifici O, 08193, Bellaterra, Spain
[3] Computer Science, Multimedia, and Telecommunications Dept, Universitat Oberta de Catalunya. Rambla del Poblenou 156, 08018 Barcelona
mbautista@cvc.uab.es, oriol@maia.ub.es, xbaro@uoc.edu, sergio@maia.ub.es

Abstract. Error Correcting Output Codes (ECOC) have demonstrate to be a powerful tool for treating multi-class problems. Nevertheless, predefined ECOC designs may not benefit from Error-correcting principles for particular multi-class data. In this paper, we introduce the Separability matrix as a tool to study and enhance designs for ECOC coding. In addition, a novel problem-dependent coding design based on the Separability matrix is tested over a wide set of challenging multi-class problems, obtaining very satisfactory results.

Keywords: Error Correcting Output Codes, Problem-dependent designs, Separability matrix, Ensemble Learning.

1 Introduction

Multi-class classification tasks are problems in which a set of N classes, categories or namely brands are categorized. Most of state-of-the-art multi-class methodologies need to deal with the categorization of each class either by modelling its probability density function, or by learning a classification boundary and using some kind of aggregation/selection function to obtain a final decision. Another way to deal with multi-class problems is to use a divide-and-conquer approach. Instead of extending a method to cope with the multi-class case, one can divide the multi-class problem into smaller binary problems and then combine their responses using some kind of strategy, such as voting.

In the ensemble learning field, Error Correcting Output Codes (ECOC) have demonstrated to be a powerful tool to solve multi-class classification problems [CS02, DB95]. This methodology divides the original problem of N classes in n binary problems (2-class problems). Commonly, the step of defining n binary partitions of the N classes is known as *coding*. At this step, a coding matrix $M_{N \times n} \in \{-1, +1\}$ is generated. The columns of M denote the n bi-partitions of the original problem, and the rows of M, known as *codewords*, identify each

C. Sansone, J. Kittler, and F. Roli (Eds.): MCS 2011, LNCS 6713, pp. 227–236, 2011.

one of the N classes of the problem uniquely. Once M is defined, a set of n base classifiers $\{h_1, \ldots, h_n\}$ learn the n binary problems coded in M.

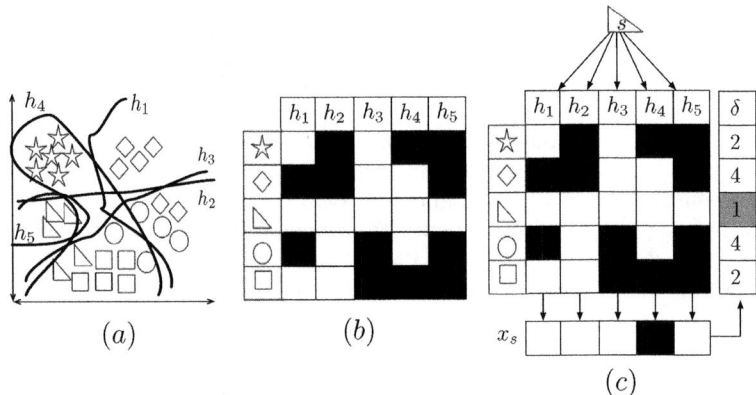

Fig. 1. (a) Feature space and trained boundaries of base classifiers. (b) Coding matrix M, where black and white cells correspond to $\{+1, -1\}$, denoting the two partitions to be learnt by each base classifier (white cells vs. black cells). (c) Decoding step, where the predictions of classifiers, $\{h_1, \ldots, h_5\}$ for sample s are compared to the codewords $\{y_1, \ldots, y_N\}$ and s is labelled as the class codeword at minimum distance.

At the *decoding* step, a new sample s is tested by each base classifier $\{h_1, \ldots, h_n\}$, obtaining a set of label predictions. The set of predictions x_s is compared to each codeword of M using a decoding measure δ and sample s is labelled as the class c_i with codeword y_i at minimum distance (i-th row of M). In Figure 1, an example for coding and decoding steps is shown for a 5−class toy problem. Note that though classifier h_4 fails its prediction, s is correctly classified.

The coding step has been widely studied in literature [TR98, RK04, ASS02], proposing either predefined [TR98, RK04] or random [ASS02] coding designs always following the trend of reducing the number of used dichotomizers. Nevertheless, predefined strategies may not be suitable for a given problem because they do not take into account the underlying distribution of the classes. In this scope, one can roughly find works on problem-dependent strategies for coding designs [EOR08, PRV06].

In this paper we introduce the Separability matrix as a way to analyse and study the properties of a certain ECOC coding matrix. Although the concept of separability has always been in the heart of all ECOC studies, up to this moment there has not been the need of defining explicitly a matrix of this kind. This is mainly due to the fact that predefined strategies assume that the coding matrix must have equidistant codewords. However, with the introduction of problem-dependent and sub-lineal coding designs this general assumption does not hold and more concise tools are needed for their analysis. The Separability matrix explicitly shows the pairwise separation between all pairs of classes. With this tool in mind, we also propose a new compact problem-dependent coding design that shows the benefits of applying the separability criteria in a problem-dependent manner.

This paper is organized as follows: Section 2 introduces the Separability matrix, in Section 3 the novel problem-dependent coding design is proposed and, Section 4 shows the experimental results. Finally, Section 5 concludes the paper.

2   The Separability Matrix

One of the main concerns of the ECOC framework is to correct as many base classifiers errors as possible. In literature, the correction capability ρ of a coding matrix M is defined as $\rho = \frac{\min(\delta(y_i, y_j)) - 1}{2}$, $\forall i, j \in \{1, \ldots, N\}$, $i \neq j$. Therefore, distance between codewords and correction capability are directly related. Given this close relationship between distance and correction capability, we define the Separability matrix S, as follows:

Given an ECOC coding matrix $M_{N \times n}$, the Separability matrix $S_{N \times N}$ contains the distances between all pairs of codes in M. Let $\{y_i, y_j\}$ be two codewords, the Separability matrix S at position (i, j), defined as $S_{i,j}$, contains the distance between the codewords $\{y_i, y_j\}$, defined as $\delta(y_i, y_j)$. An example of Separability matrix estimation for two coding designs is shown in Figure 2.

Usually, the increment in the correcting capability problem has been tackled by enlarging the codeword length, and thus, the distance between codewords [TR98]. However, Rifkin et al. show in [RK04] that if a classifier with high capacity is well optimized, small codes such as *One vs. All* are also suitable for solving the problem. Recently, following the same principle as Rifkin et al., in [BEB10] the authors propose to use a Compact ECOC matrix, with a code length of $\lceil \log_2(N) \rceil$, where $\lceil . \rceil$ round to the upper integer, which is optimized by a Genetic Algorithm in a problem-dependent manner.

If we analyse the Separability matrix S of predefined ECOC coding designs [TR98, RK04], we find that $S_{i,j} = \varsigma$ $\forall i, j \in \{1, \ldots, N\}, i \neq j$, where ς is a constant separation value. This means that codewords are equidistant, as shown in Figure 2(d). In fact, when dealing with predefined codings, the Separability matrix makes little sense and has been overlooked since all non-diagonal values are constant. Nevertheless, in problem-dependent coding strategies the Separability matrix acquires a great value, since it shows which codewords are prone to have more errors due to the lack of error correction capability. For example, if we analyse the Compact ECOC coding matrix M we find that codewords are not equidistant and the distribution of separability is not constant. An example of Compact ECOC coding and its Separability is shown in Figure 2(a) and 2(b), respectively.

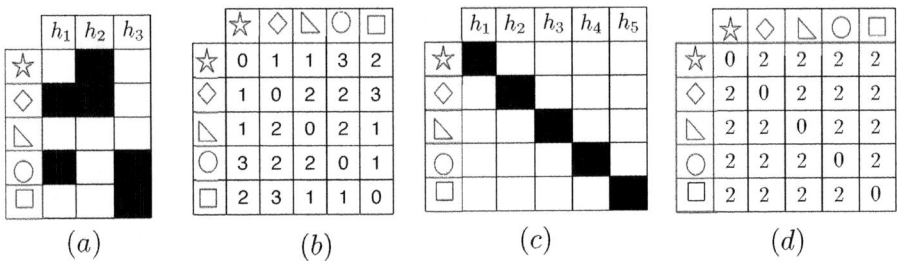

Fig. 2. (a) Compact ECOC coding matrix. (b) Separability Matrix of a Compact ECOC. (c) *One vs. All* coding matrix. (d) Separability matrix of *One vs. All* coding.

3   Application of Separability Matrix for Extension Coding

Problem-dependent coding strategies have not been thoroughly studied in literature [EOR08, PRV06]. In contrast to classical approaches [RK04, TR98, ASS02], problem-dependent coding designs combine the error correcting principles with a guided coding procedure which takes into account the distribution of the data. In this work we define a problem-dependent coding design based on the Separability matrix to enhance the error correcting capabilities of the design. Moreover, we also take profit of the Confusion matrix to define the partitions of classes of each binary classifier.

In [BEB10] the authors propose a problem-dependent Compact ECOC coding matrix of length $\lceil \log_2 N \rceil$. However, the computational cost of optimizing this coding matrix is very expensive and in every case the resultant matrix M has null correction capability since $\rho = 0$. On the other hand, one would like to have at least $\min(S) \geq 3$, to correct one error. This could be done by extending the codewords $\{y_1, \ldots, y_N\}$ of the coding matrix M until $S_{i,j} = 3$ $\forall i, j \in \{1, \ldots, N\}, i \neq j$. However, we have to take into account that confusion is not equally distributed among all the classes, and thus separability might not have to be also equally distributed. Let $\{c_i, c_j, c_k, c_l\}$ be four classes of our N-class problem, then, if $(C_{i,j} + C_{j,k}) > (C_{k,l} + C_{l,k})$ (where $C_{i,j}$ is the number of samples of class c_i classified as class c_j), it will be more probable to misclassify a sample between classes c_i and c_j than between classes c_k and c_l. Thus, it will be more efficient to increase $\delta(y_i, y_j)$ than $\delta(y_k, y_l)$.

Therefore, following the idea of Compact ECOC coding, we propose to extend the codewords of a non-optimized Compact ECOC coding (Binary ECOC), which is the binary representation of the N classes of our problem. This means that the codeword y_i of class c_i is the binary representation of a decimal value $i \ \forall i \in \{1, \ldots, N\}$. This extension is calculated in order to increase the distance δ between the most confused codes, computing a problem-dependent extension still with a reduced code length. The proposed algorithm uses both Separability $S_{N \times N}$ and Confusion $C_{N \times N}$ matrices of a Binary ECOC to compute an extension of its coding matrix M, defined as $E_{N \times k}$ where k is the number of columns (base classifiers) of the extension.

The Confusion-Separability-Extension (CSE) coding algorithm is an iterative algorithm that looks for the most confused classes in C, i.e $\{c_i, c_j\}$ and codes an Extension matrix E that increases its separability $S_{i,j}$ until a certain user-defined separability value ϱ is achieved. In addition, the Extension matrix E also increments the separability for all the classes confused with c_i or c_j. This extension is performed in order to increase the separability with all the classes that are prone to confuse with classes c_i or c_j. When no classes are confused with $\{c_i, c_j\}$ the coding is performed taking into account the overall confusion with all classes $\{c_1, \ldots, c_N\}$. Once E is completely coded, the algorithm checks if any column in E was previously on M. In that case, the algorithm changes specific codewords. Let t be an iteration of the algorithm, which codes E_t, then at iteration $t + 1$, $M_{t+1} = M_t \cup E_t$, the algorithm will stop when in M, $n \geq N$, this stop condition is defined to upper bound the code length of the design to N, though smaller codes may be suitable. In addition, we consider that if $\delta(y_i, y_j) \geq \varrho$, then $C_{i,j} = 0$. Therefore, another stop condition for the algorithm is that $\forall i, j C_{i,j} = 0$, because that means that no confusion is left to treat. Note that CSE coding algorithm only requires the C and S matrices generated by a Binary ECOC. In addition, no retraining or testing of classifiers is needed trough the extension process. Algorithm 1 shows the CSE coding algorithm, which is illustrated in the toy example of Figure 3.

Data: $M_{N \times n}$, $C_{N \times N}$, $S_{N \times N}$, ϱ
Result: $E_{N \times k}$
k // separability increment needed
$Y^E_{N \times k} \in \{-1, +1\}$// set of unused generated codewords
$S^m_{1 \times 1} \in \{0, \ldots, \infty\}$// minimum separability value
$S^c_{p \times q} \in \{0, 1\}, p \leq N, q = 2$// classes at minimum separability with $\{c_i, c_j\}$
while $k + n < N$ and $\exists\, i, j\ C_{i,j} \geq 0$ **do**

    $(i, j) := \arg\max_{i,j}(C)$ // look for the pair of classes $\{c_i, c_j\}$ with maximum
        confusion in C
    $k := \varrho - S_{i,j}$;
    Y^E :=generateCodes(k,N) // generate 2^k codes κ times until N codes are generated
    $y^E_i := Y^E_1$ // assign random code to one of the classes with maximum confusion
    $(E, Y^E) :=$findCode(y^E_i, k, Y^E) // find a code at $\delta = k$ with the code Y^E_i
    while $S_m < \varrho$ **do**
        $(S^c, S^m) :=$findMinSepClasses(E, S, C);
        $(E, Y^E) :=$codifyMinSep(S^c, E, Y^E) // look for a suitable code for S^c
        $S^m = S^m + 1$;
    end
    if $\exists\{i, j\} : E_{i,j} = 0$ **then**
        $E :=$codifyZero(E, S, C, Y^E) // codify the undefined codes in E taking into
        account confusion with $\{c_1, ..., c_N\}$
    end
    $E :=$checkExtension(M, E) // check if some column in E was previously in M
    $(C, S, M) :=$updateMatrices(M, E, S, C) // update confusion, separability and coding
    matrices
end

Algorithm 1. CSE coding algorithm.

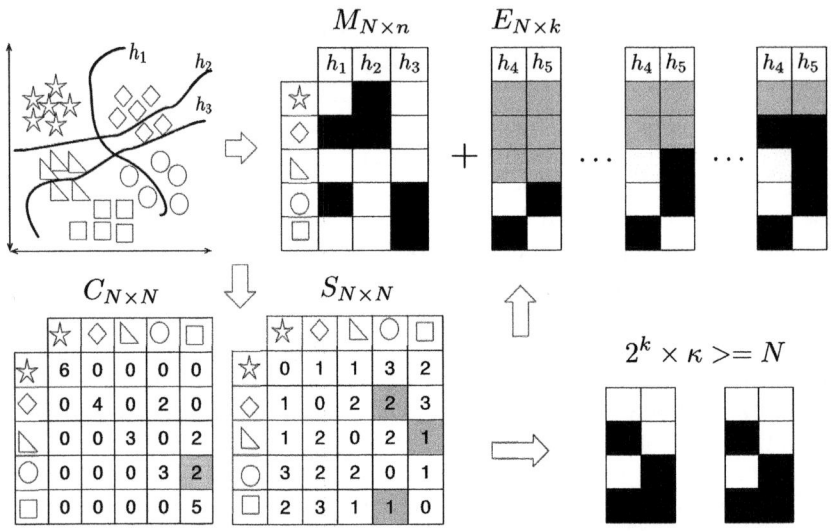

Fig. 3. CSE example in a 5−class toy problem

The CSE coding algorithm codifies an Extension matrix E based on the Separability and Confusion matrices of a certain coding matrix M. Note that though in this paper this Extension matrix is applied over a Binary ECOC, the CSE coding algorithm is independent of the initial coding matrix M, and thus it could be applied to extend any coding design.

The confusion matrix C of Figure 3 has it maximum confusion value at $C_{4,5}$ (circle and square classes). Therefore, in the first iteration, an extension to split those classes and increment its separability will be coded. For this example, let the user-defined value of ϱ be 3. Thus, the length k of the Extension matrix at the current iteration E is $k = \varrho - S_{4,5}$. To increment the distance $\delta(y_4, y_5)$ to ϱ, we have to find two codes $\{y_4^E, y_5^E\}$ so that $\delta(y_4^E, y_5^E) = k$. In fact, the algorithm generates the 2^k codes κ times until N codewords are generated, and then, searches for two codes at $\delta = k$. Once this codes are defined in E, the algorithm looks for all the classes with $\min S_{i,j}$, $i \in \{4,5\}$, $j \in \{1, \ldots, N\}$ and $\max C_{i,j}$, $i \in \{4,5\}$, $j \in \{1, \ldots, N\}$ in order to increment its distance δ. If no confusion positions are found and the codes in E are left empty, then the algorithm applies $\min(S_{i,j})$, $\forall i, j \in \{1, \ldots, N\}$ and $\max(C_{i,j})$, $\forall i, j \in \{1, \ldots, N\}$.

Once the Extension matrix E is coded the algorithm checks if $E \cap M = \emptyset$ column-wise, if not, then the codeword corresponding to the class with $\min C_{i,j}$, $i \in \{4,5\}$, $j \in \{1, \ldots, N\}$ and an opposite with $\min S_{i,j}$, $i \in \{4,5\}$, $j \in \{1, \ldots, N\}$ are interchanged, and E is checked again. When E is completely coded and checked, M, S, and C are updated. That means that for the next iteration $M = M \cup E$. In addition, S is re-estimated with the new M.

3.1   Training the Base Classifiers

In [RK04] the author concludes that if the base classifier is properly tuned, the *One vs. All* may be used without loss of generalization capability. Following this idea, our coding design is upper bounded by N classifiers and thus, we need to use powerful dichotomizers in order to reduce possible classification errors.

In literature, Support Vector Machines with a RBF-Gaussian kernel have demonstrated to be powerful dichotomizers. Nevertheless, they need some parameters to be optimized. In this case, parameters ζ, which is the regularizer, and γ, which has a relationship with the smoothness of the boundary, have to be optimized. A common way to optimize this parameters is to perform a grid search with cross-validation. Recently, in [BEB10] the authors have shown that Genetic Algorithms (GA) can be introduced in this optimization problem with good results.

For each binary problem, defined by a column of M, we use Genetic Algorithms in order to estimate values for ζ and γ. For this task, we use the same settings than in [LdC08], where individuals correspond to a pairs of genes, and each gene corresponds to the binary codification of a floating point value. This parameter estimation is performed under a 2-fold cross-validation measurement in order to avoid over-fitting bias and improve generalization.

4   Experimental Results

In order to present the results, first, we discuss the data, methods, and evaluation measurements of the experiments.

- **Data:** The first bench of experiments consists of seven muti-class problems extracted from the UCI Machine Learning Repository [AN07], showed in Table 1. In addition, we test our methodology over 3 challenging Computer Vision multi-class problems. First, we classify 70 visual object categories from the MPEG dataset [MP]. Then, 50 classes of the ARFace database [MB98] are classified. Finally, we test our method in a real traffic sign categorization problem consisting of 36 traffic sign classes [CMP$^+$04].

Table 1. UCI repository data sets characteristics

Problem	#Training samples	#Features	#Classes
Dermathology	366	34	6
Ecoli	336	8	8
Vehicle	846	18	4
Segmentation	2310	19	7
Glass	214	9	7
Vowel	990	10	11
Yeast	1484	8	10

- **Methods:** We compare the *One vs. All* [RK04] ECOC approach with the CSE coding design with separability value $\varrho = \{3, 5\}$. In addition, we also compare our results with the Dense Random coding scheme [ASS02] using N classifiers. The ECOC base classifier is the libsvm implementation of a SVM with Radial Basis Function kernel [CC01a]. The SVM ζ and γ parameters are tuned via Genetic Algorithms for all the methods, minimizing the classification error of a two-fold evaluation over the training sub-set. Furthermore, the same experiments were run with Real AdaBoost as base classifier [FS95].
- **Evaluation Measurements:** The classification performance is obtained by means of a stratified ten-fold cross-validation.

The classification results obtained for all the data sets considering the different ECOC configurations are shown in Table 2 and Table 3, with SVM an Adaboost as base classifier, respectively. In order to compare the performances provided for each strategy, the table also shows the mean rank of each ECOC design considering the twelve different experiments. The rankings are obtained estimating each particular ranking r_i^j for each problem i and each ECOC configuration j, and computing the mean ranking R for each design as $R_j = \frac{1}{N} \sum_i r_i^j$, where N is the total number of data sets. We also show the mean number of classifiers ($\#$) required for each strategy.

Table 2. UCI classification results with SVM as base classifier

Data set	*One vs. All* ECOC		CSE ECOC $\varrho = 3$		CSE ECOC $\varrho = 5$		Dense Random ECOC	
	Perf.	Classif.	Perf.	Classif.	Perf.	Classif.	Perf.	Classif.
Vowel	55.0±10.5	11	66.9±7.8	9.2	**69.8±6.3**	10.6	67.9±8.3	11
Yeast	41.0±7.3	10	54.7±11.8	5.7	53.0±9.3	9.5	**54.9±6.4**	10
Ecoli	**78.9±3.5**	8	76.4±4.4	7	78.6±3.9	7.4	72.1±2.7	8
Glass	51.6±10.2	7	**55.5±7.6**	6	52.7±8.4	3	42.8±11.02	7
Segment	**97.3±0.7**	7	96.9±0.8	6.6	96.6±1.0	6.2	96.6±1.3	7
Derma	97.1±1.2	6	**97.1±0.9**	5.2	95.9±1.2	3	95.7±0.8	6
Vehicle	80.1±4.0	4	**81.1±3.5**	3	70.6±3.4	3	81.1±3.6	4
MPEG7	83.2±5.1	70	88.5±4.5	15	89.6±4.9	20.4	**90.0±6.4**	70
ARFaces	76.0±7.22	50	80.7±5.2	13.8	84.6±5.3	20.2	**85.0±6.3**	50
Traffic	91.3±1.1	36	95.7±0.92	12.2	**96.6±0.8**	19	93.3±1.0	36
Rank & #	3.0	20.8	**2.2**	**8.8**	2.3	10.3	2.5	20.8

Table 3. UCI classification results with Real AdaBoost as base classifier

Data set	*One vs. All* ECOC		CSE ECOC $\varrho = 3$		CSE ECOC $\varrho = 5$		Dense Random ECOC	
	Perf.	Classif.	Perf.	Classif.	Perf.	Classif.	Perf.	Classif.
Vowel	40.6±1.3	11	44.7±0.8	10	46.5±1.2	10.6	**47.0±1.2**	11
Yeast	36.8±1.1	10	**45.6±0.4**	9.6	42.9±1.0	9.5	40.8±1.3	10
Ecoli	71.5±10.9	8	68.1±8.3	7.4	63.3±9.2	7.4	**75.0±7.8**	8
Glass	**53.8±12.1**	7	52.8±13.5	6	44.5±10.8	6	49.5±10.9	7
Segment	**96.4±0.7**	7	95.0±0.3	6.8	94.8±0.9	6.2	95.3±1.0	7
Derma	**89.3±4.9**	6	77.6±6.3	5.4	76.0±5.3	3	76.7±5.3	6
Vehicle	**73.6±1.3**	4	72.7±1.9	4	62.9±1.4	3	72.7±1.5	4
MPEG7	54.4±7.2	70	65.5±9.5	15	73.7±8.3	24.3	**86.5±6.4**	70
ARFaces	36.3±7.2	50	53.8±5.2	13.8	62.8±8.3	20.4	**81.5±6.3**	50
Traffic	80.6±6.2	36	81.3±8.1	12.2	87.4±7.9	20.6	**91.2±5.3**	36
Rank & #	2.6	20.8	2.4	9.16	3.0	10.89	**1.9**	**20.8**

Results show that the proposed method outperforms the *One vs. All* standard coding design in most cases, using far less number of dichotomizers. This is caused by the fact that the proposed algorithm focus the correcting capability in those classes more prone to be confused, and thus, less redundancy is needed. However, one has to notice that if designing a coding matrix with $n = N$ classifiers, Dense Random coding seems to be a suitable choice that also outperforms the standard *One vs. All* coding.

Nevertheless when comparing Dense Random coding with our method in terms of performance, no significance is found since both methods have a comparable rank. In fact, Dense Random coding seems to perform better than our proposal in the Computer Vision problems, where the number of classes was large.

This situation was expectable since Dense Random coding was bounded to code N dichotomies. In fact, since Dense Random tend construct coding matrices of equidistant codes, we can approximate the correction capability of Dense Random coding by dividing the number of classes between the minimum number of classifiers needed to increase, at least, one unit the distance between codes $(\rho_{est} = \frac{N}{\lceil \log_2(N) \rceil})$. For example, in the MPEG7 experiment [MP], the estimation of the correction capability of Dense Random coding tends to be $\rho_{est} = \frac{70}{\lceil \log_2(70) \rceil} = 10$. While for the CSE algorithm proposed with $\varrho = 5$ the estimation for the correcting capability is $\rho = 2$.

5   Conclusions

In this paper, we introduce the Separability matrix as a tool to enhance and analyse ECOC coding designs. Although separability issues have been always in the core of all ECOC coding proposals, until now there was no explicit need to define such a matrix. Nevertheless, in problem-dependent strategies and in sublinear coding designs, acquires great value since it shows which codes are prone to be confused due to the lack of correction capability, and thus, more precise and compact codes can be defined. Moreover, a novel ECOC coding design based on the Separability matrix is, focusing the correction capability on those codes which are more prone to be confused.

Results show that the proposed coding design obtains comparable or even better results than predefined compact coding designs using far less number of dichotomizers.

Acknowledgments. This work has been supported by projects TIN2009-14404-C02 and CONSOLIDER-INGENIO CSD 2007-00018.

References

[AN07]   Asuncion, A., Newman, D.J.: UCI machine learning repository. University of California, Irvine, School of Information and Computer Sciences (2007),
http://www.ics.uci.edu/$\sim$mlearn/MLRepository.html

236   M.Á. Bautista et al.

[ASS02]   Allwein, E., Schapire, R., Singer, Y.: Reducing multiclass to binary: A
           unifying approach for margin classifiers. In: JMLR, vol. 1, pp. 113–141
           (2002)
[CMP+04]   Casacuberta, J., Miranda, J., Pla, M., Sanchez, S., Serra, A., Talaya, J.:
           On the accuracy and performance of the GeoMobil system. In: Interna-
           tional Society for Photogrammetry and Remote Sensing (2004)
[CS02]   Crammer, K., Singer, Y.: On the learnability and design of output codes
           for multi-class problems. Machine Learning 47, 201–233 (2002)
[DB95]   Dietterich, T., Bakiri, G.: Solving multiclass learning problems via error-
           correcting output codes. In: JAIR, vol. 2, pp. 263–286 (1995)
[Dem06]   Demsar, J.: Statistical comparisons of classifiers over multiple data sets.
           In: JMLR, vol. 7, pp. 1–30 (2006)
[DK95]   Dietterich, T., Kong, E.: Error-correcting output codes corrects bias and
           variance. In: Prieditis, S., Russell, S. (eds.) ICML, pp. 313–321 (1995)
[Hol75]   Holland,J.H.: Adaptation in natural and artificial systems: An analysis
           with applications to biology, control, and artificial intelligence. University
           of Michigan Press (1975)
[BEB10]   Bautista, M.A., Escalera, S., Baro, X.: Compact Evolutive Design of
           Error-Correcting Output Codes. In: Supervised and Unsupervised En-
           semble methods and applications - European Conference on Machine
           Learning, pp. 119–128 (2010)
[LdC08]   Lorena, A.C., de Carvalho., A.C.P.L.F.: Evolutionary tuning of svm pa-
           rameter values in multiclass problems. Neurocomputing 71(16-18), 3326–
           3334 (2008)
[MB98]   Martinez, A., Benavente, R.: The AR face database. In: Computer Vision
           Center Technical Report # 24 (1998)
[MP]   http://www.cis.temple.edu/latecki/research.html
[PRV06]   Pujol, O., Radeva, P., Vitrià, J.: Discriminant ECOC: A heuristic method
           for application dependent design of error correcting output codes. Trans.
           on PAMI 28, 1001–1007 (2006)
[RK04]   Rifkin, R., Klautau, A.: In defense of one-vs-all classification. In: JMLR,
           vol. 5, pp. 101–141 (2004)
[TR98]   Hastie, T., Tibshirani, R.: Classification by pairwise grouping. In: NIPS,
           vol. 26, pp. 451–471 (1998)
[EOR08]   Escalera, S., Pujol, O., Radeva, P.: Sub-class error-correcting output
           codes. In: Proceedings of the 6th international conference on Computer
           vision systems, pp. 494–504 (2008)
[BEV09]   Baro, X., Escalera, S., Vitria, J., Pujol, O., Radeva, P.: Traffic Sign
           Recognition Using Evolutionary Adaboost Detection and Forest-ECOC
           Classification. IEEE Transactions on Intelligent Transportation Sys-
           tems 10(1), 113–126 (2009)
[BGV92]   Boser, B.E., Guyon, I.M., Vapnik, V.N.: A training algorithm for opti-
           mal margin classifiers. In: Proceedings of the Fifth Annual Workshop on
           Computational Learning Theory (COLT 1992), pp. 144–152. ACM Press,
           New York (1992)
[FS95]   Freund, Y., Schapire, R.E.: A decision-theoretic generalization of on-line
           learning and an application to boosting. In: Proceedings of the Second
           European Conference on Computational Learning Theory, UK pp. 23–37
           (1995)
[CC01a]   Chang, C.-C., Lin, C.-J.: LIBSVM: a library for support vector machines.,
           http://www.csie.ntu.edu.tw/~cjlin/libsvm

Improving Accuracy and Speed of Optimum-Path Forest Classifier Using Combination of Disjoint Training Subsets

Moacir P. Ponti-Jr.[1,*] and João P. Papa[2]

[1] Institute of Mathematical and Computer Sciences,
University of São Paulo (ICMC/USP)
13560-970 São Carlos, SP, Brazil
moacir@icmc.usp.br
http://www.icmc.usp.br/~moacir

[2] Department of Computing, UNESP — Univ Estadual Paulista
Bauru, SP, Brazil
papa@fc.unesp.br
http://wwwp.fc.unesp.br/~papa

Abstract. The Optimum-Path Forest (OPF) classifier is a recent and promising method for pattern recognition, with a fast training algorithm and good accuracy results. Therefore, the investigation of a combining method for this kind of classifier can be important for many applications. In this paper we report a fast method to combine OPF-based classifiers trained with disjoint training subsets. Given a fixed number of subsets, the algorithm chooses random samples, without replacement, from the original training set. Each subset accuracy is improved by a learning procedure. The final decision is given by majority vote. Experiments with simulated and real data sets showed that the proposed combining method is more efficient and effective than naive approach provided some conditions. It was also showed that OPF training step runs faster for a series of small subsets than for the whole training set. The combining scheme was also designed to support parallel or distributed processing, speeding up the procedure even more.

Keywords: Optimum-Path Forest classifier, distributed combination of classifiers, pasting small votes.

1 Introduction

The combination of classifiers was demonstrated to be effective, under some conditions, for several pattern recognition applications. These methods were widely explored, for example, to stabilize results of random classifiers and to improve performance of weak ones.

---

* Supplementary material for this paper can be found at http://www.icmc.usp.br/~moacir/project/MCS

C. Sansone, J. Kittler, and F. Roli (Eds.): MCS 2011, LNCS 6713, pp. 237–248, 2011.

If the classifiers to be combined provide only class labels as output, the majority voting is the approach commonly used to decide over an ensemble of them. The limits of vote fusion schemes were investigated by Kuncheva et al. [9], and the diversity aspect that is often claimed to produce successful combinations is currently under discussion, with the study of patterns of failure and success, and "good" and "bad" diversity [4].

In order to produce a classifier ensemble, bagging [1], boosting [7] and random subspace methods [8] were shown to be useful under some conditions [15]. They are based on the creation of classifiers trained with multiple training sets using different samples or features. The majority voting is used to produce the final decision.

To deal with large data sets, pasting small votes was proposed by Breiman [2]. The original idea was to build many classifiers from "bites", bags of small size, of data. Each one of the M bites yields one classifier. The M decisions are combined by voting.

From pasting small votes, two algorithms, Rvotes and Ivotes, were developed. While Rvote performs a simple random selection of samples to create each bite, Ivote creates consecutive training datasets based on the "importance" of the instances. It uses an algorithm similar to boosting-based algorithms to create classifiers trained with more representative samples [15]. These methods uses an evaluation set, a separate set with labeled objects, to test the accuracy of the current bite and change samples with ones that can improve the subset classification accuracy. Although Rvote is faster, it is not competitive in accuracy when compared with Ivote. Later, distributed versions for Rvote and Ivote were proposed by Chawla et al. [5] in order to reduce training time.

In this study, we propose a strategy similar to pasting small votes aiming to combine decisions of Optimum-Path Forest (OPF) classifiers [13]. The OPF technique models the feature space as a graph, using optimum-path algorithms to perform training and classification. It outputs only class labels. The proposed combination is performed by training classifiers with disjoint training subsets and it is designed to improve accuracy and reduce running time. To improve speed, the method was developed so that it could be processed using parallel or distributed processors. Since multicore and multiprocessing systems are widely available, it is an interesting feature of the proposed method.

Many papers have studied the effects of combination schemes applied to classifiers based on statistical pattern recognition, trees [8], neural-networks [3], nearest neighbors and support-vector machines [15]. Graph-based combination methods were also investigated with focus on structural pattern recognition [14,10]. In this context, as far as we know, we are the first to present a combination system for OPF classifiers. Therefore, this paper presents novel material both on the study of a new classifier and on the development of a fast combination algorithm based on this classifier.

The paper is organized as follows. Sections 2 and 3 introduces the OPF classifier and the combination method, respectively. Section 4 and 5 describe the experiments, results and discussion. Finally, the conclusions are presented in Section 6.

2 Optimum-Path Forest Classifier (OPF)

Papa et al. [13] introduced the idea of designing pattern clasifiers based on
optimum-path forest. Given a training set with samples from distinct classes,
we wish to design a pattern classifier that can assign the true class label to
any new sample, where each sample is represented by a set of features and a
distance function measures their dissimilarity in the feature space. The training
samples are then interpreted as the nodes of a graph, whose arcs are defined by
a given adjacency relation and weighted by the distance function. It is expected
that samples from a same class are connected by a path of nearby samples.
Therefore, the degree of connectedness for any given path is measured by a
connectivity (path-value) function, which exploits the distances along the path.
In supervised learning, the true label of the training samples is known and so it is
exploited to identify key samples (prototypes) in each class. Optimum paths are
computed from the prototypes to each training sample, such that each prototype
becomes the root of an optimum-path tree (OPT) composed by its most strongly
connected samples. The labels of these samples are assumed to be the same of
their root.

The OPF approach proposed by Papa et al. [13] computes prototypes as the
nearest samples from different classes in the training set. For that, a Minimum
Spanning Tree (MST) is computed over that set, and the connected samples with
different labels are marked as prototypes. A path-cost function that calculates
the maximum arc-weight along a path is used, together with a full connectedness graph. Therefore, the training phase of OPF consists, basically, of finding
prototypes and execute OPF algorithm to determine the OPTs rooted at them.
This training procedure is implemented below.

Algorithm 1 – OPF_TRAIN: OPF TRAINING ALGORITHM

INPUT: A λ-labeled training set T.
OUTPUT: Optimum-path forest P_1, cost map C_1, label map L_1, and ordered set
              T'.
AUXILIARY: Priority queue Q, set S of prototypes, and cost variable cst.

1. *Set $T' \leftarrow \emptyset$ and compute by MST the prototype set $S \subset T$.*
2. *For each $s \in T \backslash S$, set $C_1(s) \leftarrow +\infty$.*
3. *For each $s \in S$, do*
4. $\quad$ $\llcorner$ *$C_1(s) \leftarrow 0$, $P_1(s) \leftarrow nil$, $L_1(s) \leftarrow \lambda(s)$, and insert s in Q.*
5. *While Q is not empty, do*
6. $\quad$ *Remove from Q a sample s such that $C_1(s)$ is minimum.*
7. $\quad$ *Insert s in T'.*
8. $\quad$ *For each $t \in T$ such that $t \neq s$ and $C_1(t) > C_1(s)$, do*
9. $\quad\quad$ *Compute $cst \leftarrow \max\{C_1(s), d(s,t)\}$.*
10. $\quad\quad$ *If $cst < C_1(t)$, then*
11. $\quad\quad\quad$ *If $C_1(t) \neq +\infty$, then remove t from Q.*
12. $\quad\quad\quad$ *$P_1(t) \leftarrow s$, $L_1(t) \leftarrow L_1(s)$, $C_1(t) \leftarrow cst$.*
13. $\quad\quad\quad$ *Insert t in Q.*
14. *Return a classifier $[P_1, C_1, L_1, T']$.*

Further, the test phase essentially evaluates, for each test sample, which training node offered the optimum-path to it. In [13], the classification of each new sample t from test set O is carried out based on the distance $d(s,t)$ between t and each training node $s \in T$ and on the evaluation of the following equation:

$$C_2(t) = \min\{\max\{C_1(s), d(s,t)\}\}, \; \forall s \in T. \tag{1}$$

Let $s^* \in T$ be the node s that satisfies this equation. It essentially considers all possible paths from S in the training graph T extended to t by an arc (s,t), and label t with the class of s^*. Algorithm 2 shows the implementation of the OPF classification phase.

Algorithm 2 – OPF_CLASSIFY: OPF CLASSIFICATION ALGORITHM

INPUT: Classifier $[P_1, C_1, L_1, T']$, evaluation set E (or test set O), and the pair
             (v, d) for feature vector and distance computations.
OUTPUT: Label L_2 and predecessor P_2 maps defined for E.
AUXILIARY: Cost variables tmp and $mincost$.

1.   *For each $t \in E$, do*
2. $i \leftarrow 1, \; mincost \leftarrow max\{C_1(k_i), d(k_i, t)\}$.
3. $L_2(t) \leftarrow L_1(k_i) \; and \; P_2(t) \leftarrow k_i$.
4.   *While $i < |T'|$ and $mincost > C_1(k_{i+1})$, do*
5.   *Compute $tmp \leftarrow \max\{C_1(k_{i+1}), d(k_{i+1}, t)\}$.*
6.   *If $tmp < mincost$, then*
7. $mincost \leftarrow tmp$.
8. $L_2(t) \leftarrow L(k_{i+1}) \; and \; P_2(t) \leftarrow k_{i+1}$.
9. $i \leftarrow i + 1$.
10. *Return $[L_2, P_2]$.*

2.1   Learning

Large datasets usually present redundancy, so at least in theory it should be possible to estimate a reduced training set with the most relevant patterns for classification. The use of a training and an evaluation set has allowed us to learn relevant training samples from the classification errors in the evaluation set, by swapping misclassified samples of the evaluating set and non-prototype samples of the training one during a few iterations (usually less than 10 iterations) [13]. In this learning strategy, the training set remains with the same size and the classifier instance with the highest accuracy is selected to be used for classification of new objects. Algorithm 3 below implements this idea.

Algorithm 3 – OPF_LEARN: OPF LEARNING ALGORITHM

INPUT: A labeled training and evaluating sets T and E, respectively and the
             number T of iterations.
OUTPUT: Optimum-path forest P_1, cost map C_1, label map L_1, and ordered set
             T'.
AUXILIARY: Arrays FP and FN of sizes c for false positives and false negatives,
             set S of prototypes, and list LM of misclassified samples.

1.   *Set $MaxAcc \leftarrow -1$.*
2.   *For each iteration $I = 1, 2, \ldots, T$, do*
3.   *$LM \leftarrow \emptyset$ and compute the set $S \subset T$ of prototypes.*
4.   *$[P_1, C_1, L_1, T'] \leftarrow OPF\_train(T, S, (v, d))$.*
5.   *For each class $i = 1, 2, \ldots, c$, do*
6.   *$FP(i) \leftarrow 0$ and $FN(i) \leftarrow 0$.*
7.   *$[L_2, P_2] \leftarrow OPF\_classify(T', E, (v, d))$*
8.   *For each sample $t \in E$, do*
9.   *If $L_2(t) \neq \lambda(t)$, then*
10.   *$FP(L_2(t)) \leftarrow FP(L_2(t)) + 1.$*
11.   *$FN(\lambda(t)) \leftarrow FN(\lambda(t)) + 1.$*
12.   *$LM \leftarrow LM \cup t.$*
13.   *Compute accuracy Acc according to [13].*
14.   *If $Acc > MaxAcc$ then save the current instance $[P_1, C_1, L_1, T']$*
15.   *of the classifier and set $MaxAcc \leftarrow Acc$.*
16.   *While $LM \neq \emptyset$*
17.   *$LM \leftarrow LM \backslash t.$*
18.   *Replace t by a non-prototype sample, randomly selected from T.*
19. *Return the classifier instance $[P_1, C_1, L_1, T']$ with the highest accuracy in E.*

3 Combination of Classifiers Trained with Disjoint Subsets

The OPF classifier outputs only class labels, without ranking or probability information. Therefore, only abstract-level options are available in order to create ensembles and combine this kind of classifier. A method based on bagging or boosting would be interesting in this case. However, since the OPF classifier has the advantage to perform a fast training, it is interesting to design an also fast combining scheme.

We propose a fast combination method based on the idea of pasting small votes. It creates an ensemble of OPF classifiers using disjoint training subsets. A fixed number D of subsets is set, and the algorithm chooses random samples, without replacement, from the original training set T. The samples are taken so that each subset will contain approximately the same number of objects per class. The complete procedure, called OPFcd, is described in Algorithm 4. As described in Section 2.1, the OPF has a fast learning algorithm to change the training set using an evaluation set. It has a behavior similar to a boosting algorithm. This learning algorithm is used to improve accuracy of each classifier, and, therefore, is expected to improve accuracy of the final decision. Moreover, the OPF training algorithm has computational complexity of $\Theta(N^2)$, where N denotes the training set size, and therefore, it is expected to run faster for k training sets of size X then for a larger one with size $N = k \times X$.

The OPFcd was designed in order to take advantage of new (and widely available) multicore and multiprocessor systems. With use of this algorithm, training, learning and classification procedures for each classifier E_j can be assigned to a different processor. It can be carried out by assigning each call of lines 5 and 6 (see Algorithm 4) to different processors. In section 5.1 it will be shown that the use of a multicore processor can reduce the overall running time.

Algorithm 4 – OPFCD: OPF COMBINATION OF DISTRIBUTED DISJOINT SETS

INPUT: Training data set T of size N with correct labels $\omega_j \in \Omega$, $j = \{1,..,C\}$
           for C classes, the evaluation set V, the set of objects to be classified
           O (test set), the number of disjoint subsets D, the number of samples
           of each class $P = \{p_1,..,p_C\}$, and the OPF algorithm *OPF_learn* and
           OPF_classify
OUTPUT: Set of labels after final decision, L
AUXILIARY: The number of objects on each subset M, training subsets K_i, classi-
           fiers E_i, and objects labeled by each classifier I_i, where $i = \{1,..,D\}$

1. $M \leftarrow \lfloor N/D \rfloor$
2. *For each subset i, $(\forall i = 1..D)$, do*
3.   *For each class j, $(\forall j = 1..C)$, do*
4. $\llcorner$ *Select randomly $(p_j/D) \times M$ samples of class j from T without*
5.   *replacement and store them in K_i*
6. $E_i \leftarrow OPF\_learn(K_i, V)$
7. $\llcorner$ $I_i \leftarrow OPF\_classify(O, E_i)$
8. $L \leftarrow Vote(I_i)$
9. *Return L*

3.1   Nearest Neighbor Tie-Breaking

In multiclass problems and decisions by vote, it is possible to observe ties. Tie-breaking methods are often arbitrary. However, using object local information [16,14] it is possible to obtain a better solution for tie-breaking.

We propose the use of a nearest neighbor tie-breaking for multi class problems as follows. Let $\left[\omega_1^i, ..., \omega_D^i\right]$ be the class outputs of an ensemble of D classifiers for a given object i. If the voting procedure results in a tie, the class labels of the nearest previously classified object j are joined with the current object, so that the new voting will be applied at $\left[\omega_1^i, ..., \omega_D^i, \omega_1^j, ..., \omega_D^j\right]$. If a new tie is obtained, the object is assigned to a random class, chosen from the set of tied classes.

4   Experiments

4.1   Data

Several experiments were carried out to test the effectiveness of OPF combination using training subsets. The data used in the experiments includes six synthetic and four real data sets. The simulated data sets were built to have partially overlapped classes. Different covariance matrices were used to generate the 4-class Gaussian data sets. The project web page contains details and the code used to generate the simulated data using PRTools (v.4.1.4)[1] Table 1

---

[1] http://www.prtools.org/

shows the simulated data sets characteristics and Table 2 displays the real data sets used to show whether the proposed technique is effective when solving real problems.

Table 1. Synthetic databases details

Name	size	classes	features	type
B2-2d	1,000	2	2	Banana-shape
B4-2d	2,000	4	2	Banana-shape
G2-3d	1,000	2	3	Gaussian
G4-3d	2,000	4	3	Gaussian
G2-4d	100,000	2	4	Gaussian
G4-4d	100,000	4	4	Gaussian

Table 2. Synthetic databases details

Name	size	classes	features
NTL	8,067	2	4
COREL	1,000	10	150
Activity	164,860	11	6
KDD-1999	380,271	3	9

The *NTL* is a real database obtained from an electric power company. It is frequently used to identify thefts in power distribution systems, and it is composed by legal and illegal industrial profiles. The idea is to classify each profile as a fraud or not.

COREL is a subset of an image database including the classes: African people and villages, beach, buildings, buses, dinosaurs, elephants, flowers, horses, mountains and glaciers, food. The 150 features were extracted using the SIFT algorithm [11].

KDD-1999 is the data set used for "The Third International Knowledge Discovery and Data Mining Tools Competition"[6]. The task was to build a network intrusion detector, capable of distinguishing intrusions or attacks, and normal connections. It includes a variety of intrusions simulated in a military network environment.

Activity is the "Localization Data for Person Activity Data Set" [6]. The data set contains recordings of people wearing four sensors while performing one of the activities: walking, falling, lying down, lying, sitting down, sitting, standing up from lying, on all fours, sitting on the ground, standing up from sitting, standing up from sitting on the ground. We used a subset of 6 features (the original has 8), without the date and time stamp features.

4.2   Settings and Implementation

We conducted the experiments as follows: the data sets were partitioned in three sets, 15% for training, 10% for evaluating and 75% for testing. The settings for *KDD-1999* are an exception due to the size of the data set, and the high performance obtained by the single classifier, so for this data set we used 1% for training and 0.5% for evaluating.

All samples from training set was used to train the "single classifier", and disjoint partitions of the training set were used to train several classifiers and built the ensembles. We performed experiments with 3, 5, 7, 9, 11 and 13 partitions (disjoint training subsets). For multi class problems, the nearest neighbor tie-breaking described in Section 3.1 was also included in the experiments and identified by the name OPFcd+NN.

All experiments were repeated 10 times on independent training sets. The average results are presented in tables and graphics. A two-tailed t-test for samples with unequal variance was computed to verify the significant differences between the results of single classifier and the combination methods.

Regarding OPF implementation we used the LibOPF [12], a free library for the implementation of optimum-path forest-based classifiers. All experiments were carried out on a quad-core machine Intel® Core^{TM}2 Quad 3GHz with 6MB cache size, and a Linux 64-bits operational system (kernel v.2.6.32-27). To investigate the use of multicore processing we assigned the affinity of processes related to the training, learning and classification to the four different core processors using taskset routine. We are going to present the results of the running time for the data sets G2-3d and G4-3d, both with 100,000 samples.

5   Results and Discussion

The average accuracy results for each data set is shown in Table 3 for simulated and Table 4 for real data sets. The use of combined OPF classifiers significantly increased the performance for almost all data sets.

To analyze when the combination is useful, we studied the classification error when individual disjoint sets (without combination) were used to train and classify the data sets. In Figure 1 we present examples of the three different "patterns" found on the data sets when the different sizes of partitions were used to train a classifier. The best results were observed with the second pattern, as observed for the *B4-2d* data set. In this case when less training samples are taken there is a slight increase in the error, and the combination improves the performance using almost any number of partitions. The third pattern was also shown to improve results when using a particular range of partition sizes — for the *NTL* data set we observed an improvement using from 3 to 7 partitions. However, when the performance decreases as the number of partitions increases, the combination is useless or increases the error, as observed for the *Activity* data set.

For the *B4-2d* data set a significant improvement was achieved using from 5 to 13 partitions in OPFcd method and for all partitions using also the nearest

Table 3. Synthetic databases results: classification average error and standard deviation for single classifier and combination, with significant improvements in boldface for $p < 0.01$, using the best number of partitions, indicated by #part

Data set	Single classifier	OPFcd (#part)	OPFcd+NN (#part)
B2-2d	22.6±1.7	**16.8±1.2 (9)**	—
B4-2d	31.0±1.7	**26.0±1.5 (11)**	**23.8±1.9 (11)**
G2-3d	26.0±1.1	**21.0±2.5 (11)**	—
G4-3d	14.1±0.7	**12.3±0.6 (11)**	**11.9±0.6 (11)**
G2-4d	18.9±1.2	**16.6±1.8 (7)**	—
G4-4d	13.2±0.8	13.5±1.0 (7)	12.9±0.7 (5)

Table 4. Real databases results: classification average error and standard deviation for single classifier and combination, with significant improvements in boldface for $p < 0.01$, using the best number of partitions, indicated by #part

Data set	Single classifier	OPFcd (#part)	OPFcd+NN (#part)
NTL	9.2±0.5	**8.2±0.3 (5)**	—
COREL	31.0±1.7	**26.0±1.5 (11)**	**24.8±1.9 (11)**
Activity	30.1±1.1	31.3±2.5 (7)	30.6±2.3 (5)
KDD-1999	0.21±0.05	**0.14±0.02 (9)**	**0.12±0.02 (11)**

Table 5. Classification results for the data set B4-2d using combination of classifiers trained with 3–13 disjoint sets, with significant improvement in boldface for $p < 0.01$ when compared to the single classifier (31.0±1.7)

	# training set partitions					
	3	5	7	9	11	13
OPFcd	29.7±1.2	**26.8±1.1**	**26.9±1.0**	**26.5±1.5**	**26.0±1.6**	**26.1±0.9**
p values		.003	.004	.002	.002	.004
with NN	**28.3±1.2**	**24.9±1.3**	**25.8±0.7**	**24.6±1.6**	**23.8±1.4**	**24.7±1.7**
p values	.022	.001	.001	.001	.001	.001

Table 6. Classification results for the data set NTL using combination of classifiers trained with 3–13 disjoint sets, with significant improvement in boldface for $p < 0.01$ when compared to the single classifier (9.2±0.5)

	# training set partitions					
	3	5	7	9	11	13
OPFcd	**8.1±0.2**	**8.0±0.3**	**8.3±0.3**	9.1±0.2	9.4±0.6	9.5±0.6
p values	.003	.002	.008	.312	.419	.457

neighbor tie-breaking, as displayed in Table 5. The results for the *NTL* are shown in Table 6, where improvement was achieved using from 3 to 7 partitions.

The nearest neighbor tie-breaking method improved the results of multi class problems in most experiments. However, since this method is more time consuming, one must analyze the benefits of using it to possibly reduce the error with an increase in overall running time.

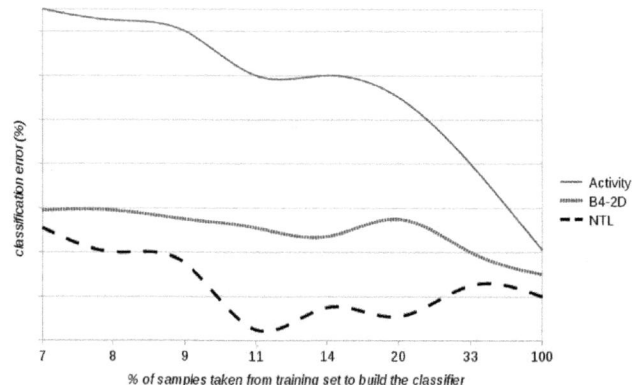

Fig. 1. Examples of patterns of classification error as a function of the percentage of samples used from the training set of three data sets: *Activity*, *B4-2d* and *NTL*

5.1   Improvement in Running Time and Multicore Processing

One of the main contributions of the proposed combination scheme is the potential to be less time consuming. The observed average running times are shown in Figure 2. Each procedure was timed during all the experiments with the *G2-4d*

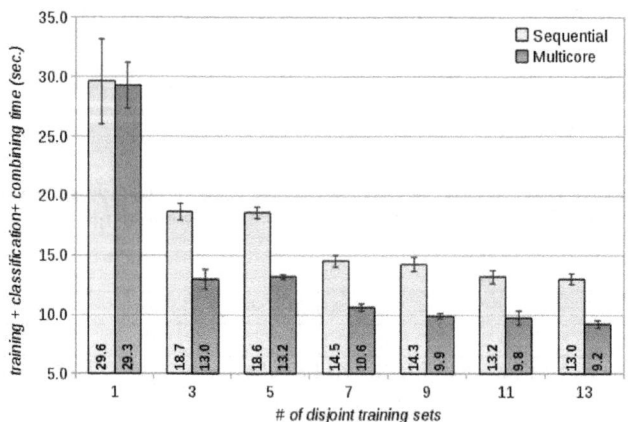

Fig. 2. Average running time results for trainining, classification and combination using single classifier and combination of classifiers using from 3 to 15 partitions, with comparison between multicore and sequential processing

data set using OPFcd and with the *G4-4d* data set using OPFcd+NN. It is possible to see a significant decrease of the overall running time as a higher number of disjoint partitions are used. This effect is due to the quadratic nature of the training and learning OPF algorithms.

It can be seen also in Figure 2 that even lower running times were observed when assigning each process to a different processor core.

6   Conclusions

The combination of OPF classifiers trained with disjoint subsets has showed to be effective under some conditions, since we have found three patterns of classification behavior. The proposed method reduced the classification error for different data sets specially when the error does not increases as the size of the partitions decrease. We also showed that the running time results for OPF can be improved by training series of smaller training sets and combining them. This improvement is due to the quadratic nature of the OPF algorithm. The training and learning algorithms can also be processed in multicore or multiprocessor machines to achieve a larger improvement in speed.

The OPF classifier is a recent and promising method for pattern recognition. Therefore, reporting an effective combining method for this classifier can be important for many applications. In this paper, we focused on the development of a fast and robust method, capable of be implemented on a parallel or distributed system for OPF classifiers. Further projects can analyze the behavior of OPF classifiers when combined using methods such as boosting, bagging and random subspace with focus on more accurate results. A deepest analysis of the proposed combination system is also a point left for future studies.

References

1. Breiman, L.: Bagging predictors. Machine Learning Journal 2(24), 123–140 (1996)
2. Breiman, L.: Pasting small votes for classification in large databases and on-line. Machine Learning 36, 85–103 (1999)
3. Breve, F.A., Ponti Jr, M.P., Mascarenhas, N.D.A.: Multilayer perceptron classifier combination for identification of materials on noisy soil science multispectral images. In: XX Brazilian Symposium on Computer Graphics and Image Processing (SIBGRAPI 2007), pp. 239–244. IEEE, Belo Horizonte (2007)
4. Brown, G., Kuncheva, L.I.: "Good" and "Bad" Diversity in Majority Vote Ensembles. In: El Gayar, N., Kittler, J., Roli, F. (eds.) MCS 2010. LNCS, vol. 5997, pp. 124–133. Springer, Heidelberg (2010)
5. Chawla, N.V., Hall, L.O., Bowyer, K.W., Moore Jr., T.E.: Distributed pasting of small votes. In: Roli, F., Kittler, J. (eds.) MCS 2002. LNCS, vol. 2364, pp. 52–62. Springer, Heidelberg (2002)
6. Frank, A., Asuncion, A.: UCI machine learning repository (2010), http://archive.ics.uci.edu/ml
7. Freund, T.: Boosting: a weak learning algorithm by majority. Information and Computation 121(2), 256–285 (1995)

8. Ho, T.: The random subspace method for constructing decision forests. IEEE Trans. Pattern Analysis and Machine Intelligence 20(8), 832–844 (1998)
9. Kuncheva, L., Whitaker, C., Shipp, C.A., Duin, R.: Limits on the majority vote accuracy in classifier fusion. Pattern Analysis and Applications 6, 22–31 (2003)
10. Lee, W.J., Duin, R.: A labelled graph based multiple classifier system. In: Benediktsson, J.A., Kittler, J., Roli, F. (eds.) MCS 2009. LNCS, vol. 5519, pp. 201–210. Springer, Heidelberg (2009)
11. Li, J., Wang, J.Z.: Automatic linguistic indexing of pictures by a statistical modeling approach. IEEE Trans Pattern Analysis and Machine Intelligence 25(9), 1075–1088 (2003)
12. Papa, J.P., Falcão, A.X., Suzuki, C.T.N.: LibOPF: a library for optimum-path forest (OPF) classifiers. (2009), http://www.ic.unicamp.br/~afalcao/libopf/
13. Papa, J.P., Falcão, A.X., Suzuki, C.T.N.: Supervised pattern classification based on optimum-path forest. Int. J. Imaging Systems and Technology 19(2), 120–131 (2009)
14. Schenker, A., Bunke, H., Last, M., Kandel, A.: Building graph-based classifier ensembles by random node selection. In: Roli, F., Kittler, J., Windeatt, T. (eds.) MCS 2004. LNCS, vol. 3077, pp. 214–222. Springer, Heidelberg (2004)
15. Skurichina, M., Duin, R.P.W.: Bagging, boosting and the random subspace method for linear classifiers. Pattern Analysis and Applications 5, 121–135 (2002)
16. Woods, K., Kegelmeyer-Jr., W., Bowyer, K.: Combination of multiple classifiers using local accuracy estimates. IEEE Trans Pattern Analysis and Machine Intelligence 19(4), 405–410 (1997)

Analyzing the Relationship between Diversity and Evidential Fusion Accuracy

Yaxin Bi

School of Computing and Mathematics, University of Ulster, Newtownabbey,
Co. Antrim, BT37 0QB, UK
y.bi@ulster.ac.uk

Abstract. In this paper, we present an empirical analysis on the relationship between diversity and accuracy of classifier ensembles in the context of the theory of belief functions. We provide a modelling for formulating classifier outputs as triplet mass functions and a unified notation for defining diversity measures and then assess the correlation between the diversity obtained by four pairwise and non-pairwise diversity measures and the improvement of accuracy of classifiers combined in decreasing and mixed orders by Dempster's rule, Proportion and Yager's rules. Our experimental results reveal that the improved accuracy of classifiers combined by Dempster's rule is positively correlated with the diversity obtained by the four measures, but the correlation between the diversity and the improved accuracy of the ensembles constructed by Proportion and Yager's rules is negative, which is not in favor of the claim that increasing diversity could lead to reduction of generalization error of classifier ensembles.

1 Introduction

The combination of multiple classifiers/ensemble approach is a powerful decision making and classification technique that has been used successfully for modelling many practical problems, such as text categorization and remote sensing. In the modelling of classifiers combination, many researchers believe the diversity being inherent in the classifiers plays an important role in constructing successful classifier ensembles. Unfortunately to date there exists no general accepted theoretical framework for capturing diversity in multiple classifier systems. Although many statistics have been employed to measure diversity among classifiers with the intension to determine whether it correlates with ensemble performance in the literature, results are varied. In particular there is little effort concerning how diversity measured by statistics imparts ensemble performance in the framework of the Dempster-Shafer (DS) theory of evidence [6], where classifier outputs are modeled as pieces of evidence that are combined by Dempster's rule of combination and its alternatives. In this paper, we present our studies on measuring diversity among classifiers and then experimentally examine the relationship between diversity obtained by four pairwise and non-pairwise diversity measures and the improved accuracy of classifiers combined by Dempster's rule of combination [6], Yager's rule [10] and Proportion rule [11].

C. Sansone, J. Kittler, and F. Roli (Eds.): MCS 2011, LNCS 6713, pp. 249–258, 2011.

Previous studies on the relationship between diversity and ensemble performance have stimulated considerable interest and they can be categorized into the contexts of regression and classification. In the context of classification, Kuncheva et al. carried out an experimental study on relationship between diversity and accuracy [5]. Their results show although there are proven connections between diversity and accuracy in some special cases, there is no strong linear and non-linear correlation between diversity and accuracy. In [8], Tang, et al. conducted a follow-up comprehensive study. They investigate the correlation among the six statistical measures used in [5] and relate these measures to the concept of margin, which is explained as a key success factor of Boosting algorithms. The experimental results indicate that large diversity may not consistently correspond to better ensemble performance and the information gained by varying diversity among classifiers cannot provide a consistent guidance for making an ensemble of classifiers to achieve good generalization performance.

In [1], we have developed new evidence structures called a *triplet* and *quartet* and a formalism for modelling classifier outputs as triplet and quartet mass functions, and we also established a range of formulae for combining these mass functions in order to arrive at a consensus decision. In [2], we formulated the problem of measuring diversity among classifiers and addressed the impact of diversity on the performance of classifiers combined only by Dempster's rule of combination. In this paper, we extend our previous study and carry out an analysis of diversity effects on the quality of classifier ensembles, in which the component classifier are independently generated by 13 machine learning algorithms and are combined in decreasing and mixed orders using three evidential combination rules. We use the triplet as an underlying evidence structure for representing classifier outputs and study the correlation between diversity and the ensemble accuracy by the Spearmans rank analysis over 12 benchmark data sets. The experimental results demonstrate that the positive correlation between the diversity and the ensemble accuracy made by Dempster's rule is stronger than the negative correlation made by the other two rules in mixed order, however the positive correlation made by Dempster's rule is weaker than the negative one made by Yager's rule. Moreover our results conjecture that the order of combining classifiers could be regarded as a useful factor in constructing successful classifier ensembles.

2   Representation of Classifier Outputs

In ensemble approaches, a learning algorithm is provided with a training data set made up of $D \times C = \{\langle d_1, c_1 \rangle, \cdots, \langle d_{|D|}, c_q \rangle\}$ $(1 \leq q \leq |C|)$ for deriving some unknown function f such that $f(d) = c$. Instance d_i is characterized by a vector in the form of $(d_{i_1}, \cdots, d_{i_n})$ where d_{i_j} is either a nominal or ordinal value, and c_i is typically drawn from a set of categorical classes in terms of class labels. Given a set of training data $D \times C$, a learning algorithm is aimed at learning a function φ in terms of classifier from the training data set, where classifier φ is an approximation to an unknown function f.

Given a new instance d, a classification task is to make decision for d using φ about whether d belongs to class c_i. Instead of single-class assignment, we regard such a classification process as a mapping:

$$\varphi : D \to C \times [0, 1] \tag{1}$$

where $C \times [0, 1] = \{(c_i, s_i) \mid c_i \in C, 0 \le s_i \le 1\}$, s_i is a numeric value.The greater the value of class s_i, the greater the possibility of the instance belonging to that class. Simply we denote a classifier output by $\varphi(d) = \{s_1, \cdots, s_{|C|}\}$ and the accuracy of the classifier by $F(\varphi)$. Given a group of classifiers, $\varphi_1, \varphi_2, \cdots, \varphi_M$, all the classifier outputs on instance d can be organized into a matrix as illustrated in formula (2).

$$\begin{pmatrix} \varphi_1(d) \\ \varphi_2(d) \\ \vdots \\ \varphi_M(d) \end{pmatrix} = \begin{pmatrix} s_{11} & s_{12} & \cdots & s_{1|C|} \\ s_{21} & s_{22} & \cdots & s_{2|C|} \\ \vdots & \vdots & \cdots & \vdots \\ s_{M1} & s_{M2} & \cdots & s_{M|C|} \end{pmatrix} \tag{2}$$

3 Basics of the Dempster-Shafer (DS) Theory of Evidence

Definition 1. Let Ω be a frame of discernment. Let m be a mass function, which is defined as a assignment function assigning a numeric value in $[0,1]$ to $X \in 2^{\Omega}$ with two conditions below [6].

$$1) \; m(\emptyset) = 0, \quad 2) \sum_{X \subseteq \Omega} m(X) = 1$$

where X is called a focal element or focus if $m(X) > 0$ and a singleton if $|X| = 1$.

Given the general representation of classifier outputs in formula (2) and Definition 1, we denote $\Omega = \{c_1, \ldots c_{|\Omega|}\}$ where $\{c\} \subseteq \Omega$ represents a proposition of interest.

Definition 2. Let Ω be a frame of discernment and let $\varphi(d)$ be a list of scores as before, an application-specific mass function is defined a mapping function, $m : 2^{\Omega} \to [0, 1]$ as follows:

$$m(\{c_i\}) = \frac{s_i}{\sum_{j=1}^{|\Omega|} s_j} \tag{3}$$

where $c_i \in \Omega$ and $1 \le i \le |\Omega|$.

Definition 3. Let Ω be a frame of discernment. Let m_1 and m_2 be two mass functions defined for $X, Y \subseteq \Omega$. Dempster's rule of combination (or Dempster's rule) is, denoted by $\oplus$, defined as

$$(m_1 \oplus m_2)(A) = \frac{\sum_{X \cap Y = A} m_1(X) m_2(Y)}{\sum_{X \cap Y \ne \emptyset} m_1(X) m_2(Y)} \tag{4}$$

where $A \subseteq \Omega$ and $\oplus$ is also called the *orthogonal sum*. $N = \sum_{X \cap Y \neq \emptyset} m_1(X) m_2(Y)$ is the *normalization constant*. $E = 1 - N$ is called the *conflict factor*. This rule strongly emphasizes the agreement between multiple independent sources and ignores all the conflicting evidence through a normalization factor.

The issue associated with Dempster's rule of combination is the normalization process, which may produce counter-intuitive results when evidence sources are in highly conflict. In the past decades there has been a great deal of debate about overcoming counter-intuitive results. Much of research has been devoted to develop alternatives to Dempster's rule of combination. Due to the limited space, we confine our study in the classical combination rules: Dempster's rule, Yager's rule and Proportion rule and formulating pieces of evidence derived from classifier outputs in terms of triplet mass functions below.

3.1   Triplet Mass Function and Computation

Given the formulation of classifier outputs in formula (2), by formula (3), we can rewrite $\varphi(d)$ as $\varphi(d) = \{m(\{c_1\}), m(\{c_2\}), \cdots, m(\{c_\Omega\})\}$, referred to as a list of decisions — a piece of evidence. By formula (4) two or more pieces of evidence can then be combined to make the final classification decision. To improve the efficiency of computing the orthogonal sum operation and the accuracy of the final decision on the basis of the combined results, a new structure, called a triplet, has been developed in [1]. A brief introduction of the structure is given below.

Definition 4. Let Ω be a frame of discernment and $\varphi(d) = \{m(\{c_1\}), m(\{c_2\}), \ldots, m(\{c_\Omega\})\}$ where $|\Omega| \geq 2$, the expression of $Y = \langle \{u\}, \{v\}, \Omega \rangle$ is defined as a *triplet*, where $\{u\}, \{v\}$ are singletons and they satisfy

$$m(\{u\}) + m(\{v\}) + m(\Omega) = 1$$

Based on the number of singleton decisions, we also refer to a triplet as a structure of *two-point focuses*, and call the associated mass function a *two-point mass function*. To obtain triplet mass functions, we define a focusing operation in terms of the *outstanding rule* and denote it by m^σ as follows:

$$\{u\} = arg \max(\{m(\{c_1\}), m(\{c_2\}), ..., m(\{c_\Omega\})\}) \tag{5}$$

$$\{v\} = arg \max(\{m(\{c\}) \mid c \in \{c_1, ..., c_\Omega\} - \{u\}\}) \tag{6}$$

$$m^\sigma(\Omega) = 1 - m^\sigma(\{u\}) + m^\sigma(\{v\}) \tag{7}$$

We refer to m^σ as a *triplet mass function* or as a *two-point mass function*, simply m. By applying formulas (3), (5), (6), (7), formula (2) is simply rewritten as formula (8) below.

$$\begin{pmatrix} \varphi_1(d) \\ \varphi_2(d) \\ \vdots \\ \varphi_M(d) \end{pmatrix} = \begin{pmatrix} m_1(\{u_1\}) & m_1(\{v_1\}) & m_1(\Omega) \\ m_2(\{u_2\}) & m_2(\{v_2\}) & m_2(\Omega) \\ \vdots & \vdots & \vdots \\ m_M(\{u_M\}) & m_M(\{v_M\}) & m_M(\Omega) \end{pmatrix} \tag{8}$$

Definition 5. Suppose m_1 and m_2 are two triplet mass functions on the frame of discernment Ω. Let X and Y be two triplets. Yager's combination rule is defined as:

$$(m_1 \,\textcircled{y}\, m_2)(A) = \begin{cases} 0 & \text{if } A = \emptyset \\ \sum_{X \cap Y = A} m_1(X) m_2(Y) & \text{if } \emptyset \subset A \subset \Omega \\ \sum_{X \cap Y = \Omega} m_1(X) m_2(Y) + & \\ \sum_{X \cap Y = \emptyset} m_1(X) m_2(Y) & \text{if } A = \Omega \end{cases} \tag{9}$$

Yager's rule keeps the condition of $m(\emptyset) = 0$ and adds masses allocated to both the empty set and the frame of discernment together into Ω.

Definition 6. Suppose m_1 and m_2 are two triplet mass functions on the frame of discernment Ω. Let X and Y be two triplets. Proportion combination rule is defined as:

$$(m_1 \,\textcircled{a}\, m_2)(A) = \begin{cases} 0 & \text{if } A = \emptyset \\ (\sum_{X \cap Y = A} m_c(X) + & \\ \sum_{X \cap Y = A} m_r(Y))/2 & \text{if } A \neq \Omega \end{cases} \tag{10}$$

where $m_c(X)$ and $m_r(Y)$ are the average of mass functions based on columns and rows in the intersection table, respectively [11]. The detailed calculation of combining triplet mass functions and incorporating the outstanding rule to formulas (4), (9) and (10) is omitted due to the limited space.

4   Diversity Measures

Since there has been no convincing theory or experimental study, suggesting which of statistical measures can be reliably used to improve ensemble performance [8]. In this study, based on the way of measuring agreement and disagreement we employ three pairwise and one non-pairwise methods to measure diversity being inherent in binary classifier outputs as discussed in most studies in the literature [5].

Formally suppose we are given M classifiers denoted by $\varphi_1, \cdots, \varphi_M$, a set of classes $\Omega = \{c_1, \cdots, c_\Omega\}$ and a test set $T = \{x_1, \cdots, x_{|T|}\}$. For any instance $x \in T$, each classifier output $\varphi_i(x)$ is modeled as an binary output, i.e. $\varphi_i(x) = 1$ if φ_i correctly classifies x, $\varphi_i(x) = 0$ if φ_i incorrectly classifies x. For the classifiers that correctly classify instances, we denote them by $\widehat{\varphi}(x) = \{\varphi(x) | \varphi_i(x) = 1, 1 \leq i \leq M, x \in T\}$.

With this notation, we implemented four statistical diversity measures for our experiments, including Kappa (κ) statistic [3], Disagreement (dis) measure [7], Q-statistic (qs) [5] [8] and Kohavi-Wolpert variance [4].

5   Experimental Evaluation

5.1   Experimental Settings

In our experiments, we used 12 data sets, including *anneal, audiology, balance, car, glass, autos, iris, letter, segment, soybean, wine* and *zoo*. All the data sets have at least three or more classes as required by the triplet structure.

For base (individual) classifiers, we used thirteen learning algorithms which all are taken from the Waikato Environment for Knowledge Analysis (Weka) version 3.4, including AOD, NaiveBayes, SOM, IB1, IBk, KStar, DecisionStump, J48, RandomForest, DecisionTable, JRip, NNge, and PART. These algorithms were simply chosen on the basis of their performance on three randomly picked data sets. Parameters used for each algorithm was at the Weka default settings described in [9].

Given 13 classifiers $\varphi_1, \cdots, \varphi_{13}$ that are generated by the 13 learning algorithms, for the combination of classifiers in decreasing order, we first rank all the classifiers as $F(\varphi_{k_1}) \geq F(\varphi_{k_2}) \geq \cdots \geq F(\varphi_{k_{12}}) \geq F(\varphi_{k_{13}})$ $(1 \leq k_i \leq 13)$, and then we combine φ_{k_1} with φ_{k_2} as a classifier ensemble, denoted by $2C$, and combine the resulting $2C$ with φ_{k_3} as another classifier ensemble, denoted by $3C$, and so forth, until combine $12C$ with $\varphi_{k_{13}}$, denoted by $13C$. We also denote the groups of classifiers that make up the classifier ensembles iC by $i\Phi$ $(1 \leq i \leq 13)$. With respect to the combination of classifiers in mixed order, the order of classifiers is random. We first pick up two classifiers to combine them as a classifier ensemble, denoted by $2C$, and then combine the resulting $2C$ with the third classifier that is randomly chosen, denoted by $3C$, until combine the resulting ensembles with the last classifier, denoted by $13C$.

To assess how the diversity $div(i\Phi)$ (detailed in Section 4) is actually correlated the accuracy $F(iC)$, we carried out a Spearmans rank correlation analysis on each pair of $\langle div(2\Phi), F(2C) \rangle$, $\langle div(3\Phi), F(3C) \rangle$, ..., $\langle div(13\Phi), F(13C) \rangle$ over the 12 data sets, resulting in 12 pairs of correlation coefficient $r \in [-1, 1]$ and $p\text{-}value \in [0, 1]$. A positive correlation coefficient r indicates a positive correlation between the diversity and accuracy, whereas a negative r indicates a negative correlation between them. $p\text{-}value$ indicates the degree of that the correlation is statistically significant.

To further quantify the relationship between the diversity $div(i\Phi)$ and the improvement of the accuracy $F(i\Phi)$, we calculate the mean accuracy of $i\Phi$, denoted by $F_M(i\Phi)$ with $[F(\varphi_{K_1}) + F(\varphi_{K_2}) + \ldots + F(\varphi_{K_i})]/i$ and then calculate each difference between $F(iC)$ and $F_M(i\Phi)$ in terms of the improvement of the accuracy, denoted by $F_{IM}(i\Phi)$ $(1 \leq i \leq 13)$.

5.2   Results in Decreasing Order

In this experiment we constructed three groups of classifier ensembles by Dempter's rule, Proportion rule and Yager's rule, respectively, each of which shares the same group of classifiers in the form of 2Φ, 3Φ, ..., 13Φ, thus 3×13 classifier ensembles have been generated in total. For $i\Phi$, we calculate diversity $\kappa(i\Phi)$, $qs(i\Phi)$, $dis(i\Phi)$ and $kw(i\Phi)$ $(1 \leq i \leq 13)$, resulting in four groups of diversity collectively 4×13 pieces of diversity. We scaled these calculations up over the 12 data sets, and the results are graphed in Fig. 1.

According to the behaviors of the curves and the nature of the four diversity measures in Fig.1, the diversity curves of 12 groups of classifiers from $2c$ to $13c$ over the 12 data sets can be characterized into two groups: one is measured by qs and κ, and the other is measured by kw and dis. It can be observed

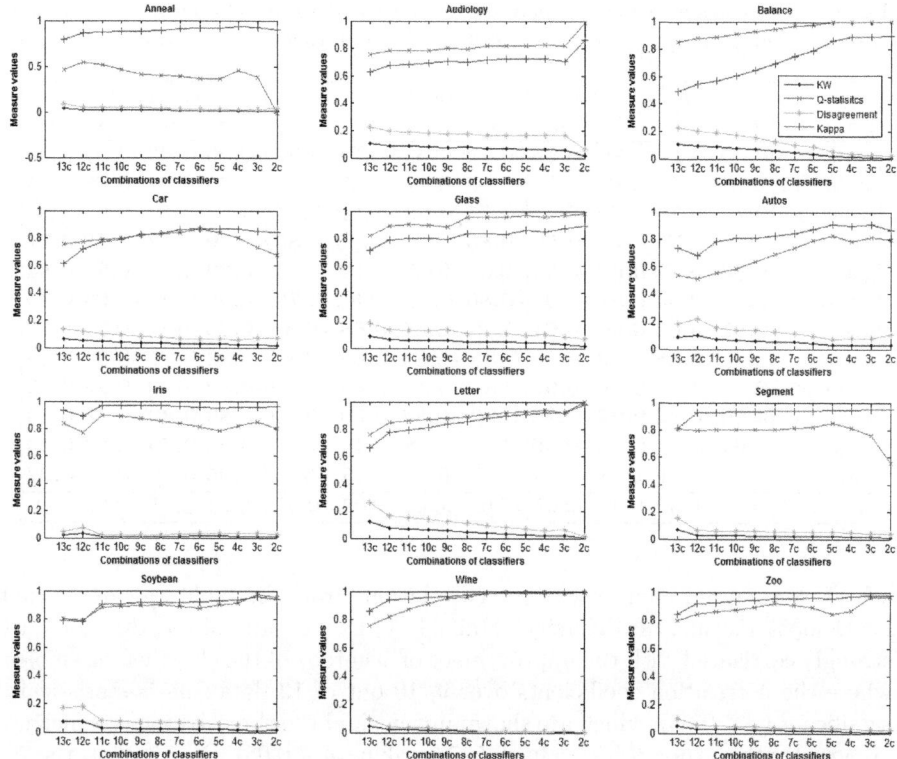

Fig. 1. Diversity and accuracy of the corresponding combinations of 13 classifiers over the 12 data sets in decreasing order

that the fitness between kw and dis is better than that between qs and κ, and the curve margins between qs and κ are larger than those between kw and dis. Roughly speaking, for the former group the curves decrease as more classifiers are added, whereas for the latter group, the curves increase with the addition of more classifiers, i.e. both of them go towards the closer from $2c$ to $13c$. These results suggest that the number of classifiers has an impact on the agreement among the classifiers – the more classifiers are added into the groups of classifiers, the more diversity appears among the groups.

We further calculated the improvement of the accuracy $F_{IM}(i\Phi) = F(iC)$ - $F_M(i\Phi)$, and the correlation coefficients between pairs of $\langle div(2\Phi), F_{IM}(2C) \rangle$, $\langle div(3\Phi), F_{IM}(3C) \rangle$, ..., $\langle div(13\Phi), F_{IM}(13C) \rangle$ $(1 \leq i \leq 13)$ for the three combination rules and the four diversity measures over the 12 data sets. The resulting correlation coefficients are presented in Table 1. In the table, each cell value is a correlation coefficient between the diversity obtained by one measure and the accuracy improvement on one data set. Based on the properties of the diversity measures, when the qs and κ coefficients are positive and kw and dis are negative, they represent a positive correlation between the diversity and accuracy,

Table 1. Correlation coefficients between diversity and improved accuracy of combined classifiers using Dempster's rule, Proportion rule and Yager's rule in decreasing order

Dataset	Dempster's rule				Proportion rule				Yager's rule			
	kw	qs	dis	κ	kw	qs	dis	κ	kw	qs	dis	κ
anneal	**0.95**	0.42	**0.96**	**-0.96**	**-0.79**	-0.33	**-0.81**	**0.81**	**-0.96**	-0.42	**-0.96**	**0.96**
audiology	0.18	-0.19	0.12	-0.11	**-0.63**	0.45	-0.56	0.54	**-0.73**	0.56	**-0.66**	**0.64**
balance	**0.79**	**-0.70**	**0.79**	**-0.77**	**-0.71**	**0.63**	**-0.72**	**0.71**	0.41	-0.51	0.42	-0.45
car	-0.25	0.17	-0.32	0.44	**-0.83**	0.24	**-0.86**	**0.85**	**-0.89**	-0.17	**-0.78**	**0.71**
glass	**0.63**	**-0.77**	**0.67**	**-0.68**	-0.01	0.05	-0.02	-0.01	**-0.90**	**0.81**	**-0.85**	**0.84**
autos	**0.99**	**-0.95**	**0.97**	**-0.98**	**-0.80**	**0.88**	**-0.80**	**0.79**	**-0.96**	**0.97**	**-0.96**	**0.96**
iris	**0.91**	**-0.63**	**0.90**	**-0.91**	**0.70**	**-0.61**	**0.68**	**-0.68**	**-0.65**	-0.12	-0.55	0.54
letter	**0.89**	**-0.88**	**0.89**	**-0.88**	**0.85**	**-0.84**	**0.85**	**-0.84**	**-0.83**	**0.77**	**-0.81**	**0.83**
segment	**0.99**	0.31	**0.99**	**-0.99**	**0.81**	0.21	**0.83**	**-0.83**	**-0.89**	-0.40	**-0.83**	**0.83**
soybean	**0.99**	**-0.93**	**0.99**	**-0.99**	**0.90**	**-0.84**	**0.90**	**-0.89**	**-0.89**	**0.80**	**-0.88**	**0.89**
wine	**0.64**	-0.43	**0.66**	**-0.66**	0.44	-0.38	0.48	-0.48	**-0.94**	**0.79**	**-0.94**	**0.94**
zoo	0.39	-0.11	0.55	-0.55	0.11	0.09	0.25	-0.25	**-0.95**	**0.76**	**-0.97**	**0.97**
Av	0.64	-0.34	0.60	-0.67	-0.02	-0.04	0.02	0.03	-0.72	0.27	-0.65	0.68

otherwise they represent a negative correlation. From the table we can see that for Dempster's rule, the diversity obtained by the measures of kw, dis and κ are strongly correlated with the improvement of accuracy of the classifier ensembles, where the correlation coefficients over 9, 10 out of 12 data sets is statistically significant ($p \leq 0.05$), which are shown in *bold*, and the three average coefficients are greater than the critical value 0.577 (making $p \leq 0.05$). By contrast, the relationship between the diversity measured by qs and the improved accuracy is rather weak. From the results of Proportion rule, we were surprised to find that the correlation between the diversity and the improved accuracy is negatively strong where the kw and κ coefficients over 8, 9 out of 12 data sets is statistically significant ($p \leq 0.05$), and even stronger negative correlation occurred in the results of yager's rule, where the correlation coefficients over 10, 11 out of 12 data sets are statistical significant. Interestingly, for Proportion rule and Yager's rule, the diversity measured by qs is weakly correlated to the improvement of accuracy achieved by the three combination rules, which is similar to the case of Dempster's rule. From this experiment we develop an understanding of that Dempster's rule of combination is better to avail of the diversity among the classifiers to improve the performance of the individual classifiers than Proportion and Yager's rules, and the diversity measured by qs is less sensitive to the improved accuracy of the classifiers.

5.3 Results in Mixed Order

In this experiment we took the same way as detailed in Subsections 5.1 and 5.2, but classifiers $\varphi_1, \ldots, \varphi_i$ ($1 \leq i \leq 13$) within each group $i\Phi$ combined by the three combination rules are in a mixed order. The diversity curves of 12 groups of classifiers can be similarly characterized into two groups and the graph is omitted. It has been observed that when more classifiers are added into

Table 2. Correlation coefficients between diversity and improved accuracy of combined classifiers using Dempster's rule, Proportion rule and Yager's rule in mixed order

Dataset	Dempster's rule				Proportion rule				Yager's rule			
	kw	qs	dis	κ	kw	qs	dis	κ	kw	qs	dis	κ
anneal	**0.89**	**-0.59**	**0.87**	**-0.87**	**0.65**	**-0.64**	**0.63**	**-0.63**	**-0.92**	0.50	**-0.90**	**0.90**
audiology	**0.96**	**-0.94**	**0.94**	**-0.92**	**0.97**	**-0.95**	**0.98**	**-0.98**	0.17	-0.13	0.31	-0.40
balance	**0.64**	**-0.74**	0.48	-0.49	-0.19	0.12	-0.27	0.28	**-0.59**	**0.60**	-0.50	0.51
car	**0.98**	**-0.60**	**0.90**	**-0.91**	-0.17	0.01	-0.21	0.20	**-0.78**	0.23	**-0.63**	**0.73**
glass	**0.61**	-0.30	0.50	-0.53	0.44	-0.31	0.38	-0.36	**-0.59**	0.25	-0.45	0.46
autos	**0.82**	**-0.77**	**0.87**	**-0.83**	**0.60**	-0.44	**0.66**	**-0.60**	**-0.76**	-0.15	**-0.66**	**0.63**
iris	**0.95**	**-0.89**	**0.97**	**-0.97**	**0.82**	**-0.92**	**0.86**	**-0.86**	**-0.97**	**0.80**	**-0.97**	**0.97**
letter	**0.96**	**-0.91**	**0.97**	**-0.96**	**0.76**	**-0.73**	**0.71**	**-0.67**	-0.20	0.21	-0.01	0.02
cleveland	-0.25	0.28	-0.15	-0.06	-0.32	0.28	-0.25	0.13	-0.19	0.13	-0.15	0.30
segment	**1.00**	-0.29	**1.00**	**-1.00**	**0.94**	-0.14	**0.96**	**-0.96**	**-0.99**	0.30	**-0.98**	**0.98**
soybean	**0.97**	**-0.82**	**0.97**	**-0.97**	**0.84**	**-0.70**	**0.85**	**-0.85**	-0.56	0.42	-0.54	0.54
wine	**0.97**	**-0.82**	**0.98**	**-0.98**	**0.68**	-0.37	**0.67**	**-0.67**	**-0.96**	**0.59**	**-0.95**	**0.95**
zoo	**0.99**	**-0.76**	**0.99**	**-0.99**	**0.59**	-0.47	**0.60**	**-0.59**	**-0.88**	0.46	**-0.82**	**0.82**
Av	0.81	-0.63	0.79	-0.81	0.51	-0.40	0.50	-0.51	-0.63	0.32	-0.56	0.57

the groups of classifiers, the first group of the curves roughly decreases, while the second group increases but both of them have some fluctuations, there is no consistent trend that can be visually identified from them. Compared with the diversity curves in two orders, although the curves in both of the orders share a similar shape, there are more fluctuations in mixed order than in decreasing order, which could be caused by the order of classifiers.

Following the same way as in decreasing order, we calculated the correlation coefficients between the diversity and the improved accuracy of the different groups of classifiers, and the resulting coefficients are presented in Table 2. The table clearly shows the strong correlation between the diversity obtained by kw, qs, dis and κ and the improved accuracy of the different groups of classifiers by Dempster's rule of combination, where the correlation coefficients over 10, 12 of the 12 data sets are statistically significant. However the diversity measured by the four measures is negatively correlated to the improved accuracy of the classifiers combined by Proportion rule and Yager's rule, and the correlation of Proportion rule is stronger than that of Yager's rule and qs is less sensitive to the improved accuracy.

6   Summary and Future Work

This study reports a range of experiments to assess the relationship between diversity and accuracy over the 12 benchmark data sets with the 12 groups of classifiers. The experimental results reveal that

- Dempster's rule can better utilize the diversity among classifiers to improve the performance of individual classifiers than Proportion and Yager's rules;

- the diversity measured by qs is less sensitive to the improved accuracy of the classifiers;
- the order of classifiers has an impact not only on the diversity among the classifiers, but also on the ensemble accuracy.

However the impact of the orders is varied on the combination rules and data sets. For Dempster's rule, the average accuracy of the classifier ensembles built in decreasing order is 77.89%, which is 1.10% better than that of mixed order; for Yager's rule, the average ensemble accuracy in decreasing order is 65.58%, 2.68% better than that of mixed order; for Proportion rule, the average ensemble accuracy in decreasing order is 74.81%, 0.05% better than that of mixed order. Nevertheless we conclude that the decreasing order could be a better way for combining classifiers. Meanwhile the negative correlation between the improvement of ensemble accuracy made by Proportion and Yager's rules and the diversity obtained by the four measures raises the issue of developing a sound theoretical understanding of effectiveness of the alternative combination rules in this context, in particular how to explain such negative correlations between the diversity and accuracy warrants a further investigation.

References

1. Bi, Y., Guan, J., Bell, D.: The combination of multiple classifiers using an evidential approach. Artificial Intelligence 17, 1731–1751 (2008)
2. Bi, Y., Wu, S.: Measuring Impact of Diversity of Classifiers on the Accuracy of Evidential Ensemble Classifiers. In: Hüllermeier, E., Kruse, R., Hoffmann, F. (eds.) IPMU 2010. Communications in Computer and Information Science, vol. 80, pp. 238–247. Springer, Heidelberg (2010)
3. Fleiss, J.L., Cuzick, J.: The reliability of dichotomous judgments: unequal numbers of judgments per subject. Applied Psychological Measurement 3, 537–542 (1979)
4. Kohavi, R., Wolpert, D.: Bias plus variance decomposition for zero-one loss functions. In: Proc 13th International Conference of Machine Learning, pp. 275–283 (1996)
5. Kuncheva, L., Whitaker, C.J.: Measures of diversity in classifier ensembles and their relationship with the ensemble accuracy. Machine Learning 51, 181–207 (2003)
6. Shafer, G.: A Mathematical Theory of Evidence, 1st edn. Princeton University Press, Princeton (1976)
7. Skalak, D.: he sources of increased accuracy for two proposed boosting algorithms. In: Proc. American Association for Artificial Intelligence, AAAI-1996, Integrating Multiple Learned Models Workshop (1996)
8. E., Tang, E.K., Suganthan, P.N., Yao, X.: An analysis of diversity measures. Machine Learning 65(1), 247–271 (2006)
9. Witten, I.H., Frank, E.: Data Mining: Practical machine learning tools and techniques, 2nd edn. Morgan Kaufmann, San Francisco (2005)
10. Yager, R.R.: On the dempster-shafer framework and new combination rules. Information Science 41, 93–137 (1987)
11. Anand, S.S., Bell, D., Hughes, J.G.: EDM: A General Framework for Data Mining Based on Evidence Theory. Data Knowl. Eng 18(3), 189–223 (1996)

Classification by Cluster Analysis:
A New Meta-Learning Based Approach

Anna Jurek, Yaxin Bi, Shengli Wu, and Chris Nugent

School of Computing and Mathematics University of Ulster,
Jordanstown, Shore Road, Newtownabbey, Co. Antrim, UK, BT37 0QB
jurek-a@email.ulster.ac.uk,
{y.bi,s.wu1,cd.nugent}@ulster.ac.uk

Abstract. Combination of multiple classifiers, commonly referred to as an classifier ensemble, has previously demonstrated the ability to improve classification accuracy in many application domains. One popular approach to building such a combination of classifiers is known as stacking and is based on a meta-learning approach. In this work we investigate a modified version of stacking based on cluster analysis. Instances from a validation set are firstly classified by all base classifiers. The classified results are then grouped into a number of clusters. Two instances are considered as being similar if they are correctly/incorrectly classified to the same class by the same group of classifiers. When classifying a new instance, the approach attempts to find the cluster to which it is closest. The method outperformed individual classifiers, classification by a clustering method and the majority voting method.

Keywords: Combining Classifiers, Stacking, Ensembles, Clustering, Meta-Learning.

1   Introduction

A classifier ensemble is a group of classifiers whose individual decisions are combined to provide, as an output, a consensus opinion during the process of classification. The key to producing a successful ensemble can be viewed as an approach which applies both, an appropriate combination scheme along with the careful selection of base classifiers that are going to be combined. Two key issues must therefore be considered: firstly, which and how a group of base classifiers should be selected (learning methods), and secondly, how to combine individual results into one final decision (combination methods) [9]. Very often diversity [6] and individual classifier accuracy are used as criterion for the selection of a good collection of base classifiers. Many approaches have been investigated to this problem. They are based either on using different learning methods and the same training set [1], or the same learning method and different training samples [8]. Different methods use different types of base classifier outputs during the combination phase, for instance class label or class probability distribution [1, 2]. Alternatively, another approach is to use predictions as a set of attributes to train a combining function in terms of meta-learning [12].

C. Sansone, J. Kittler, and F. Roli (Eds.): MCS 2011, LNCS 6713, pp. 259–268, 2011.
© Springer-Verlag Berlin Heidelberg 2011

In our work, we investigate a new technique of combining multiple classifiers, which can be reviewed as meta-learning based approach. We apply a framework of clusters built on the outputs of the base models. Base classifiers are used for generating clusters, however, are not reused in the testing phase. In such a way, the computational complexity can be reduced. Based on this concept, classification of unseen instances is reduced to a simple clustering task.

The remainder of the paper is organized as follows: Section 2 presents a brief review of previously investigated meta-learning techniques. A description of the new approach is introduced in Section 3. Section 4 presents the experimental results, which are discussed in Section 5. Conclusions and future work are presented in Section 6.

2   Related Work

Stacking, an approach based on the Meta-Learning technique, usually refers to the combination of models that are built using different learning methods, all with the same single data set [3]. Firstly, a collection of base-level classifiers is generated (level-0). Secondly, all instances from the validation set are classified by all base classifiers. The results of this process compose a training data set for a meta model. In the next step, a meta-learner model is built for combining the decisions of the base-level models (level-1) [3]. In the testing phase, the meta-level classifier uses the base classifiers' predictions as the input attributes to the model.

In a study by Ting and Witten [12] four different learning algorithms were applied in level-1: C4.5, IB1, NB and multi response linear regression (MLR). Results demonstrated that MLR was the only good candidate. For classification problems with n classes, n regression problems were defined. For each class a linear equation was formulated. Given a new test pattern, the equations were calculated for all the classes and the class with the greatest value was nominated as the final answer. In the same work, it was shown that using class probability distributions was more successful than applying class labels.

In [11], the concept of Meta Decision Trees (MDT) was proposed. The difference in comparison with an ordinary decision tree is that instead of predicting a class label directly, the MDT approach predicts the model which should be used to make the final decision. C4.5 was employed as the learning algorithm at the meta-level (MLC4.5). In the study, probability distribution characteristics were used as attributes. The method outperformed the stacking algorithm using an ordinary classification tree built with C4.5. Re-implementation of the MLC4.5 referred to as MLJ4.8 was proposed in [14]. It outperformed other techniques such as bagging, boosting or stacking with ordinary classification trees, however, it did not provide better results in comparison to stacking with MLR.

A modified version of Stacking with MLR, referred to as StackingC, was proposed in [10]. In StackingC, as opposed to Stacking, for each linear model assigned to a specific class, only the partial probability distribution related to that class is applied during training and testing phases. Results showed that for multi-class problems, StackingC outperformed Stacking. In 2-class cases, the performance of the two approaches were similar with accuracies of 85.51% and 85.54%, respectively.

Two further extensions to Stacking with MLR were introduced in [3]. In the first extension beside probability distribution the entropies and the probability distributions multiplied by the maximum probability were applied as a meta-level attributes. In the second extension model tree induction was applied instead of linear regression with the remainder of the process remaining the same. Therefore, instead of n linear equations, n model trees were inducted. The second approach was found to be successful and outperformed many other methods including Majority Voting, Select Best, Stacking with MDT, Stacking with MLR and StackingC.

In stacking, if the number of classes increases, the dimensionality of the meta-level attributes increases proportionally. This has the effect of increasing the complexity of the problem. StackingC is more effective in multi-class problems since only the probability related to specific classes are considered in the learning phase. This approach may, however, fail in the situations when the class distribution is non-symmetric [4]. An approach called Troika, was proposed in [7] to address multi-class problems. The architecture of this method contains 3 layers: Level-0 has a structure where all base classifiers provide class probability distributions. In Level-1, base classifiers are combined to obtain a group of specialists, where each one can distinguish between different pairs of classes. Predictions obtained at this level are used by the meta-classifiers in Level-2, which are trained with a one-against-all binarization method. This has the result that each classifier is a specialist being responsible for one class. Level-3 is the last layer, and contains just one model: the super classifier, which outputs a vector of probabilities as the final decision of the ensemble. Troika aimed to avoid the dimensionality problem by applying more than one combining classifier in Level-1. To avoid the effects of a skewed class distribution, a one-against-one instead of one-against-all binarization training method was employed. This method outperformed Stacking and StackingC in terms of classification accuracy.

In our work we investigate a new meta-learning based approach that is less complex than all existing meta models. In the proposed algorithm we take only class labels as the output of the base classifiers. Consequently the efficiency of the model is not affected by the number of classes and non-symmetric class distribution. In addition, we do not require the application of base classifiers in the testing phase. They are applied only to the training process of the meta model, that decreases the computational cost of the testing process.

3   Classification by Cluster Analysis (CBCA)

The general concept of the proposed approach is to use cluster analysis in the classification process. It is applied with different learning methods and the same training dataset to obtain a set of base classifiers. The main aim is to generate a collection of clusters, each of which contains similar instances according to their classification results. This means that one cluster should contain objects that were correctly/incorrectly classified to the same class by the same group of base classifiers. Additionally, to improve the quality of the clusters, the most significant features were considered during the clustering process. In the next step centroids of the cluster and their class labels were calculated. Upon presentation of an unseen instance, the centroid to which it is closest is identified. The class label of the selected centroid informs us to which class the instance belongs.

We notice that, similar to stacking, the base classifiers' outputs are used to generate a meta-model, which in our case is composed of a collection of clusters. The difference between our method and stacking is, that during the testing phase, our method does not require the further use of the base classifiers. New patterns are classified based only on their features. The following 2 sub-sections describe the general process of clustering and classification in more detail. All technical details of the implementation of the proposed algorithm are provided in Section 3.3.

3.1 Clustering Process

The clustering approach is applied with N different learning methods denoted by $L_1,...,L_N$. In the first step, the data set D is divided into 3 disjoint subsets: $D = D_1 \cup D_2 \cup D_3$. D_1 is used for the training of the base classifiers, D_2 is used for the purpose of building the clusters, which will be used as a meta classifier and D_3 is the testing set. Fig. 1 presents the entire process of generating a collection of clusters.

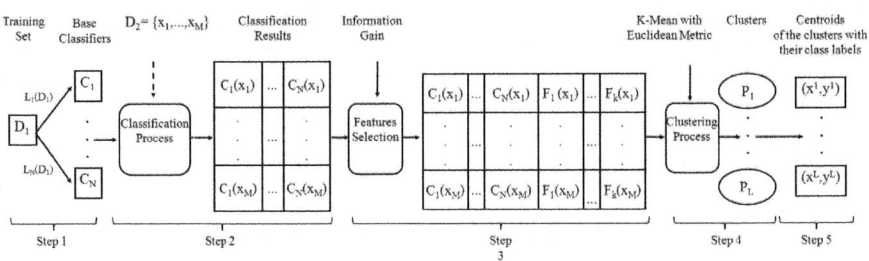

Fig. 1. Outline of the clusters generation process

In the first step, we apply N different learning methods and D_1 as a training set, to obtain N different base classifiers: $C_1,...,C_N$. All models are meant to provide a class label as an output. In the second step, all instances in D_2, denoted by $x_1,..., x_M$, are classified by all base classifiers. As a result of this process, we obtain an $M \times N$ matrix containing classification results. Rows represent instances from D_2, and columns represent base classifiers. For example, cell $C_j(x_i)$ of the matrix, contains the label of the class, where x_i was classified by C_j. Next, we intended to cluster all rows according to the values in the columns to obtain a collection of clusters denoted by $P_1,...,P_L$. Instances in each cluster were expected to belong to the same class. Nevertheless, given that the matrix does not contain information about real class labels of the instances, it may occur that in one cluster there will be instances from many different classes. This situation may happen, for example, if one instance will be misclassified by all the models to the same class given by c_k. Consequently, it will be located in the cluster with the instance that really belongs to the class c_k and were correctly classified by all base classifiers since the 2 equivalent rows in the matrix will be identical. In order to alleviate this problem, we decided to apply a subset of the most significant features in the clustering process. This helps to ensure that instances in one cluster generally belong to the same class. In step 3 K most significant features were selected. K new columns, labeled as $F_1,...,F_k$, were subsequently added to the matrix. Each of them represents one attribute. In the 4 step, the rows are clustered according to values in $N+K$ columns. As a result we obtain a collection of clusters. In the last step centroids of all the clusters and their class labels are calculated. The whole classification process is described in the following section.

3.2 Classification Process

In our method, classification of a new instance differs from the stacking approach. In stacking, all base classifiers are re-used to provide meta data for each pattern being considered. Based on this data, the meta classifier makes a final decision. In our method, the final classifier only uses the features of an instance to make the decision. The entire process is presented in Fig. 2.

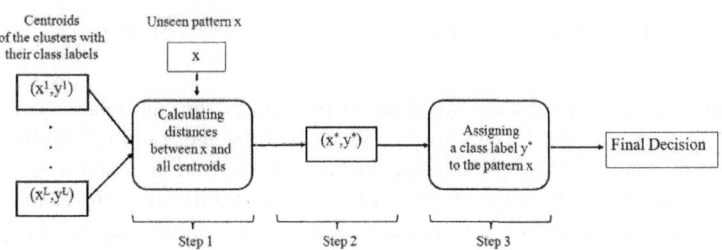

Fig. 2. Outline of new instance classification process

As a result of the training process we obtain a collection of L centroids. In step 1, the distance between the unseen instance, represented by the K most significant features, and the centroids are calculated using the Euclidean distance. In step 2, the centroid that is most similar to the unseen instance, is identified. In the final step, the class where the selected centroid belongs is considered as the final decision. Section 3.3 presents all technical details of the described algorithm.

3.3 Technical Details of CBCA

Figure 3 presents the pseudo code of the proposed CBCA approach.

Training process	
Input:	$D_1, D_2 =(x_1,..,x_M)$; $L_1,...,L_N$ - learning methods; K–no. of features to be selected
Process:	//Build Base Classifiers
	1) for i=1...N
	2) $C_i=L_i(D_1)$;
	3) end
	//Select K most significant features
	4) Information Gain $(D_1,D_2)\rightarrow F_1,...,F_K$;
	//Build matrix A
	5) For i=1$\rightarrow$M
	6) For j=1$\rightarrow$N $A(i,j)=C_j(x_i)$;
	7) For k=1$\rightarrow$ K $A(i,N+k)=F_k(x_i)$;
	//Determine the number of clusters
	8) Z=number of classes;
	9) Do
	10) { Z++;
	11) Calculate Accuracy (Z,D_1); }
	12) While (Accuracy(Z,D_1)> Accuracy $(Z-1,D_1)$)
	13) Z--;
	//Build final collection of clusters
	14) Kmeans$([A(1,1),...,A(1,N+K)],...,[A(M,1)...A(M,N+K)], Z)\rightarrow P_1,...,P_Z$;
	15) For i=1$\rightarrow$Z
	16) { Identify centroid of $P_i \rightarrow x_i^*$;
	17) Read class label of instance $x_i^* \rightarrow y_i^*$; }
Outputs:	$(x_1^*, y_1^*),...,(x_Z^*, y_Z^*)$-centroids of the cluster with their class labels
Testing process	
Input:	x - testing pattern
Process:	18) Select (x',y') such that d(x,x',K)=min$\{$d$(x, x_1^*,K),...,$d$(x, x_Z^*,K)\}$, where d is the Euclidean distance, and assign y' to the x

Fig. 3. Outline of CBCA method for training and testing process

As an input for the method we have to provide a training data sets, that is later divided in 2 parts: D_1 for training base classifiers and D_2 for building clusters. Followed by that, there are N learning methods given that are used to generate base classifiers. The third input is a number of the most significant features that are taken into account during the clustering process (K). The initial step in the process is building N classifiers using N different learning methods and a training set D_1. For the selection of learning methods and value of K we tried a number of different combinations and chose the one that gave the best results on the training data. We applied rather low values of N, 2 or 3 in most cases, to not increase drastically dimension of the data. Following this, the Information Gain[1] is calculated for each attribute. This method discretizes all numeric attributes. All attributes are then ranked[2] from the highest to the lowest. We applied for this both training and validation data sets. The first K attributes in the ranking are selected. In the next step a matrix A with size $Nx(M+K)$ is constructed. All rows of the matrix are clustered by the K-Means algorithm[3], that automatically handles numerical and categorical attributes. The

---

[1] weka.InfoGainAttributeEval.buildEvaluator
[2] weka.attributeSelection.Ranker
[3] weka.clusterers.SimpleKMeans

method uses the Euclidean distance measure[4]. For categorical data the distance is either 0 or 1. Numerical data are normalized into 0-1 range to make them equal influence in the distance calculation. To determine the number of clusters required we implemented a straightforward searching algorithm. Initially the number of clusters Z equals the number of classes. In each iteration we increase Z by one and calculate the accuracy of our method on the training data set. The procedure is repeated until the performance improves when we increase the number of clusters. Following determining the parameter Z, all instances from D_2 are divided into Z clusters. For each cluster we calculate the centre and its class label[5]. Class labels of the centre is calculated as a mode - value that appears most often within the cluster. As an output of the training process we obtain a collection of centroids with their class labels. Following presentation of an unseen pattern the distance between it and the all centers are calculated using the Euclidean distance. Only the K number of the most significant features are considered in this process. The class label of the closest centre is considered as the final decision.

4 Experiments

To evaluate the method, experiments were conducted with 15 data sets taken from the UCI Machine Learning Repository [13]. A brief description is presented in Table 1.

Table 1. Data Sets used in Experiments taken from the UCI Machine Learning Repository

Data Set	No. of Instances	No. of classes	No. of attributes	Data Set	No. of Instances	No. of classes	No. of attributes
Breast Cancer	286	2	9	Spect	187	2	22
Colic	368	2	23	Tytanic	2201	2	6
Hepatitis	155	2	20	Credit-g	1000	2	4
Heart-c	303	5	14	Iris	150	3	4
Breast-w	699	2	10	Omin	479	2	65
Haberman	306	2	4	Waveform	600	2	15
Heart-h	294	5	13	Heart-statlog	270	2	14
Glass	214	7	10				

We have used 6 base classifiers implemented in Weka [5]: Naive Bayes[6] (NB), NB Tree[7] (DT), Simple Logistic Regression[8] (SL), J48 Tree[9] (J), Multilayer Perceptron[10](MP), and BF Tree[11] (BF). For each data set, the best combination of classifiers was selected, based on the performance on the training data.

---

[4] weka.core.EuclideanDistance
[5] weka.clusterers.SimpleKMeans.GetClustersCentoids
[6] weka.classifiers.bayes.NaiveBayes
[7] weka.classifiers.trees.NBTree
[8] weka.classifiers.functions.SimpleLogistic -I 0 -M 500 -H 50 -W 0.0
[9] weka.classifiers.trees.J48 -C 0.25 -M 2
[10] weka.classifiers.functions.MultilayerPerceptron -L 0.3 -M 0.2 -N 500 -V 0 -S 0 -E 20 -H a
[11] weka.classifiers.trees.BFTree -S 1 -M 2 -N 5 -C 1.0 -P POSTPRUNED

4.1 Implementation Details

To evaluate the proposed approach, we have implemented an experimental system that is composed of 6 learning methods in addition to the proposed meta learning approach and a majority voting (MV) method. MV is tested twice, one with all 6 base classifiers and the other with the same combination of base classifiers that was selected and applied with the CBCA approach. All base classifiers were implemented in Weka. Since some parameters of the system were established by validation on the training set, to reduce over fitting problem, a 5-fold cross-validation was performed for all experiments. The original data set was divided into 5 folds. One fold was used for testing and the remaining 4 folds were divided into 2 parts. The first part was used for training and the second part was used for building the clusters. To investigate the relationship between the size of the validation/training set and the performance of the CBCA approach we performed three types of division with relations 1:1, 1:3 and 3:1. The process is repeated for each fold and the final accuracy is calculated as an average over the 5 folds.

4.2 Results

The experiments were carried out with 15 data sets. The proposed method was tested with each of the data sets and then compared with individual classifiers and the MV approach. For each data set, the CBCA approach was considered with different combinations of the base models. Combinations which provided the best final results on the training data were identified and applied to build a final classifier (to generate the collection of clusters). For the different data sets, the different numbers of the most significant attributes were applied in the clustering process. Additionally, we implemented and compared with our approach the standard "classification by clustering" method (CL). Clusters in this method are built by using only the attributes from the instances. In this way we want to consider if the first N columns in the matrix (attributes presenting classification results) are significant in the whole approach. Table 2 presents

Table 2. Accuracy for the different methods considered following evaluation for 15 data sets

Data Set	NB	DT	SL	J	MP	BF	MV	MV(*)	CL	CBCA(*)
BC	72.6	70.2	73.3	73.7	69.5	68.6	72.3	72.3	72.1	**76.5**
Colic	79.7	84.9	82.2	85.8	79.7	86.3	86.6	81.4	77.0	**86.8**
Hepatitis	84.0	77.3	80.7	77.3	83.3	78.0	82.0	82.0	78.7	**86.0**
Heart-c	82.7	81.3	81.0	74.0	80.0	75.7	81.3	81.3	82.0	**84.6**
Breast-w	95.8	96.1	96.0	94.1	94.0	94.7	96.1	93.8	97.0	**97.4**
Haberman	75.1	72.1	71.5	70.8	69.2	71.1	73.4	72.5	73.1	**76.5**
Heart-h	**84.5**	81.7	**84.5**	78.3	81.0	79.3	81.7	80.0	80.3	**84.5**
Spect	66.5	71.9	66.5	61.1	59.5	68.1	68.1	61.1	73.5	**74.1**
Titanic	77.9	**79.1**	77.9	79.0	78.8	**79.1**	77.9		77.5	**79.1**
Credit-g	**75.5**	71.0	75.1	71.2	72.1	70.0	**75.5**	71.9	72.1	73.2
Iris	95.9	93.8	95.2	91.7	94.5	93.8	94.5	95.1	92.4	**97.3**
Omin	65.3	71.6	68.8	60.9	69.7	71.0	69.7	60.6	70.1	**72.2**
Waveform	88.7	89.4	91.1	84.5	88.7	85.7	90.4	90.4	84.5	**91.1**
Heart-s	84.1	81.1	83.0	76.2	78.5	76.6	81.9	76.6	82.3	**84.5**
Glass	50.0	64.3	65.7	68.6	66.7	68.6	68.6	**69.1**	64.8	**69.1**

the accuracy of all the methods considered for all data sets. The values in the table present the mean percentages of the instances which were correctly classified for all 5 folds. The last column in the table presents the results of the proposed CBCA approach. The shaded cells in Table 2 indicate which base classifiers were applied with our method in each data set. The MV column represents the accuracy of the MV method applied with all base classifiers. MV(*) stands for MV applied with the same combination of base classifiers as the CBCA approach. The CL column presents the results of the standard classification accuracy attained by the clustering method.

5   Discussion

Table 2 shows that the CBCA approach outperformed all base classifiers, MV and MV(*) in 11 out of the 15 data sets considered. In 3 datasets the CBCA approach attained the same best result as a number of the other algorithms. The CBCA was only outperformed by one of the other approaches in one of the datasets.

Based on the results we can see, that in 14 cases the performance of the base classifiers was improved. The only case where no improvement over base models was achieved was when considering the Credit-g dataset. For Heart-h CBCA gave the same result as NB and SL. Nevertheless, in this case only J and BF were considered as base classifiers, and both methods were still significantly improved. Similar situations arose for the Titanic dataset.

During the experiments we compared our method with MV applied with all 6 base classifiers (MV) and with only selected models (MV(*)). CBCA outperformed MV in 14 out of 15 cases. We performed a T-test with the confidence level $\alpha=0.1$, and the results showed that the difference was statistically significant for 8 out of 14 cases. Given that for each data set we only have 5 samples obtained from 5 folds, we decided to run another T-test on all data together hence providing us with $15*5=75$ samples. According to the results attained, CBCA is significantly more effective than MV with the confidence level of $\alpha=99\%$. MV(*) provided improved results in comparison with the MV in only two of the data sets. CBCA outperformed MV(*) in 13 out of 15 cases and in 10 the difference was statistically significant with the confidence level $\alpha=0.1$. We performed the T-test for all 75 samples and the result showed that CBCA is more effective than MV(*) with the confidence level $\alpha=99\%$.

The final comparison we performed was between CBCA and the standard CL. We can see from Table 2 that CL was outperformed by the CBCA in all data sets. This confirms that the information contained in the attributes representing classification results is significant in the entire process.

6   Conclusion and Future Work

In this paper we have investigated a new approach to combining multiple classifiers based on cluster analysis. A collection of base classifiers was applied to generate a group of clusters, which were subsequently used to compose the final classifier. Instances were grouped according to their classification results together with the number of the most significant attributes. The proposed method is a modified Meta-Learning approach since the training set of the final model is composed of the base classifier outputs. The approach differs from the stacking approach, since we do not reuse base classifiers in the testing phase. The final model makes decisions based only on the features of the instances. This significantly reduces the complexity of the classification process. After obtaining the framework of clusters, classification of unseen patterns turns into a relatively simple clustering task. The proposed method appears to be an effective combination tool. For 14 out of 15 data sets it improved performance of all base classifiers combined. CBCA appeared more effective than the MV method. It has been observed as well that cluster analysis based only on the features of the instances is less effective than the one proposed by us.

Given the results of the experiments are promising, we wish to continue our work on a number of related issues. The key limitation of the proposed system is that some of the parameters are selected based on trial-and-error. It cost the problem of low generality of the system. This is the first issue we want to consider in the future work. Beside this, we would like to investigate more carefully an influence of random initialization of the K-Means on the proposed system.

References

1. Bi, Y., McClean, S., Anderson, T.: On Combining Multiple Classifiers Using an Evidential Approach. Artificial Intelligence, 324–329 (2006)
2. Danesh, A., Moshiri, B., Fatemi, O.: Improve Text Classification Accuracy based on Classifier Fusion Methods, pp. 1–6 (2007)
3. Dzeroski, S., Zenko, B.: Is Combining Classifiers with Stacking Better than Selecting the Best One? Machine Learning, pp. 255–273 (2004)
4. Furnkranz, J.: Pairwise Classification as an Ensemble Technique. In: 13th European Conference on Machine Learning, UK, pp. 97–110. Springer, London (2002)
5. Hall, M., E. Frank, G. Holmes, B. Pfahringer, P. Reutemann, I. H. Witten. The WEKA Data Mining Software: An Update SIGKDD Explorations, Volume 11, Issue 1 (2009).
6. Kuncheva, L.I., Whitaker, C.J.: Measures of Diversity in Classifier Ensembles and Their Relationship with the Ensemble Accuracy. Machine Learning, 181–207 (2004)
7. Menahem, E., Rokach, L., Elovici, Y.: Troika - An improved stacking schema for classification tasks. Information Sciences, 4097–4122 (2009)
8. Rodriguez, J., Maudes, J.: Boosting recombined weak classifiers. Pattern Recognition Letters, 1049–1059 (2008)
9. Rokach, L.: Ensemble-based classifiers. Artificial Intelligence Review, 1–39 (2010)
10. Seewald, A.K.: How to Make Stacking Better and Faster While Also Taking Care of an Unknown Weakness. In: 19th International Conference on Machine Learning, pp. 554–561. Morgan Kaufmann Publishers, San Francisco (2002)
11. Todorovski, L., Dzeroski, S.: Combining Multiple Models with Meta Decision Trees. In: 4th European Conference on Principles of Data Mining and Knowledge Discovery, pp. 54–64. Springer, Berlin (2000)
12. Ting, K.M., Witten, I.F.: Issues in stacked generalization. Artificaial Intelligence Research, 271–289 (1999)
13. UC Irvine Machine Learning Repository,
    http://repository.seasr.org/Datasets/UCI/arff
14. Zenko, B., Todorovski, L., Dzeroski, S.: A comparison of stacking with MDTs to bagging, boosting, and other stacking methods. In: Int. Conference on Data Mining, pp. 669–670 (2001)

C³E: A Framework for Combining Ensembles of Classifiers and Clusterers

A. Acharya[1], E.R. Hruschka[1,2], J. Ghosh[1], and S. Acharyya[1]

[1] University of Texas (UT) at Austin, USA
[2] University of Sao Paulo (USP) at Sao Carlos, Brazil

Abstract. The combination of multiple classifiers to generate a single classifier has been shown to be very useful in practice. Similarly, several efforts have shown that cluster ensembles can improve the quality of results as compared to a single clustering solution. These observations suggest that ensembles containing both classifiers and clusterers are potentially useful as well. Specifically, clusterers provide supplementary constraints that can improve the generalization capability of the resulting classifier. This paper introduces a new algorithm named **C³E** that combines ensembles of classifiers and clusterers. Our experimental evaluation of **C³E** shows that it provides good classification accuracies in eleven tasks derived from three real-world applications. In addition, **C³E** produces better results than the recently introduced Bipartite Graph-based Consensus Maximization (**BGCM**) Algorithm, which combines multiple supervised and unsupervised models and is the algorithm most closely related to **C³E**.

Keywords: Ensembles, Classification, Clustering.

1 Introduction

The combination of multiple classifiers to generate a single classifier has been an active area of research for the last two decades [8,7]. For instance, an analytical framework to quantify the improvements in classification results due to combining multiple models has been addressed in [13]. More recently, a survey of traditional ensemble techniques — including applications of them to many difficult real-world problems such as remote sensing, person recognition, one vs. all recognition, and medicine — has been presented in [9]. In brief, the extensive literature on the subject has shown that from independent, diversified classifiers, the ensemble created is usually more accurate than its individual components. Analogously, several research efforts have shown that cluster ensembles can improve the quality of results as compared to a single clustering solution — *e.g.*, see [6]. Actually, the potential motivations and benefits for using cluster ensembles are much broader than those for using classifier ensembles, for which improving the predictive accuracy is usually the primary goal. More specifically, cluster ensembles can be used to generate more robust and stable clustering results (compared to a single clustering approach), perform distributed computing under privacy or sharing constraints, or reuse existing knowledge [12].

C. Sansone, J. Kittler, and F. Roli (Eds.): MCS 2011, LNCS 6713, pp. 269–278, 2011.

In this paper, an algorithm that combines ensembles of classifiers and clusterers is introduced. As far as we know, this topic has not been addressed in the literature. Most of the motivations for combining ensembles of classifiers and clusterers are similar to those that hold for the standalone use of either classifier ensembles or cluster ensembles. However, some additional nice properties can emerge from such a combination — *e.g.*, unsupervised models can provide a variety of supplementary constraints for classifying new data [11]. From this viewpoint, the underlying assumption is that similar new objects in the target set are more likely to share the same class label. Thus, the supplementary constraints provided by the cluster ensemble can be useful for improving the generalization capability of the resulting classifier, specially when labelled data is scarce. Also, they can be useful for designing learning methods that are aware of the possible differences between training and target distributions, thus being particularly interesting for applications in which concept drift may take place.

The remainder of this paper is organized as follows. The proposed algorithm — named $\mathbf{C^3E}$, from Consensus between Classification and Clustering Ensembles — is described in the next section. Related work is addressed in Section 3. An experimental study is reported in Section 4. Finally, Section 5 concludes the paper and describes ongoing work.

Notation. Vectors and matrices are denoted by bold faced lowercase and capital letters, respectively. Scalar variables are written in italic font. A set is denoted by a calligraphic uppercase letter. The effective domain of a function $f(y)$, i.e., the set of all y such that $f(y) < +\infty$ is denoted by $dom(f)$, while the interior and the relative interior of a set $\mathcal{Y}$ are denoted by $int(\mathcal{Y})$ and $ri(\mathcal{Y})$, respectively. Also, for $\mathbf{y}_i, \mathbf{y}_j \in \mathbb{R}^k$, $\langle \mathbf{y}_i, \mathbf{y}_j \rangle$ denotes their inner product.

2   Description of $\mathbf{C^3E}$

The proposed framework that combines classifier and cluster ensembles to generate a more consolidated classification is depicted in Fig. 1. It is assumed that an ensemble of classifiers has been previously induced from a training set. Such an ensemble is part of the framework that will be used for classifying new data — *i.e.*, objects from the target set[1] $\mathcal{X} = \{\mathbf{x}_i\}_{i=1}^n$. The ensemble of classifiers is employed to estimate initial class probabilities for every object $\mathbf{x}_i \in \mathcal{X}$. These probability distributions are stored as a set of vectors $\{\boldsymbol{\pi}_i\}_{i=1}^n$ and will be refined with the help of a cluster ensemble. From this point of view, the cluster ensemble provides supplementary constraints for classifying the objects of $\mathcal{X}$, with the rationale that similar objects are more likely to share the same class label. Each of $\boldsymbol{\pi}_i$'s is of dimension k so that, in total, there are k classes denoted by $C = \{C_\ell\}_{\ell=1}^k$. In order to capture the similarities between the objects of $\mathcal{X}$, $\mathbf{C^3E}$ also takes as input a similarity (co-association) matrix $\mathbf{S}$, where each entry corresponds to the relative co-occurrence of two objects in the same cluster [6,12] — considering all the data partitions that form the cluster ensemble induced

---

[1] The target set is a test set that has not been used to build the ensemble of classifiers.

from $\mathcal{X}$. To summarize, $\mathbf{C^3E}$ receives as inputs a set of vectors $\{\pi_i\}_{i=1}^n$ and the similarity matrix $\mathbf{S}$. After processing these inputs, $\mathbf{C^3E}$ outputs a consolidated classification — represented by a set of vectors $\{\mathbf{y}_i\}_{i=1}^n$, where $\mathbf{y}_i = \hat{P}(C \mid \mathbf{x}_i)$ — for every object in $\mathcal{X}$. This procedure is described in more detail below, where r_1 classifiers, indexed by q_1, and r_2 clusterers, indexed by q_2, are employed to obtain a consolidated classification.

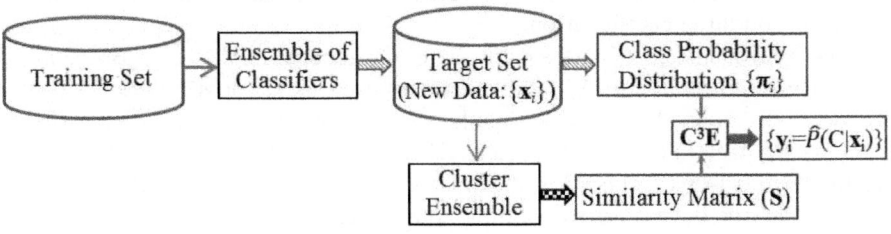

Fig. 1. Combining Ensembles of Classifiers and Clusterers

Step A: Obtain Input From Classifier Ensemble. The output of classifier q_1 for object $\mathbf{x}_i$ is a k-dimensional class probability vector $\pi_i^{(q_1)}$. From the set of such vectors $\{\pi_i^{(q_1)}\}_{q_1=1}^{r_1}$, an average vector can be computed for $\mathbf{x}_i$ as:

$$\pi_i = \frac{1}{r_1} \sum_{q_1=1}^{r_1} \pi_i^{(q_1)}. \tag{1}$$

Step B: Obtain Input From Cluster Ensemble. After applying r_2 clustering algorithms (clusterers) to $\mathcal{X}$, a similarity (co-association) matrix $\mathbf{S}$ is computed. Assuming each clustering is a hard data partition, the similarity between two objects is simply the fraction of the r_2 clustering solutions in which those two objects lie in the same cluster[2]. Note that such similarity matrices are byproducts of several cluster ensemble solutions, e.g., CSPA algorithm in [12].

Step C: Obtain Consolidated Results from $\mathbf{C^3E}$. Having defined the inputs for $\mathbf{C^3E}$, namely the set $\{\pi_i\}_{i=1}^n$ and the similarity matrix, $\mathbf{S}$, the problem of combining the ensembles of classifiers and clusterers can be posed as an optimization problem whose objective is to minimize J in (2) w.r.t. the set of probability vectors $\{\mathbf{y}_i\}_{i=1}^n$, where $\mathbf{y}_i = \hat{P}(C \mid \mathbf{x}_i)$, i.e., $\mathbf{y}_i$ is the new and hopefully improved estimate of the aposteriori class probability distribution for a given object in $\mathcal{X}$.

$$J = \sum_{i \in \mathcal{X}} \mathcal{L}(\pi_i, \mathbf{y}_i) + \alpha \sum_{(i,j) \in \mathcal{X}} s_{ij} \mathcal{L}(\mathbf{y}_i, \mathbf{y}_j) \tag{2}$$

---

[2] A similarity matrix can also be defined for soft clusterings — e.g., see [10].

The quantity $\mathcal{L}(\cdot, \cdot)$ denotes a loss function. Informally, the first term in Eq. (2) captures dissimilarities between the class probabilities provided by the ensemble of classifiers and the output vectors $\{\mathbf{y}_i\}_{i=1}^n$. The second term encodes the cumulative weighted dissimilarity between all possible pairs $(\mathbf{y}_i, \mathbf{y}_j)$. The weights to these pairs are assigned in proportion to the similarity values $s_{ij} \in [0, 1]$ of matrix $\mathbf{S}$. The coefficient $\alpha \in \mathbb{R}_+$ controls the relative importance of classifier and cluster ensembles. Therefore, minimizing the objective function over $\{\mathbf{y}_i\}_{i=1}^n$ involves combining the evidence provided by the ensembles in order to build a more consolidated classification.

The approach taken in this paper is quite general in that any Bregman divergence (defined in the Appendix) can be used as the loss function $\mathcal{L}(\cdot, \cdot)$ in Eq. (2). Bregman divergences include a large number of useful loss functions such as the well-known squared loss, KL-divergence, logistic loss, Mahalanobis distance, and I-divergence. A specific Bregman Divergence (e.g. KL-divergence) can be identified by a corresponding convex function ϕ (e.g. negative entropy for KL-divergence), and hence be written as $d_\phi(\mathbf{y}_i, \mathbf{y}_j)$. Using this notation, the optimization problem can be rewritten as:

$$\min_{\{\mathbf{y}_i\}_{i=1}^n} J = \min_{\{\mathbf{y}_i\}_{i=1}^n} \left[\sum_{i \in \mathcal{X}} d_\phi(\boldsymbol{\pi}_i, \mathbf{y}_i) + \alpha \sum_{(i,j) \in \mathcal{X}} s_{ij} d_\phi(\mathbf{y}_i, \mathbf{y}_j) \right]. \tag{3}$$

All Bregman divergences have the remarkable property that the single best (in terms of minimizing the net loss) representative of a set of vectors, is simply the expectation of this set (!) provided the divergence is computed with this representative as the second argument of $d_\phi(\cdot, \cdot)$ — see Th. 1 in the Appendix for a more formal statement of this result. Unfortunately this simple form of the optimal solution is not valid if the variable to be optimized occurs as the first argument. In that case, however, one can work in the (Legendre) dual space, where the optimal solution has a simple form (see [1] for details). Re-examining Eq. (3), we notice that the $\mathbf{y}_i$'s to be minimized over occur both as first and second arguments of a Bregman divergence. Hence optimization over $\{\mathbf{y}_i\}_{i=1}^n$ is not available in closed form. We circumvent this problem by creating two copies for each $\mathbf{y}_i$ — the left copy, $\mathbf{y}_i^{(l)}$, and the right copy, $\mathbf{y}_i^{(r)}$. The left(right) copies are used whenever the variables are encountered in the first(second) argument of the Bregman divergences. The right and left copies are updated iteratively, and an additional constraint is used to ensure that the two copies of a variable remain close during the updates. First, keeping $\{\mathbf{y}_i^{(l)}\}_{i=1}^n$ and $\{\mathbf{y}_i^{(r)}\}_{i=1}^n \setminus \{\mathbf{y}_j^{(r)}\}$ fixed, the part of the objective function that only depends on $\mathbf{y}_j^{(r)}$ can be written as:

$$J_{[\mathbf{y}_j^{(r)}]} = d_\phi(\boldsymbol{\pi}_j^{(r)}, \mathbf{y}_j^{(r)}) + \alpha \sum_{i^{(l)} \in \mathcal{X}} s_{i^{(l)} j^{(r)}} d_\phi(\mathbf{y}_i^{(l)}, \mathbf{y}_j^{(r)}). \tag{4}$$

Note that the optimization of $J_{[\mathbf{y}_j^{(r)}]}$ in (4) w.r.t. $\mathbf{y}_j^{(r)}$ is constrained by the fact that the left and right copies of $\mathbf{y}_j$ should be equal. Therefore, a soft constraint is added in (4), and the optimization problem now becomes:

$$\min_{\mathbf{y}_j^{(r)}} \left[d_\phi(\boldsymbol{\pi}_j^{(r)}, \mathbf{y}_j^{(r)}) + \alpha \sum_{i^{(l)} \in \mathcal{X}} s_{i^{(l)} j^{(r)}} d_\phi(\mathbf{y}_i^{(l)}, \mathbf{y}_j^{(r)}) + \lambda_j^{(r)} d_\phi(\mathbf{y}_j^{(l)}, \mathbf{y}_j^{(r)}) \right], \quad (5)$$

where $\lambda_j^{(r)}$ is the corresponding Lagrange multiplier. It can be shown (see the Appendix) that there is a unique minimizer for the optimization problem in (5):

$$\mathbf{y}_j^{(r)*} = \left[\left[\boldsymbol{\pi}_j^{(r)} + \gamma_j^{(r)} \sum_{i^{(l)} \in \mathcal{X}} \delta_{i^{(l)} j^{(r)}} \mathbf{y}_i^{(l)} + \lambda_j^{(r)} \mathbf{y}_j^{(l)} \right] / \left[1 + \gamma_j^{(r)} + \lambda_j^{(r)} \right] \right], \quad (6)$$

where $\gamma_j^{(r)} = \alpha \sum_{i^{(l)} \in \mathcal{X}} s_{i^{(l)} j^{(r)}}$ and $\delta_{i^{(l)} j^{(r)}} = s_{i^{(l)} j^{(r)}} / \left[\sum_{i^{(l)} \in \mathcal{X}} s_{i^{(l)} j^{(r)}} \right]$. The same optimization in (5) is repeated over all the $\mathbf{y}_j^{(r)}$'s. After the right copies are updated, the objective function is (sequentially) optimized w.r.t. all the $\mathbf{y}_i^{(l)}$'s. Like in the first step, $\{\mathbf{y}_j^{(l)}\}_{j=1}^n \setminus \{\mathbf{y}_i^{(l)}\}$ and $\{\mathbf{y}_j^{(r)}\}_{j=1}^n$ are kept fixed, and the equality of the left and right copies of $\mathbf{y}_i$ is added as a soft constraint, so that the optimization w.r.t. $\mathbf{y}_i^{(l)}$ can be rewritten as:

$$\min_{\mathbf{y}_i^{(l)}} \left[\alpha \sum_{j^{(r)} \in \mathcal{X}} s_{i^{(l)} j^{(r)}} d_\phi(\mathbf{y}_i^{(l)}, \mathbf{y}_j^{(r)}) + \lambda_i^{(l)} d_\phi(\mathbf{y}_i^{(l)}, \mathbf{y}_i^{(r)}) \right], \quad (7)$$

where $\lambda_i^{(l)}$ is the corresponding Lagrange multiplier. As mentioned earlier, one needs to work in the dual space now, using the convex function ψ (Legendre dual of ϕ) which is defined as:

$$\psi(\mathbf{y}_i) = \langle \mathbf{y}_i, \nabla \phi^{-1}(\mathbf{y}_i) \rangle - \phi(\nabla \phi^{-1}(\mathbf{y}_i)). \quad (8)$$

One can show that $\forall \mathbf{y}_i, \mathbf{y}_j \in int(dom(\phi))$, $d_\phi(\mathbf{y}_i, \mathbf{y}_j) = d_\psi(\nabla \phi(\mathbf{y}_j), \nabla \phi(\mathbf{y}_i))$ (see [1] for more details). Thus, the optimization problem in (7) can be rewritten in terms of the Bregman divergence associated with ψ as follows:

$$\min_{\nabla \phi(\mathbf{y}_i^{(l)})} \left[\alpha \sum_{j^{(r)} \in \mathcal{X}} s_{i^{(l)} j^{(r)}} d_\psi(\nabla \phi(\mathbf{y}_j^{(r)}), \nabla \phi(\mathbf{y}_i^{(l)})) + \lambda_i^{(l)} d_\psi(\nabla \phi(\mathbf{y}_i^{(r)}), \nabla \phi(\mathbf{y}_i^{(l)})) \right]. \quad (9)$$

The unique minimizer of the problem in (9) can be computed using Corollary 1 (see the Appendix). $\nabla \phi$ is monotonic and invertible for ϕ being strictly convex and hence the inverse of the unique minimizer for problem (9) is unique and equals to the unique minimizer for problem (7). Therefore, the unique minimizer of problem (7) w.r.t. $\mathbf{y}_i^{(l)}$ is given by:

$$\mathbf{y}_i^{(l)*} = \nabla \phi^{-1} \left[\left[\gamma_i^{(l)} \sum_{j^{(r)} \in \mathcal{X}} \delta_{i^{(l)} j^{(r)}} \nabla \phi(\mathbf{y}_j^{(r)}) + \lambda_i^{(l)} \nabla \phi(\mathbf{y}_i^{(r)}) \right] / \left[\gamma_i^{(l)} + \lambda_i^{(l)} \right] \right], \quad (10)$$

where $\gamma_i^{(l)} = \alpha \sum_{j^{(r)} \in \mathcal{X}} s_{i^{(l)} j^{(r)}}$ and $\delta_{i^{(l)} j^{(r)}} = s_{i^{(l)} j^{(r)}} / \left[\sum_{j^{(r)} \in \mathcal{X}} s_{i^{(l)} j^{(r)}} \right]$.

For the experiments reported in this paper, the generalized I-divergence, defined as:

$$d_\phi(\mathbf{y}_i, \mathbf{y}_j) = \sum_{\ell=1}^{k} y_{i\ell} \log(\frac{y_{i\ell}}{y_{j\ell}}) - \sum_{\ell=1}^{k} (y_{i\ell} - y_{j\ell}), \forall \mathbf{y}_i, \mathbf{y}_j \in \mathbb{R}^k, \tag{11}$$

has been used. The underlying convex function is given by $\phi(\mathbf{y}_i) = \sum_{\ell=1}^{k} y_{i\ell} \log(y_{i\ell})$ so that $\nabla\phi(\mathbf{y}_i) = (1 + \log(y_{i\ell}))_{\ell=1}^{k}$. Thus, Eq. (10) can be rewritten as:

$$\mathbf{y}_i^{(l)*,I} = \exp\left(\left[\gamma_i^{(l)} \sum_{j^{(r)} \in \mathcal{X}} \delta_{i^{(l)} j^{(r)}} \nabla\phi(\mathbf{y}_j^{(r)}) + \lambda_i^{(l)} \nabla\phi(\mathbf{y}_i^{(r)})\right] / \left[\gamma_i^{(l)} + \lambda_i^{(l)}\right]\right) - 1. \tag{12}$$

Optimization over the left and right arguments of all the data points constitutes one pass (iteration) of the algorithm. These two steps are repeated till convergence. Since, at each step, the algorithm minimizes the objective in (3) and the minimizer is unique due to the strict convexity of ϕ, the algorithm is guaranteed to converge. On convergence, all $\mathbf{y}_i$'s are normalized to unit L_1 norm, to yield the individual class probability distributions for every object $\mathbf{x}_i \in \mathcal{X}$. The main steps of $\mathbf{C^3E}$ are summarized in Algorithm 1.

Algorithm 1 - C³E

Input: $\{\pi_i\}, \mathbf{S}$
Output: $\{\mathbf{y}_i\}$
    Step 0: Initialize $\{\mathbf{y}_i^{(r)}\}, \{\mathbf{y}_i^{(l)}\}$ so that $y_{i\ell}^{(r)} = y_{i\ell}^{(l)} = \frac{1}{k} \ \forall i \in \{1, 2, \cdots, n\}, \forall \ell \in \{1, 2, \cdots, k\}$
    Loop until convergence
        Step 1:
            Update $\mathbf{y}_j^{(r)}$ using equation (6) $\forall j \in \{1, 2, \cdots, n\}$
        Step 2:
            Update $\mathbf{y}_i^{(l)}$ using equation (10) $\forall i \in \{1, 2, \cdots, n\}$
    End Loop
    Step 4: Compute $\mathbf{y}_i = 0.5[\mathbf{y}_i^{(l)} + \mathbf{y}_i^{(r)}] \ \forall i \in \{1, 2, \cdots, n\}$
    Step 5: Normalize $\mathbf{y}_i \ \forall i \in \{1, 2, \cdots, n\}$

---

3   Related Work

Of late there has been substantial interest in exploiting both labeled and unlabeled data for a variety of learning scenarios, including works on semi-supervised learning and transductive learning [11,2,14]. In almost all cases, these approaches use a *single* (clustering, classification or regression) model, which is then tempered by additional data or other constraints. A notable exception is a recent work by Gao et al. [5], in which the outputs of *multiple* supervised and unsupervised models are combined. Here, it is assumed that each model partitions the target dataset $\mathcal{X}$ into groups, so that the objects in the same group share either the same predicted class label or the same cluster label. The data, models and outputs are summarized by a bipartite graph. In this graph, on one side the nodes denote the groups output by the models, whereas on the other side the nodes denote objects. A group node and an object node are connected if the object is assigned to the group — no matter if it comes from a supervised or unsupervised model. From the resulting graph, the goal is to predict the class labels so that they agree with the supervised models and also satisfy the constraints enforced by the clustering models, as much as possible. In other words, the authors in [5] aim at consolidating a classification solution by maximizing the consensus among both supervised predictions and unsupervised constraints, casting it as an optimization problem on a bipartite graph. The objective function is designed to maximize such a consensus by promoting smoothness of label assignment over the graph and consistency with the initial labeling. To solve the optimization problem, they introduce the Bipartite Graph-based Consensus Maximization (**BGCM**) Algorithm.

The **C^{3}E** algorithm can also be viewed as a semi-supervised ensemble working at the output level. Unlike **BGCM**, however, it does not receive as input several supervised and unsupervised models. Instead, **C^{3}E** ultimately processes only two fused models, namely: (i) an ensemble of classifiers that delivers a class probability vector for every object in $\mathcal{X}$; and (ii) an ensemble of clusterers that provides a similarity matrix, where each entry corresponds to the relative co-ocurrence of two objects in the same cluster of $\mathcal{X}$ (considering all the available data partitions). Contrary to **BGCM**, which is based on hard classification inputs from supervised models, **C^{3}E** can deal with class probability distributions obtained by the ensemble of classifiers, and caters to both hard and soft clusterings. Moreover, **C^{3}E** avoids solving a difficult correspondence problem — *i.e.*, aligning cluster labels to class labels — implicitly tackled by **BGCM**.

4   Experimental Evaluation

The **C^{3}E** algorithm has been evaluated on the same classification datasets employed by Gao *et al.* [5] to assess their **BGCM** algorithm[3]. Following Gao *et al.* [5], eleven classification tasks from three real-world applications (20 Newsgroups, Cora, and DBLP) have been used. In each task, there is a target set on which the class labels should be predicted. In [5], two supervised models ($\mathbf{M}_1$ and $\mathbf{M}_2$) and two unsupervised models ($\mathbf{M}_3$ and $\mathbf{M}_4$) were used to obtain (on the target sets) class and cluster labels, respectively. These same labels have been used as inputs to **C^{3}E**. In doing so, comparisons between **C^{3}E** and **BGCM** are performed using exactly the same base models, which were trained in the same datasets. In other words, both **C^{3}E** and **BGCM** receive the same inputs

---

[3] Datasets available at http://ews.uiuc.edu/ jinggao3/nips09bgcm.htm.

w.r.t. the components of the ensembles, from which consolidated classification solutions for the target sets are generated.

For the sake of compactness, the description of the datasets and learning models used in [5] are not reproduced here, and the interested reader is referred to that paper for further details. However, the results achieved by Gao *et al.* [5] from their four base models ($\mathbf{M_1}$, $\mathbf{M_2}$, $\mathbf{M_3}$, and $\mathbf{M_4}$), from **BGCM** [5], and from two well-known cluster ensemble approaches — **MCLA** [12] and **HBGF** [4] — are reproduced here for comparison purposes. Being cluster ensemble approaches, **MCLA** [12] and **HBGF** [4] ignore the class labels, considering that the four base models provide just cluster labels. Therefore, to evaluate classification accuracy obtained by these ensembles, the cluster labels are matched to the classes through an Hungarian method that favors the best possible class predictions. In order to run $\mathbf{C^3E}$, the supervised models ($\mathbf{M_1}$ and $\mathbf{M_2}$) have been fused to obtain class probability estimates for every object in the target set. Also, the co-association matrix used by $\mathbf{C^3E}$ was achieved by fusing the unsupervised models ($\mathbf{M_3}$ and $\mathbf{M_4}$). The parameters of $\mathbf{C^3E}$ have been manually optimized for better performance in each dataset. In particular the following pairs of $(\alpha, \lambda)^4$ have been respectively used for the datasets News, Cora, and DBLP: $(4 \times 10^{-2}, 10^{-2})$; $(10^{-4}, 10^{-2})$; $(10^{-7}, 10^{-3})$.

Table 1. Classification Accuracies — Best Results in Boldface

Method	News1	News2	News3	News4	News5	News6	Cora1	Cora2	Cora3	Cora4	DBLP
M_1	0.7967	0.8855	0.8557	0.8826	0.8765	0.8880	0.7745	0.8858	0.8671	0.8841	0.9337
M_2	0.7721	0.8611	0.8134	0.8676	0.8358	0.8563	0.7797	0.8594	0.8508	0.8879	0.8766
M_3	0.8056	0.8796	0.8658	0.8983	0.8716	0.9020	0.7779	0.8833	0.8646	0.8813	0.9382
M_4	0.7770	0.8571	0.8149	0.8467	0.8543	0.8578	0.7476	0.8594	0.7810	0.9016	0.7949
MCLA	0.7592	0.8173	0.8253	0.8686	0.8295	0.8546	0.8703	0.8388	0.8892	0.8716	0.8953
HBGF	0.8199	0.9244	0.8811	0.9152	0.8991	0.9125	0.7834	0.9111	0.8481	0.8943	0.9357
BGCM	0.8128	0.9101	0.8608	0.9125	0.8864	0.9088	0.8687	0.9155	0.8965	0.9090	0.9417
C^3E	**0.8501**	**0.9364**	**0.8964**	**0.9380**	**0.9122**	**0.9180**	**0.8854**	**0.9171**	**0.9060**	**0.9149**	**0.9438**

The classification accuracies achieved by the studied methods are summarized in Table 1, where one can see that the proposed $\mathbf{C^3E}$ has shown the best accuracies for all datasets. In order to provide some reassurance about the validity and non-randomness of the obtained results, the outcomes of statistical tests, following the study of Demsar [3], are also reported. In brief, multiple algorithms have been compared on multiple datasets by using the Friedman test, with a corresponding post-hoc test. The adopted statistical procedure indicates that the null hypothesis of equal accuracies — considering the results obtained by the ensembles — can be rejected at $\alpha = 0.05$. In pairwise comparisons, significant statistical differences have only been observed between $\mathbf{C^3E}$ and the other ensembles, *i.e.*, there is no evidence that the accuracies of **MCLA**, **HBGF**, and **BGCM** are statistically different from one to another.

---

4 $\lambda_i^{(r)} = \lambda_i^{(l)} = \lambda$ has been set for all i.

5 Conclusions

The **C^{3}E** algorithm, which combines ensembles of classifiers and clusterers, was introduced. **C^{3}E** has shown better accuracies than the recently proposed **BGCM** Algorithm [5], which combines the outputs of multiple supervised and unsupervised models and is the most closely related algorithm to **C^{3}E**. The asymptotic time complexity of **C^{3}E** is quadratic with the number of objects in the target set and linear with the number of ensemble components, whereas **BGCM** has cubic time complexity with respect to these input sizes.

There are several aspects that can be investigated in future work. For example, the impact of the number of classifiers and clusterers in **C^{3}E** deserves further investigations. Also, the relative relevance of each component of the ensemble can be straightforwardly incorporated into **C^{3}E**. Finally, a more comprehensive experimental evaluation, specially considering comparisons with other semi-supervised algorithms, is in order.

Acknowledgments

This work has been supported by NSF Grants (IIS-0713142 and IIS-1016614) and by the Brazilian Research Agencies FAPESP and CNPq.

References

1. Banerjee, A., Merugu, S., Dhillon, I., Ghosh, J.: Clustering with Bregman divergences. In: JMLR (2005)
2. Schlkopf, B., Zien, A., Chapelle, O.: Semi-Supervised Learning. MIT Press. Cambridge (2006)
3. Demšar, J.: Statistical comparisons of classifiers over multiple data sets. J. Mach. Learn. Res. 7, 1–30 (2006)
4. Fern, X.Z., Brodley, C.E.: Solving cluster ensemble problems by bipartite graph partitioning. In: Proc. of the ICML, pp. 36–43 (2004)
5. Gao, J., Liang, F., Fan, W., Sun, Y., Han, J.: Graph-based consensus maximization among multiple supervised and unsupervised models. In: Proc. of NIPS, pp. 1–9 (2009)
6. Ghosh, J., Acharya, A.: Cluster ensembles. WIREs Data Mining and Knowledge Discovery 1, 1–12 (to appear 2011)
7. Kittler, J., Roli, F. (eds.): IPSN 2003. LNCS, vol. 2634. Springer, Heidelberg (2003)
8. Kuncheva, L.I.: Combining Pattern Classifiers: Methods and Algorithms. Wiley, Chichester (2004)
9. Oza, N., Tumer, K.: Classifier ensembles: Select real-world applications. Information Fusion 9(1), 4–20 (2008)
10. Punera, K., Ghosh, J.: Consensus based ensembles of soft clusterings. Applied Artificial Intelligence 22, 109–117 (2008)
11. Davidson, I., Basu, S., Wagstaff, K.L. (eds.): Clustering with Balancing Constraints. CRC Press, Boca Raton (2008)
12. Strehl, A., Ghosh, J.: Cluster ensembles – a knowledge reuse framework for combining multiple partitions. In: JMLR, vol. 617, pp. 583–617 (2002)

13. Tumer, K., Ghosh, J.: Analysis of decision boundaries in linearly combined neural classifiers. Pattern Recognition 29, 341–348 (1996)
14. Goldberg, A., Zhu, X.: Introduction to Semi-Supervised Learning. Morgan and Claypool Publishers, San Rafael (2009)

Appendix

Definition 1. *Let $\phi : \mathcal{S} \to \mathbb{R}$ be a strictly convex, differentiable function defined on a convex set $\mathcal{S} = dom(\phi) \subseteq \mathbb{R}^k$. Then the **Bregman divergence** $d_\phi : \mathcal{S} \times ri(\mathcal{S}) \to [0, \infty)$ between $\mathbf{y}_i \in \mathcal{S}$ and $\mathbf{y}_j \in ri(\mathcal{S})$ is defined as: $d_\phi(\mathbf{y}_i, \mathbf{y}_j) = \phi(\mathbf{y}_i) - \phi(\mathbf{y}_j) - \langle \mathbf{y}_i - \mathbf{y}_j, \nabla\phi(\mathbf{y}_j) \rangle$.*

From this Definition, it follows that $d_\phi(\mathbf{y}_i, \mathbf{y}_j) \geq 0 \ \forall \mathbf{y}_i \in \mathcal{S}, \mathbf{y}_j \in ri(\mathcal{S})$ and equality holds *iff* $\mathbf{y}_i = \mathbf{y}_j$. Then, it can be shown that there is a unique minimizer for the optimization problem in (5) by considering the following theorem:

Theorem 1 (from [1]). *Let Y be a random variable that takes values in $\mathcal{Y} = \{\mathbf{y}_i\}_{i=1}^n \subset \mathcal{S} \subseteq \mathbb{R}^k$ following a probability measure v such that $\mathbb{E}_v[Y] \in ri(\mathcal{S})$. Given a Bregman divergence $d_\phi : \mathcal{S} \times ri(\mathcal{S}) \to [0, \infty)$, the optimization problem $\min_{\mathbf{s} \in ri(\mathcal{S})} \mathbb{E}_v[d_\phi(Y, \mathbf{s})]$ has a unique minimizer given by $\mathbf{s}^* = \boldsymbol{\mu} = \mathbb{E}_v[Y]$.*

Corollary 1. *Let $\{Y_i\}_{i=1}^n$ be a set of random variables, each of which takes values in $\mathcal{Y}_i = \{\mathbf{y}_{ij}\}_{j=1}^{n_i} \subset \mathcal{S} \subseteq \mathbb{R}^d$ following a probability measure v_i such that $\mathbb{E}_{v_i}[Y_i] \in ri(\mathcal{S})$. Consider a Bregman divergence d_ϕ and an objective function of the form $J_\phi(\mathbf{s}) = \sum_{i=1}^m \alpha_i \mathbb{E}_{v_i}[d_\phi(Y_i, \mathbf{s})]$ with $\alpha_i \in \mathbb{R}_+ \ \forall i$. This objective function has a unique minimizer given by $\mathbf{s}^* = \boldsymbol{\mu} = \left[\sum_{i=1}^m \alpha_i \mathbb{E}_{v_i}[Y_i] \right] / \left[\sum_{i=1}^m \alpha_i \right]$.*

Proof. The proof for this corollary is pretty straightforward and similar to that of Theorem 1 as given in [1] but omitted here for space constraints. ∎

A Latent Variable Pairwise Classification Model of a Clustering Ensemble

Vladimir Berikov

Sobolev Institute of mathematics, Novosibirsk State University, Russia
berikov@math.nsc.ru
http://www.math.nsc.ru

Abstract. This paper addresses some theoretical properties of clustering ensembles. We consider the problem of cluster analysis from pattern recognition point of view. A latent variable pairwise classification model is proposed for studying the efficiency (in terms of "error probability") of the ensemble. The notions of stability, homogeneity and correlation between ensemble elements are introduced. An upper bound for misclassification probability is obtained. Numerical experiment confirms potential usefulness of the suggested ensemble characteristics.

Keywords: clustering ensemble, latent variable model, misclassification probability, error bound, ensemble's homogeneity and correlation.

1 Introduction

Collective decision-making based on a combination of simple algorithms is actively used in modern pattern recognition and machine learning. In last decade, there is growing interest in clustering ensemble algorithms [1,2]. In the ensemble design process, the results obtained by different algorithms, or by one algorithm with various parameters settings are used. After construction of partial clustering solutions, a final collective decision is built.

Modern literature on clustering ensembles can be roughly divided into several main categories. There are a great deal of works in which the ensemble methodology is adapted to new application areas such as magnetic resonance imaging, satellite images analysis, analysis of genetic sequences etc (see, for example, [3,4,5]). Another direction aims to develop clustering ensembles methods of general usage and elaborate efficient algorithms using various optimization techniques (e.g., [6]). Other categories of works are of more theoretical nature; their purpose is to study the properties of clustering ensembles, improve measures of ensemble quality, suggest the ways to achieve the best quality (e.g., [7,8,9]).

There is a large number of experimental evidences confirming a significant raise in stability of clustering decisions for ensemble algorithms (see, for example, [2,10]). At the same time, theoretical grounds of clustering ensembles algorithms, as opposed to the pattern classifier ensembles theory (e.g., [11]), are still in

C. Sansone, J. Kittler, and F. Roli (Eds.): MCS 2011, LNCS 6713, pp. 279–288, 2011.

the early development stage. Existing works consider mainly the asymptotic properties of clustering ensembles (e.g., [7]).

Cluster analysis problems are characterized by the complexity of formalization caused by substantially subjective nature of grouping process. For the definition of clustering quality it is necessary to apply additional a priori information in terms of "natural" classification, to use data generation models etc. In the given work we attempt to corroborate clustering ensemble methodology by utilizing of a pattern recognition model with latent class labels. To avoid problems with class renumbering, a pairwise classification approach is used.

The rest of the paper is organized as follows. In the next section we give basic definitions and introduce the model of ensemble cluster analysis. In the third section we receive an upper bound for error probability (in classifying a pair of arbitrary objects according to latent variable labels) and give some qualitative consequences of the result. In the forth section we introduce the estimates of ensemble characteristics. The next section describes numerical experiment that demonstrates the usage of these notions. The conclusion summaries the work and gives possible future directions.

2   Ensemble Model

Let us consider a sample $s = \{o^{(1)}, \ldots, o^{(N)}\}$ of objects independently and randomly selected from a general population. The purpose of the analysis is to group the objects into $K \geq 2$ classes in accordance with some clustering criterion; the number of classes may be either given beforehand or not (in the latter case an optimal number of classes should be determined automatically).

Let each of the objects be characterized by variables $X_1, \ldots, X_n$. Denote by $x = (x_1, \ldots, x_n)$ the vector of these variables for an object o, $x_j = X_j(o)$, $j = 1, \ldots, n$.

In many clustering tasks it is allowable to consider that there exists a ground truth (latent, directly unobserved) variable

$$Y \in \{1, \ldots, K\}$$

that determines to which class an object belongs. Suppose that the observations of k-th class are distributed according to the conditional dencity function $p_k(x) = p(x|Y = k)$, $k = 1, \ldots, K$.

Consider the following model of data generation. Let each object be assigned to class k in accordance with a priori probabilities $P_k = \mathbb{P}(Y = k)$, $k = 1, \ldots, K$, where $\sum_{k=1}^{K} P_k = 1$. After the assignment, an observable value of x is determined with use of $p_k(x)$. This procedure is repeated independently for each object.

For an arbitrary pair of different objects $a, b \in s$, their correspondent observations are denoted by $x(a)$ and $x(b)$. Let

$$Z = \mathbb{I}(Y(a) \neq Y(b)),$$

where $\mathbb{I}(\cdot)$ is indicator function. Denote by $P_Z = \mathbb{P}[Z = 1|x(a), x(b)]$ the probability of the event "a and b belong to different classes, given $x(a)$ and $x(b)$":

$$P_Z = 1 - \mathbb{P}[Y(a) = 1|x(a)]\,\mathbb{P}[Y(b) = 1|x(b)] - \ldots$$

$$- \mathbb{P}[Y(a) = K|x(a)]\,\mathbb{P}[Y(b) = K|x(b)] = 1 - \sum_{k=1}^{K} \frac{p_k(x(a))p_k(x(b))P_k^2}{p(x(a))p(x(b))},$$

where $p(x(o)) = \sum_{k=1}^{K} p_k(x(o))P_k$, $o = a, b$.

Let a clustering algorithm μ be run to partition s into K subsets (clusters). Because the numberings of clusters do not matter, it is convenient to consider the equivalence relation, i.e. to indicate whether the algorithm μ assigns each pair of objects to the same class or to different classes. Let

$$h_\mu(a, b) = \mathbb{I}[\mu(a) \neq \mu(b)].$$

Let us consider the following model of *ensemble clustering*. Suppose that algorithm μ is randomized, i.e. it depends from a random value Ω from a given set of allowable values (parameters or more generally "learning settings" such as bootstrap samples, order of input objects etc). In addition to Ω, the algorithm's decisions are dependent from the true status of the pair a, b (i.e., from Z):

$$h_\mu(a, b) = h_{\mu(\Omega)}(Z, a, b).$$

Hereinafter we will denote $h_{\mu(\Omega)}(Z, a, b) = h(\Omega, Z)$.

Suppose that

$$\mathbb{P}[h(\Omega, Z) = 1|Z = 1] = \mathbb{P}[h(\Omega, Z) = 0|Z = 0] = q,$$

i.e. the conditional probabilities of correct decision (either partition or union of objects a,b) coincide. One can say that q reflects the *stability* of algorithm under various learning settings. We shall suppose that $q > 1/2$; it means that algorithm μ provides better clustering quality than just random guessing. In machine learning theory, such a condition is known as the condition of *weak learnability*.

Denote $P_h = \mathbb{P}[h(\Omega, Z) = 1]$. This quantity shows the *homogeneity* of algorithm's decisions: P_h close to 0 or 1; or *homogeneity index*

$$I_h = 1 - P_h(1 - P_h)$$

close to 1 means high agreement between the solutions. Note that

$$P_h = \mathbb{P}[h(\Omega, Z) = 1|Z = 1]P_Z + \mathbb{P}[h(\Omega, Z) = 1|Z = 0](1 - P_Z) =$$
$$qP_Z + (1 - q)(1 - P_Z).$$

Suppose that algorithm μ is running L times under randomly and independently chosen settings. As a result, we get random decisions $h(\Omega_1, Z)$, $\ldots$,

$h(\Omega_L, Z)$. By $\Omega_1, \ldots, \Omega_L$ we denote independent statistical copies of a random vector Ω.

For every Ω_l, algorithm μ is running independently (it does not use the results obtained with other $\Omega_{l'}$, $l' \neq l$). Suppose that the decisions are conditionally independent:

$$\mathbb{P}[h(\Omega_{i_1}, Z) = h_{i_1}, \ldots, h(\Omega_{i_j}, Z) = h_{i_j} | Z = z] =$$
$$\mathbb{P}[(h(\Omega_{i_1}, Z) = h_{i_1} | Z = z] \cdot \cdots \cdot \mathbb{P}[h(\Omega_{i_j}, Z) = h_{i_j} | Z = z],$$

where $\Omega_{i_1}, \ldots, \Omega_{i_j}$ are arbitrary learning settings, $h_{i_1}, \ldots, h_{i_j}, z \in \{0, 1\}$ (we shall assume that L is odd).

Let $P_{h,h} = \mathbb{P}[h(\Omega', Z) = 1, h(\Omega'', Z) = 1]$, where Ω', Ω'' have the same distribution as Ω, and $\Omega' \neq \Omega''$. It follows from the assumptions of independence and stability that

$$P_{h,h} = \mathbb{P}[h(\Omega', Z) = 1, h(\Omega'') = 1 | Z = 1] P_Z +$$
$$\mathbb{P}[h(\Omega', Z) = 1, h(\Omega'') = 1 | Z = 0](1 - P_Z) =$$
$$q^2 P_Z + (1-q)^2(1 - P_Z). \quad (1)$$

Denote $\bar{H} = \frac{1}{L} \sum_{l=1}^{L} h(\Omega_l, Z)$. The function

$$c(h(\Omega_1, Z), \ldots, h(\Omega_L, Z)) = \mathbb{I}[\bar{H} > \frac{1}{2}]$$

shall be called *the ensemble solution* for a and b, based on the majority voting.

For constructing a final ensemble clustering decision, various approaches can be utilized [2]. For example, it is possible to use a methodology based on the pairwise dissimilarity matrix $\mathbb{H} = (\bar{H}(o^{(i_1)}, o^{(i_2)}))$, where $o^{(i_1)}, o^{(i_2)} \in s$, $o^{(i_1)} \neq o^{(i_2)}$. This matrix can be considered as a matrix of pairwise distances between objects and used as input information for a dendrogram construction algorithm to form a sample partition on a desired number of clusters.

3   An Upper Bound for Misclassification Probability

Let us consider the *margin [11]* of the ensemble:

$$mg = \frac{1}{L}\{ \text{ number of votes for } Z - \text{ number of votes against } Z\},$$

where $Z = 0, 1$. It is easy to show that the margin equals:

$$mg = mg(\bar{H}, Z) = (2Z - 1)(2\bar{H} - 1).$$

Using the notion of margin, one can represent the probability of wrong prediction of the true value of Z:

$$P_{err} = \mathbb{P}_{\Omega_1, \ldots, \Omega_L, Z}[mg(\bar{H}, Z) < 0].$$

It follows from the Tchebychev's inequality that

$$\mathbb{P}[U < 0] < \frac{\text{Var}U}{(\text{E}U)^2},$$

where $\text{E}U$ is population mean, $\text{Var}U$ is variance of random value U (it is required that $\text{E}U > 0$). Thus,

$$\mathbb{P}_{\Omega_1,\ldots,\Omega_L,Z}[mg(\bar{H},Z) < 0] < \frac{\text{Var}\, mg(\bar{H},Z)}{(\text{E}\, mg(\bar{H},Z))^2},$$

provided that $\text{E}\, mg(\bar{H},Z) > 0$.

Theorem. The expected value and variance of the margin are:

$$\text{E}\, mg(\bar{H},Z) = 2q - 1,$$

$$\text{Var}\, mg(\bar{H},Z) = \frac{4}{L}\left(P_h - P_{h,h}\right).$$

Proof. We have:

$$\text{E}\, mg(\bar{H},Z) = \text{E}\,(2Z-1)(\frac{2}{L}\sum_l h(\Omega_l,Z) - 1) =$$

$$\frac{4}{L}\sum_l \text{E}\, Zh(\Omega_l,Z) - 2\text{E}Z - \frac{2}{L}\sum_l \text{E}\, h(\Omega_l,Z) + 1.$$

Because all $h(\Omega_l,Z)$ are distributed in the same way as $h(\Omega,Z)$, we get:

$$\text{E}\, mg(\bar{H},Z) = 4\text{E}\, Zh(\Omega,Z) - 2P_Z - 2\text{E}\, h(\Omega,Z) + 1 =$$
$$4\mathbb{P}[Z=1, h(\Omega,Z)=1] - 2P_Z - 2\mathbb{P}[h(\Omega,Z)=1] + 1.$$

As $\mathbb{P}[h(\Omega,Z)=1] = \mathbb{P}[Z=1, h(\Omega,Z)=1] + \mathbb{P}[Z=0, h(\Omega,Z)=1] =$
$$qP_Z + (1-q)(1-P_Z) = 2qP_Z + 1 - q - P_Z,$$

we obtain:

$$\text{E}\, mg(\bar{H},Z) = 4qP_Z - 2P_Z - 2(2qP_Z + 1 - q - P_Z) + 1 = 2q - 1.$$

Consider the variance of margin:

$$\text{Var}\, mg(\bar{H},Z) = \text{Var}\,(4Z\bar{H} - 2\bar{H} - 2Z) = \text{E}\,(4Z\bar{H} - 2\bar{H} - 2Z)^2 -$$
$$(\text{E}\,(4Z\bar{H} - 2\bar{H} - 2Z))^2 = \text{E}\,(16Z^2\bar{H}^2 + 4\bar{h}^2 + 4Z^2 - 16Z\bar{H}^2 - 16Z^2\bar{H} + 8Z\bar{H}) -$$
$$(\text{E}\, mg(\bar{H},Z) - 1)^2 = \text{E}\,(4\bar{H}^2 + 4Z - 8Z\bar{H}) - 4(1-q)^2$$

(we apply $Z^2 = Z$). Next, we have

$$
\mathrm{E}\,\bar{H}^2 = \frac{1}{L^2}\mathrm{E}\left(\sum_l h(\Omega_l, Z)\right)^2 = \frac{1}{L^2}\mathrm{E}\left(\sum_l h^2(\Omega_l, Z)\right) +
$$
$$
\frac{1}{L^2}\sum_{\substack{l',l'': \\ l'\neq l''}}\mathrm{E}\left(h(\Omega_{l'}, Z)h(\Omega_{l''}, Z)\right) =
$$
$$
\frac{1}{L}\mathrm{E}\,h(\Omega, Z) + \frac{L-1}{L}\sum_{\substack{\Omega',\Omega'': \\ \Omega'\neq\Omega''}}\mathrm{E}\left(h(\Omega', Z)h(\Omega'', Z)\right) = \frac{P_h}{L} + \frac{L-1}{L}P_{h,h}.
$$

From this, we obtain:

$$
\mathrm{Var}\,mg(\bar{H}, Z) = 4\frac{P_h}{L} + 4\frac{L-1}{L}P_{h,h} + 4P_Z - 8qP_Z - 4(1-q)^2.
$$

Using (1) finally we get:

$$
\mathrm{Var}\,mg(\bar{H}, Z) = 4\frac{P_h}{L} + 4\frac{L-1}{L}P_{h,h} - 4P_{h,h} = \frac{4}{L}(P_h - P_{h,h}).
$$

The theorem is proved.

Evidently, the requirement $\mathrm{E}\,mg(\bar{H}, Z) > 0$ is fulfilled if $q > 1/2$.

Let us consider the correlation coefficient ρ between $h' = h(\Omega', Z)$ and $h'' = h(\Omega'', Z)$, where $\Omega' \neq \Omega''$. We have

$$
\rho = \rho_{h',h''} = \frac{P_{h,h} - P_h^2}{P_h(1 - P_h)}.
$$

Because $P_h - P_{h,h} = P_h - P_h^2 + P_h^2 - P_{h,h}$, we obtain

$$
\mathrm{Var}\left(mg(\bar{H}, Z)\right) = \frac{4}{L}(1 - \rho)P_h(1 - P_h).
$$

Note that $P_h - P_{h,h} = q(1 - q)$, and after necessary transformations we get the following upper bound for error probability:

$$
P_{err} < \frac{1}{L}\left(\frac{1}{1 - 4(1 - \rho)P_h(1 - P_h)} - 1\right).
$$

The obtained expression allows to make some qualitative conclusions. Namely, if the model assumptions are fulfilled and $q > 1/2$, then under other conditions being equal the following statements are valid:

- the probability of error decreases with an increase in number of ensemble elements;
- an increase in homogeneity of the ensemble and raise of correlation between its outputs reduce the probability of error (note that a signed value of the correlation coefficient is meant).

4   Estimating Characteristics of a Clustering Ensemble

To evaluate the quality of a clustering ensemble, it is necessary to estimate ensemble's characteristics (in our model – homogeneity and correlation) from a finite number of ensemble elements. For an arbitrary pair of different objects a and b, the estimate of the ensemble's homogeneity can be found as follows:

$$\hat{I}_h(a,b) = 1 - \hat{P}_h(a,b)(1 - \hat{P}_h(a,b)),$$

where

$$\hat{P}_h(a,b) = \frac{1}{L} \sum_{l=1}^{L} h_l(a,b).$$

Unfortunately, the straightforward estimation of the correlation coefficient $\rho(a,b)$ is impossible: under a fixed sample, every pair of clustering algorithms give conditionally independent decisions. Let us introduce a similar notion: the averaged correlation coefficient, where the averaging is done over all pairs of different objects:

$$\bar{\rho} = \frac{cov(h', h'')}{\sigma^2(h)};$$

where the covariance

$$cov(h', h'') = \overline{h'h''} - \bar{h}^2,$$

$$\overline{h'h''} = \frac{2}{N(N-1)} \frac{2}{L(L-1)} \sum_{\substack{a,b: \\ a \neq b}} \sum_{\substack{l',l'': \\ l' \neq l''}} h_{l'}(a,b)\, h_{l''}(a,b),$$

$$\bar{h} = \frac{2}{N(N-1)} \frac{1}{L} \sum_{\substack{a,b: \\ a \neq b}} \sum_{l} h_l(a,b) = \frac{2}{N(N-1)} \sum_{\substack{a,b: \\ a \neq b}} \hat{P}_h(a,b),$$

and the variance

$$\sigma^2(h) = \overline{h^2} - \bar{h}^2 = \bar{h} - \bar{h}^2.$$

Similarly, it is possible to introduce the averaged homogeneity index:

$$\bar{I}_h = \frac{2}{N(N-1)} \sum_{\substack{a,b: \\ a \neq b}} \hat{I}_h(a,b).$$

5   Numerical Experiment

To verify the applicability of the suggested methodology for the analysis of clustering ensemble behavior, the statistical modeling approach was used. In the modeling, artificial data sets are repeatedly generated according to certain distribution class (a type of "clustering tasks"). Each data set is classified by the ensemble algorithm. The correct classification rate, averaged over a given number of trials, determines algorithm's performance for the given type of tasks.

The following experiment was performed. In each trial, two classes are independently sampled according to the Gauss multivariate distributions

$$\mathcal{N}(m_1, \Sigma), \ \mathcal{N}(m_2, \Sigma),$$

where m_1, m_2 are vectors of means,

$$\Sigma = \sigma \mathbf{I}$$

is a diagonal n-dimensional covariance matrix,

$$\sigma = (\sigma_1, \ldots, \sigma_n)$$

is a vector of variances. Variable X_i shall be called "noisy", if for some $i \in \{1, ..., n\}$, $\sigma_i = \sigma_{noise} >> 1$. In our experiment, the set of noisy variables

$$\{X_{i_1}, ..., X_{i_{n_{noize}}}\}$$

is chosen at random. For those variables that are not noisy, we set $\sigma_i = \sigma_0 = const$. Both classes have the same sample size.

The mixture of samples is classified by the ensemble of k-means clustering algorithms. Each algorithm performs clustering in the randomly chosen variable subspace of dimensionality n_{ens}. The ensemble decision for each pair of objects (i.e., either unite them or assign to different classes) is made by the majority voting procedure. The true overall performance P_{cor} of the ensemble is determined as the proportion of correctly classified pairs.

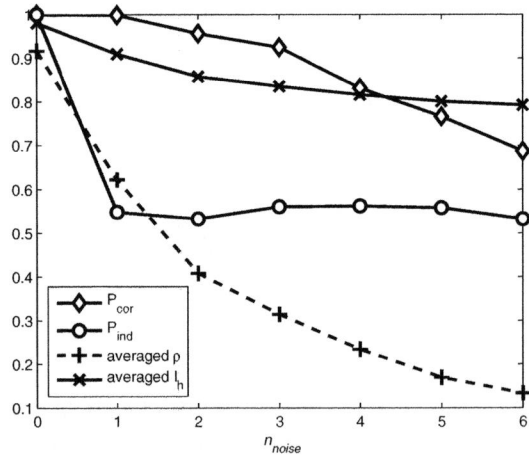

Fig. 1. Example of experiment results (averaged over 100 trials). Experiment settings: $N = 60$, $n = 10$, $m_1 = 0$, $m_2 = 1$, $\sigma_{noise} = 10$, $\sigma_0 = 0.2$, $n_{ens} = 2$, ensemble size $L = 15$.

An example of experiment results is shown in Figure 1. The graphs display the dependence of averaged values of P_{cor}, $\bar{\rho}$ and $\bar{I}_h$ from the number of noisy variables n_{noise}. To demonstrate the effectiveness of the ensemble solution in comparison with individual clustering, the averaged performance rate P_{ind} of a single k-means clustering algorithm is given (this algorithm performs clustering in the whole feature space of dimensionality n).

From this example, one can conclude that

a) in average, the ensemble algorithm has better performance than a single clustering algorithm (when a noise presents), and

b) the dynamics of estimated ensemble characteristics (averaged homogeneity and correlation) reproduces well the behavior of correct classification rate (note that this rate is directly unobserved in real clustering problems). When averaged correlation and homogeneity index are sufficiently large, one can expect good classification quality.

Conclusion

A latent variable pairwise classification model is proposed for studying non-asymptotic properties of clustering ensembles. In this model, the notions of stability, homogeneity and correlation between ensemble elements are utilized. An upper bound for probability of error is obtained. Theoretical analysis of the suggested model allows to make a conclusion that the probability of correct decision increases with an increase in number of ensemble elements. It is also found that a large degree of agreement between partial clustering solutions (expressed in our model in terms of homogeneity and correlation between ensemble elements), under condition of independence of base clustering algorithms, indicates good classification performance. Numerical experiment also confirms this conclusion.

The following possible future directions can be indicated. It is interesting to study intensional connections between the notions used in the suggested model (conditional independence, stability, homogeneity and correlation) and other known concepts such as mutual information [2] and diversity in clustering ensembles (e.g., [8,9]). Another direction could aim to improve the tightness of the obtained error bound.

Acknowledgements

This work was partially supported by the Russian Foundation for Basic Research, projects 11-07-00346a, 10-01-00113a.

References

1. Jain, A.K.: Data Clustering: 50 Years Beyond K-Means. Pattern Recognition Letters 31(8), 651–666 (2010)
2. Strehl, A., Ghosh, J.: Clustering ensembles - a knowledge reuse framework for combining multiple partitions. The Journal of Machine Learning Research 3, 583–617 (2002)

3. Kuncheva, L.I., Rodriguez, J.J., Plumpton, C.O., Linden, D.E.J., Johnston, S.J.: Random Subspace Ensembles for fMRI Classification. IEEE Transactions on Medical Imaging 29(2), 531–542 (2010)
4. Pestunov, I.A., Berikov, V.B., Kulikova, E.A.: Grid-based ensemble clustering algorithm using sequence of fixed grids. In: Proc. of the 3rd IASTED Intern. Conf. on Automation, Control, and Information Technology, pp. 103–110. ACTA Press, Calgary (2010)
5. Iam-on, N., Boongoen, T., Garrett, S.: LCE: a link-based cluster ensemble method for improved gene expression data analysis. Bioinformatics 26(12), 1513–1519 (2010)
6. Hong, Y., Kwong, S.: To combine steady-state genetic algorithm and ensemble learning for data clustering. Pattern Recognition Letters 29(9), 1416–1423 (2008)
7. Topchy, A., Law, M., Jain, A., Fred, A.: Analysis of Consensus Partition in Cluster Ensemble. In: Fourth IEEE International Conference on Data Mining, pp. 225–232. IEEE Press, New York (2004)
8. Hadjitodorov, S.T., Kuncheva, L.I., Todorova, L.P.: Moderate diversity for better cluster ensembles. Information Fusion 7(3), 264–275 (2006)
9. Azimi, J., Fern, X.: Adaptive Cluster Ensemble Selection. In: Proceedings of International Joint Conference on Artificial Intelligence, pp. 992–997 (2009)
10. Kuncheva, L.: Combining Pattern Classifiers. Methods and Algorithms. John Wiley & Sons, Hoboken (2004)
11. Breiman, L.: Random Forests. Machine Learning 45(1), 5–32 (2001)

CLOOSTING: CLustering Data with bOOSTING

F. Smeraldi[1], M. Bicego[2,3], M. Cristani[2,3], and V. Murino[2,3]

[1] School of Electronic Engineering and Computer Science,
Queen Mary University of London, UK
[2] Computer Science Department, University of Verona, Italy
[3] Istituto Italiano di Tecnologia (IIT), Italy

Abstract. We present a novel clustering approach, that exploits boosting as the primary means of modelling clusters. Typically, boosting is applied in a supervised classification context; here, we move in the less explored unsupervised scenario. Starting from an initial partition, clusters are iteratively re-estimated using the responses of one-vs-all boosted classifiers. Within-cluster homogeneity and separation between the clusters are obtained by a combination of three mechanisms: use of regularised Adaboost to reject outliers, use of weak learners inspired to subtractive clustering and smoothing of the decision functions with a Gaussian Kernel. Experiments on public datasets validate our proposal, in some cases improving on the state of the art.

1 Introduction

Boosting algorithms [9] are a class of ensemble methods that have repeatedly proved highly effective in the context of supervised classification. A somehow less explored scenario is the use of boosting techniques for unsupervised classification, namely clustering. Recently, there have been a few attempts to extend the boosting approach to the clustering domain. Some authors have proposed to combine different clustering algorithms in a boosting-like framework. For instance [20] and [10] introduced weighted re-sampling of data points according to how reliably they are classified. In [16], a general iterative clustering algorithm is presented that combines several algorithms by keeping track both of point weights and of a membership coefficient for each pair of a point and a model. Point weights are updated by a Bregman divergence optimisation procedure very similar to Adaboost [8].

All these works represent attempts at combining different clustering algorithms in an Adaboost-like framework. However, an alternative stance consists in using the Adaboost algorithm itself to describe the data, moving from the "boosting clustering methods" paradigm to "clustering *with* boosting methods" paradigm. This is in line with some other recent and very successfully approaches, based on Support Vector Machines [3,2]. The approach presented in this paper explores this direction, and is aimed at developing a clustering technique which employs a boosted classifier to describe each cluster. This may be particularly

C. Sansone, J. Kittler, and F. Roli (Eds.): MCS 2011, LNCS 6713, pp. 289–298, 2011.

advantageous for clustering since it is known that, depending on the chosen weak learners, boosting classifiers may describe non convex non connected regions in the feature space. A somehow related approach is the very recent MCBoost [11] where a noisy-or scheme is used to combine the output of several boosted classifier, one per cluster; however the weight update equations for the single boosting algorithm depend in this case on the decision of the algorithms after fusion, so that strictly speaking the strong classifiers are not obtained by Adaboost.

In this work, we explore the use of Adaboost proper to model the clusters, via a cycle of learning and classification iterations reminiscent of k-means – in line of what has been done by Camastra and Verri with Support Vector Machines [3]. Contrary to [11] we choose to model the data with a regularised Adaboost algorithm, namely Adaboost-REG [18]. The specificity of the clustering problem is dealt with at the level of the weak learners. These are balls in feature space, centred at the locations identified by a potential similar to what is done in Subtractive Clustering [4,5]. The locality of clusters is enforced by spatial smoothing of the output of the strong classifiers.

The suitability of the proposed approach has been tested on 4 different standard ML datasets, yielding results that compare favourably with the state of the art. Interestingly, the experiments suggest that our algorithm is especially advantageous in higher-dimensional spaces.

The rest of the paper is organised as follows: Section 2 presents an introduction to the Adaboost algorithm, followed by a discussion of related topics and variants of the algorithm that are relevant to its application to the clustering problem. Our proposal is detailed in Section 3, followed by experimental results in Section 4. A brief discussion in Section 5 concludes the paper.

2   Standard and Regularised Adaboost

Boosting concerns itself with the problem of combining several prediction rules with poor individual performance into a highly accurate predictor. This is commonly referred to as combining a set of "weak" learners into a "strong" classifier. Adaboost, introduced by Yoav Freund and Robert Schapire [8], differs from previous algorithms in that it does not require previous knowledge of the performance of the weak hypotheses, but it rather adapts accordingly. This adaptation is achieved by maintaining a distribution of weights over the elements of the training set.

Adaboost is structured as an iterative algorithm, that works by maintaining a distribution of weights over the training examples. At each iteration a new weak learner (classifier) is trained, and the distribution of weights is updated according to the examples it misclassifies; the underlying idea is to increase the importance of the training examples that are "difficult" to classify. The weights also determine the contribution of the each weak learner to the final strong classifier.

Unfortunately, as the number of iterations increases Adaboost tends to put higher weight on examples that are difficult to classify (see [18] and references

Table 1. The Adaboost and Adaboost-REG algorithms. For Adaboost, $C = 0$ and the minimisation (step 1) can be solved in closed form to give $\alpha_t = (\log(1 - \epsilon_t) - \log \epsilon_t)/2$.

Adaboost and Adaboost-REG:

Initialise $d_{1,i} = \frac{1}{m}$.
For $t = 1, \ldots, T$:

1. Select the weak learner h_t that minimises the training error ϵ_t, weighted by current distribution of weights $d_{t,i}$: $\epsilon_t = \sum_{i|u_{t,i}=-1} d_{t,i}$.
2. Find the coefficient α_t that minimises

$$Z(\alpha_t, \boldsymbol{\alpha}_{t-1}) = \sum_i \exp\left(-\rho(\boldsymbol{u}_{t,i}, \boldsymbol{\alpha}_t) + C\zeta(\boldsymbol{u}_{t,i}, \boldsymbol{\alpha}_t)\right) \tag{1}$$

where $\rho(\boldsymbol{u}_{t,i}, \boldsymbol{\alpha}_t) = \sum_{\tau=1}^{t} \alpha_\tau u_{\tau,i}$ and $C\zeta(\boldsymbol{u}_{t,i}, \boldsymbol{\alpha}_t)$ is a regularization term; abort if $\alpha_t = 0$ or $\alpha_t \geq \Gamma$ where Γ is a large constant.
3. Compute the new weights according to $d_{t,i} \propto \exp(-\rho(\boldsymbol{u}_{t,i}, \boldsymbol{\alpha}_t))$ and normalise.

Output the final classifier: $H(\boldsymbol{x}) = \text{sign}\left(F(\boldsymbol{x})\right)$, where

$$F(\boldsymbol{x}) = \sum_{t=1}^{T} \alpha_t h_t(\boldsymbol{x}) \tag{2}$$

---

therein). In high-noise cases these often happen to be outliers. This is of particular relevance to our clustering application because of the arbitrariness in the initial choice of the clusters (see Section 3).

A number of algorithms have been developed to alleviate this issue [7, 18, 6, 22, 21]. In the Adaboost-REG algorithm [18] this is done by substituting the *soft margin* of a point for its hard margin. This is an estimate of the hard margin corrected by a measure of how much we trust the point.

In the following, we will indicate the m training examples as $\{(\boldsymbol{x}_1, y_1), \ldots, (\boldsymbol{x}_m, y_m)\}$, where the $\boldsymbol{x}_i$ are vectors in a feature space X and the $y_i \in \{-1, +1\}$ are the respective class labels. Given a family of weak learners $\mathcal{H} = \{h_t : X \rightarrow \{-1, +1\}\}$, it is convenient to define $u_{t,i} = y_i h_t(\boldsymbol{x}_i)$, so that $u_{t,i}$ is $+1$ if h_t classifies $\boldsymbol{x}_i$ correctly, and -1 otherwise. If we now let $d_{t,i}$ be the distribution of weights at time t, with $\sum_i d_{t,i} = 1$, the cycle of iterations can be described as in Table 1 in a way that summarises both Adaboost and Adaboost-REG.

The regularization term ζ that appears in Equation 1 is used in Adaboost-REG to quantify the "mistrust" of each particular point. It is defined based on a measure of influence of the pattern on the combined hypothesis [18]:

$$\zeta(\boldsymbol{u}_{t,i}, \boldsymbol{\alpha}_t) = \left(\sum_t \alpha_t d_{t,i}\right)^2, \tag{3}$$

based on the intuition that a pattern that is often misclassified will have a high influence. Note that the minimisation in Equation 2 must be carried out numerically with some linear search algorithm, e.g. the Golden Section algorithm [17].

2.1 Using Weak Learners That Abstain

Because of the local nature of the clustering problem, we would like to be able to use weak learners that specialise on a particular region of feature space. This can be done by allowing weak learners to abstain on part of the feature vectors. To this aim, we extend to the case of Adaboost-REG one of the estimates for Alpha reported in [19], specifically:

$$\alpha = \frac{1}{2} \ln \left(\frac{W_+ + W_0/2}{W_- + W_0/2} \right). \tag{4}$$

where W_+ is the sum of the weight of the examples that the weak hypothesis classifies correctly, W_- is the misclassified weight and W_0 the weight on which the hypothesis abstains.

We extend this to Adaboost-REG by adding the following contribution to $Z_t(\alpha_t, \boldsymbol{\alpha}_{t-1})$ for each point i on which the weak learner h_t abstains:

$$\frac{1}{2} e^{-\rho_t^+(\boldsymbol{u}_{t,i}, \alpha_t) + C\zeta(\boldsymbol{u}_{t,i})} + \frac{1}{2} e^{-\rho_t^-(\boldsymbol{u}_{t,i}, \alpha_t) + C\zeta(\boldsymbol{u}_{t,i})}. \tag{5}$$

where

$$\rho_t^+(\boldsymbol{u}_{t,i}, \boldsymbol{\alpha}) = \rho_{t-1}(\boldsymbol{u}_{t-1,i}, \boldsymbol{\alpha}_{t-1}) + \alpha_t \tag{6}$$

and

$$\rho_t^-(\boldsymbol{u}_{t,i}, \boldsymbol{\alpha}) = \rho_{t-1}(\boldsymbol{u}_{t-1,i}, \boldsymbol{\alpha}_{t-1}) - \alpha_t. \tag{7}$$

Thus the penalty term for the point is calculated by treating half of its weight as misclassified and the other half as correctly classified; this is a generalisation to the regularised case of the expression for α in Equation 4.

3 Iterative Clustering with Adaboost

We propose an iterative clustering scheme outlined in Table 2. Clusters are iteratively re-estimated using one-vs-all boosted classifiers.

Within-cluster homogeneity and separation between the clusters are obtained by a combination of three mechanism:

1. use of regularised Adaboost to reject outliers, as detailed in Section 2;
2. use of weak learners (possibly localised) inspired to Subtractive Clustering, as explained in Section 3.1 below;
3. smoothing of the decision functions with a Gaussian kernel, see point 2 in Table 2.

Table 2. Iterative clustering with Adaboost

The Cloosting algorithm:

Input: points $\{\boldsymbol{x}_i\}$, number of clusters K.

Assign initial cluster labels k_i to each $\boldsymbol{x}_i$ either at random or based on a k-means run

Repeat until convergence:

1. Train K strong classifiers $F_k(\boldsymbol{x})$ to recognise the elements of each cluster, using the Adaboost-REG algorithm in Table 1
2. Compute the scores $S_k(\boldsymbol{x}_i) = \sum_j F_k(\boldsymbol{x}_j) \exp(-\beta \|\boldsymbol{x}_i - \boldsymbol{x}_j\|^2)$
3. Assign each point $\boldsymbol{x}_i$ to cluster $k_i = \arg\max_k S_k(\boldsymbol{x}_i)$

Output the final clusters $\{(\boldsymbol{x}_i, k_i)\}$

Regularization through smoothing in a boosted clustering setting also appears in [1], where however it is applied to a separate set of 2D spatial coordinates associated to the feature vectors and not to the vectors $\boldsymbol{x}_i$ themselves. The spatial weak learners used in [1] are ineffective in our context, while smoothing the final decision functions greatly improves the stability of the algorithm.

Experimentally, the algorithm is found to reliably converge within a limited number of iterations; for details, see Section 4.

3.1 "Subtractive" Weak Learners

We use spherical neighbourhoods ("balls") in feature space as weak learners. A weak learner classifies points that fall inside the ball as belonging to the cluster; the other points are rejected (i.e. assigned to any other cluster).

When training the boosted classifier for cluster k, at each iteration the centre of a ball learner is chosen as the point $\boldsymbol{c}_{k,t} = \boldsymbol{x}_{i^*(t)}$ that maximises the following weighted data density:

$$D(\boldsymbol{x}_{i^*(t)}) = \sum_{j|\boldsymbol{x}_j \in k} d_{j,t} e^{-\gamma \|\boldsymbol{x}_i - \boldsymbol{x}_j\|^2} - \Delta(\boldsymbol{x}_i, \boldsymbol{d}) \tag{8}$$

where $\gamma = 4/r_a^2$, r_a is the effective radius of the neighbourhood on which the average is computed, and $\Delta(\boldsymbol{x}_i, \boldsymbol{d})$ is a penalty term that discounts previously chosen centres:

$$\Delta(\boldsymbol{x}_i, \boldsymbol{d}) = \sum_{\tau=1}^{t-1} d_{i^*(\tau),\tau} e^{-\delta \|\boldsymbol{x}_i - \boldsymbol{x}_{i^*(\tau)}\|^2}. \tag{9}$$

This is a weighted version of the density function used for the selection of cluster centres in Subtractive Clustering [4,5]. Once a centre is chosen, the radius of the ball is optimised for the best weighted error rate, according to the Adaboost distribution of weights.

It is observed that the number of centres selected adaptively by Adaboost for representing each cluster often diminishes as better centres are chosen in later iterations of Cloosting.

3.2 Weak Learners That Abstain

In order to increase the locality of the clustering process, we tested our algorithms using weak learners that specialise on a local area in feature space. These are obtained by selecting a centre in feature space based on weighted density, as described above. Two spherical neighbourhoods are then constructed around this centre. Points inside the inner sphere are accepted; points in the spherical shell between the two surfaces are rejected, and the classifier abstains on points outside the outer sphere. The radii of the two surfaces are chosen to minimise the penalty function

$$\tilde{Z} = W_0 + 2\sqrt{W_+ W_-} \tag{10}$$

where notation is as in Section 2.1. This strikes a compromise between classification accuracy and the effective domain of the weak learner [19]. We have found it useful nevertheless to constrain the radius of the outer sphere to be larger than a multiple of the radius of the inner sphere, in order to avoid an excessive localisation of the weak learners.

Using specialist weak learners forces all decisions to be local, which might be advantageous in the case, for instance, of non-convex clusters. From the point of view of the boosting algorithm, these weak learners are treated as specified in Section 2.1

4   Experimental Evaluation

The experimental evaluation was based on four well-known real data-sets: the Iris dataset, the Wisconsin breast cancer (referred to as WBC), the Ionosphere data set – all from the USC Machine Learning Repository[1] – and the Biomed dataset[2]. Datasets differ in terms of number of patterns and dimension – details may be found in Table 3. All datasets have been standardised before applying the clustering methodology.

Table 3. Description of the datasets used for testing

Dataset	n. of clusters	n. of patterns	n. of features
Iris	3	150	4
Biomed	2	194	5
WBC	2	683	9
Ionosphere	2	351	34

---

[1] Available at http://archive.ics.uci.edu/ml/datasets.html
[2] Available at http://lib.stat.cmu.edu/datasets/

Table 4. Results for Cloosting on different publicly available datasets

Dataset	With Abstention		Without Abstention		K-Means
	RAND-INIT	KM-INIT	RAND-INIT	KM-INIT	
Iris	88.7%	**94.7%**	90.7%	90.7%	89.0%
Biomed	88.6%	88.7%	88.1%	**89.2%**	88.7%
WBC	96.8%	**97.2%**	96.6%	96.6%	96.1%
Ionosphere	**72.9%**	62.1%	69.8%	64.4%	65.2%

We compare two versions of the proposed approach, the first with weak learners that abstain, the second with the original version of the weak learners. Tests with two kinds of initialisation have been carried out: in the former, the initial labels have been chosen using the k-means algorithm (KM-INIT), while in the latter they have been randomly selected (RAND-INIT). In order to be robust against initialisation-driven fluctuations (also the k-means algorithm starts from a random assignment), all experiments have been repeated 5 times and the best result has been taken.

A preliminary evaluation of the impact of the choice of the parameters has shown that the number of Adaboost iterations T, the regularization parameter C of Adaboost-REG and the constants γ and δ for calculating the data density in subtractive weak learners are fairly insensitive; in all experiments they have been fixed to $T = 15$, $C = 20$ and $\gamma = 1.5$ respectively, with $\delta = 3\gamma/4$. On the contrary, the constant β of the smoothing kernel in Equation 2 of Table 2 and (limited to weak learners that abstain) the minimum ratio between the outer sphere and the inner sphere of the learner have proved to be dataset dependent, so that a different value has been used in each case.

Since true labels are known, clustering accuracies can be quantitatively assessed. In particular, given a specific group, an error is considered when a pattern does not belong to the most frequent class inside the group (following the protocol of [3]). The results obtained for each dataset with the different versions of the algorithm are displayed in Table 4. As a reference, in the last column, we report the results obtained with the standard k-means (for the sake of fairness, also in this case we performed 5 random initialisations, picking the best result).

As can be inferred from the table, Cloosting appears to work rather well on these datasets, that are heterogeneous in terms of the number of patterns and of dimensionality of the feature space. Concerning the different versions of the method, the use of weak learners that abstain seems to increase performance. Moreover, it may be noted that k-means initialisation seems to permit a more accurate clustering than random initialisation. Interestingly this is not true for the Ionosphere database, on which k-means performs rather poorly. In this case our method works better if started from random assignments. Finally it may be noted that Cloosting compares favourably with the k-means algorithm (for two datasets the improvement is significant), especially when the dimension of the space is rather large (as in the Ionosphere case, dimensionality 34).

Table 5. Comparative results for three datasets. We refer the reader to the cited papers for details of the experimental protocols.

Iris	
Self organising maps (SOM) [12]	81.0
Neural gas [13]	91.7
Spectral clustering [15]	84.3
Gaussian Mixture Models [14]	89.3
Kernel clustering [3]	**94.7**
Soft kernel clustering [2]	93.3
Cloosting	**94.7**

WBC	
Self organising maps (SOM) [12]	96.7
Neural gas [13]	96.1
Spectral clustering [15]	95.5
Gaussian Mixture Models [14]	94.6
Kernel clustering [3]	97.0
Soft kernel clustering [2]	97.1
Cloosting	**97.2**

Biomed	
Kernel clustering [3]	83.0
Soft kernel clustering [2]	88.2
Cloosting	**89.2**

As a further comparison with the literature, in Table 5 we report state of the art results[3] for some of the datasets – we also include our best result for comparison. It may be noticed that the performances of Cloosting are in line with those of other state of the art methods.

A final consideration concerns the convergence of Cloosting. Although we do not yet have a theoretical justification for the convergence of the iterative clustering algorithm, in all our experiments the method always converged; the number of iterations required was, on average, around ten to fifteen, depending of course on the initialisation (for k-means initialisation convergence was clearly faster).

5   Conclusions

In this work we take a novel look at clustering from a boosting perspective. Each cluster is modelled by a regularised boosted classifier composed of weak learners inspired by Subtractive Clustering, that can be localised to different areas of feature space. Learning follows a cycle of iterations similar to k-means;

---

[3] Some of the results have been computed by the authors, some others have been taken from [3] and [2].

a smoothing kernel is used to stabilise the assignment of points to the various clusters.

The convergence properties of the algorithm are at this early stage established empirically, and a smart parameter setting policy still needs to be designed. However, the experiments we report show that Cloosting achieves optimal results on several standard datasets, either matching or improving the performance of state of the art algorithms. This proves, in our view, the potential of the technique and justifies further inquiry into the use of boosting for modelling clusters.

Acknowledgements

Most of the present work was done while F. Smeraldi was visiting professor at PLUS Lab, Italian Institute of Technology, Genoa; F.S. gratefully acknowledges the financial support of IIT.

References

1. Avidan, S.: SpatialBoost: Adding spatial reasoning to adaBoost. In: Leonardis, A., Bischof, H., Pinz, A. (eds.) ECCV 2006. LNCS, vol. 3954, pp. 386–396. Springer, Heidelberg (2006)
2. Bicego, M., Figueiredo, M.: Soft clustering using weighted one-class support vector machines. Pattern Recognition 42(1), 27–32 (2009)
3. Camastra, F., Verri, A.: A novel kernel method for clustering. IEEE Trans. on Pattern Analysis and Machine Intelligence 27, 801–805 (2005)
4. Chiu, S.: Fuzzy model identification based on cluster estimation. Journal of intelligent and fuzzy systems 2, 267–278 (1994)
5. Chiu, S.L.: In: Dubois, D., Prade, H., Yager, R. (eds.) Fuzzy Information Engineering: A guided tour of applications, ch. 9, John Wiley & Sons, Chichester (1997)
6. Demiriz, A., Bennet, K.P., Shawe-Taylor, J.: Linear programming boosting via column generation. Machine Learning 46, 225–254 (2002)
7. Freund, Y.: An adaptive version of the boost by majority algorithm. In: Proceedings of the Twelfth Annual Conference on Computational Learning Theory, pp. 102–113 (2000)
8. Freund, Y., Schapire, R.: A decision-theoretic generalization of on-line learning and an application to boosting. Journal of Computer and System Science 55(1) (1997)
9. Freund, Y., Schapire, R.: A short introduction to boosting. Journal of Japanese society for Artificial Intelligence Science 14(5) (1999)
10. Frossyniotis, D., Likas, A., Stafylopatis, A.: A clustering method based on boosting. Pattern Recognition Letters 25, 641–654 (2004)
11. Kim, T.K., Cipolla, R.: MCBoost: multiple classifer boosting for perceptual co-clustering of images and visual features. In: Advances in Neural Information Processing Systems, Vancouver, Canada, pp. 841–856 (2008)
12. Kohonen, T.: Self-Organizing Maps. Springer, Heidelberg (1997)
13. Martinetz, T., Schulten, K.: Neural-gas network for vector quantization and its application to time-series prediction. IEEE Trans. on Neural Networks 4(4), 558–569 (1993)

14. McLachlan, G., Peel, D.: Finite mixture models. John Wiley and Sons, Chichester (2000)
15. Ng, A., Jordan, M., Weiss, Y.: On spectral clustering: Analysis and an algorithm. In: Advances in Neural Information Processing Systems, pp. 849–856 (2001)
16. Nock, R., Nielsen, F.: On weighting clustering. IEEE Transactions on PAMI 28(8), 1223–1235 (2006)
17. Press, W.H., Teukolsky, S.A., Vetterling, W.T., Flannery, B.P.: Numerical recipes in C++, ch.10 Cambridge University Press, Cambridge (2002)
18. Rätsch, G., Onoda, T., Muller, K.R.: Soft margins for adaboost. Machine Learning 42(3), 287–320 (2001), citeseer.ist.psu.edu/657521.html
19. Schapire, R., Singer, Y.: Improved boosting algorithms using confidence-rated predictions. Machine Learning 37(3) (1999)
20. Topchy, A., Minaei-Bidgoli, B., Jain, A.K., Punch, W.F.: Adaptive clustering ensembles. In: Proc. 17th Conf. Pattern Recognition, pp. 272–275 (2004)
21. Warmuth, M., Glocer, K.A., Rätsch, G.: Boosting algorithms for maximizing the soft margin. In: Advances in Neural Information Processing Systems 20, pp. 1585–1592. MIT Press, Cambridge (2008)
22. Warmuth, M.K., Glocer, K.A., Vishwanathan, S.V.N.: Entropy regularized lP-Boost. In: Freund, Y., Györfi, L., Turán, G., Zeugmann, T. (eds.) ALT 2008. LNCS (LNAI), vol. 5254, pp. 256–271. Springer, Heidelberg (2008)

A Geometric Approach to Face Detector Combining*

Nikolay Degtyarev and Oleg Seredin

Tula State University
n.a.degtyarev@gmail.com
http://lda.tsu.tula.ru

Abstract. In this paper, a method of combining face detectors is proposed, which is based on the geometry of the competing face detection results. The main idea of the method consists in finding groups of similar face detection results obtained by several algorithms and further averaging them. The combination result essentially depends on the number of algorithms that have fallen in each of the groups. The experimental evaluation of the method is based on seven algorithms: Viola-Jones (OpenCV 1.0), Luxand© FaceSDK, Face Detection Library, SIFinder, Algorithm of the University of Surrey, FaceOnIt, Neurotechnology© VeriLook. The paper contains practical results of their combination and a discussion of future improvements.

Keywords: combining classifiers, face detection, clustering of detector outputs, combination of face detectors, comparative test.

1 Introduction

The state-of-the-art algorithms of face detection (FD) have excellent performance for many tasks [1,2,3,8,13]. However, even the best of them still have significant error rates, e.g., $5 - 6\%$ False Rejection Rate, separating them from the desired error-free result. At the same time, it was shown by Degtyarev et al. [3] that the percentage of challenging images incorrectly processed by all tested algorithms is much smaller, only 0.13%, whereas each of the remaining 99.87% of images is correctly processed by at least one of the algorithms.

This fact reveals the possibility of reducing the error rate of face detection through harnessing several diverse algorithms in parallel. We call such principle the *combination* or *fusion of face detectors* on the analogy of the commonly adopted term of *combination/fusion of classifiers*, introduced by J. Kittler [11] and R. Duin [5].

In classifier combining, an object submitted to analysis is supposed to be indivisible, and the final output is result of jointly processing a number of elementary decisions on its class membership – voting, optimal weighting, etc. However, as

---

* This work is supported by the Russian Foundation for Basic Research, Grant No. 09-07-00394.

C. Sansone, J. Kittler, and F. Roli (Eds.): MCS 2011, LNCS 6713, pp. 299–308, 2011.

to face detector combining, images under processing are not atomic, and what is to be fused is an ensemble of diverse suggestions on the position of the face in the given image in addition to the information of its presence.

In this paper, we propose a quite naive and obvious *geometric approach* to fusing several face detectors, which takes into account only geometric properties of their outputs and ignores, for the computational simplicity sake, other individual properties like False Rejection/Acceptance Rates (FRR/FAR), Confidence Rates, etc. The main principle of combining consists in clustering of the detected face represented by the centers of the eyesand further averaging the cluster centers with respect to the portions of the detectors that have fallen in each of the clusters.

The most tangible disadvantage of the very idea of combining several diverse face detectors is the increasing computational time. However, the recent advances in the multi-core CPU technology (Central Processing Unit) allows, in principle, for a natural parallelization by the scheme "one detector – one core".

2   Models of Face Representation and Localization Accuracy

To combine or correctly compare face detectors, they should represent faces in a unified form. Coordinates of the eye centers (i.e., centers of the pupils) are the most suitable description of faces for these tasks. The reasons for this proposition are, first, the convenience of this kind of representation from the viewpoint of comparing the results, second, the necessity of matching the eye centers as an inevitable step in the majority of learning algorithms, and, third, the fact that ground-truthing eyes by a human is faster, easier and can be done more confidently than locating faces by rectangles.

Thus, we consider *all faces* as represented by their eye centers. If some FD returns a face location in the rectangular form, we first additionally estimate the coordinates of the eye centers by the eye reconstruction algorithm proposed in [3] and examined in [4].

If a detected face is represented by the centers of the eyes (Fig. 1), we consider them as correctly detected, if and only if the detected eyes belong to the pair of circles D^A around the true locations of the eyes. The common diameter of the circles $D_{Eyes} = 2\alpha \times l_{Eyes}$ depends on the distance l_{Eyes} between the centers of the eyes with coefficient α taken equal to 0.25. This criterion was originally used by Jesorsky et al. [7]).

3   A Geometric Method of Face Detectors Combining

As mentioned above, the proposed method of FD combining is based on the *geometric approach*. This means that the method takes into account only the positions of detected faces and disregards any additional information related to or provided by FD e.g. false rejection or acceptance error rate, confidence rate of detection, etc. Before describing the method, let us introduce some definitions.

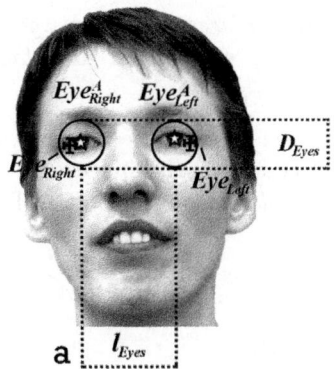

Fig. 1. Schematic face representation. Eye_{Left} and Eye_{Right} – absolute coordinates of detected left and right eyes respectively; l_{Eyes} – distance between eye centers; D_{Eyes} – diameter of the area of acceptable eyes' coordinates deviation from the true eyes location Eye_{Right}^{A} and Eye_{Left}^{A};

Definition 1. *The distance between the faces of a given pair (g,h) each represented by the centers of the eyes is the greatest of the Euclidean distances between the left eyes of the pair and the right ones:*

$$d_{Faces}(g,h) = \max\left(\left\|Eye_{Left}^{g} - Eye_{Left}^{h}\right\|, \left\|Eye_{Right}^{g} - Eye_{Right}^{h}\right\|\right). \quad (1)$$

Hereafter, Eye_{Right}^{g} and Eye_{Left}^{g} stand for the coordinates of, respectively, the left and the right eye of the given face g.

Definition 2. *The merged face is a synthetic pair of the coordinates of eye centers, averaged among the given group of K faces:*

$$Eye_{Left}^{Merge} = \frac{1}{K}\sum_{i=1}^{K} Eye_{Left}^{i}, \quad Eye_{Right}^{Merge} = \frac{1}{K}\sum_{i=1}^{K} Eye_{Right}^{i},$$
$$l_{Eyes}^{Merge} = \frac{1}{K}\left\|\sum_{i=1}^{K} Eye_{Left}^{i} - \sum_{i=1}^{K} Eye_{Right}^{i}\right\|. \quad (2)$$

The model of eyes localization accuracy described in Section 2 implies that if each algorithm of a group of algorithms has correctly detected a face, then the distances between the detected faces are smaller or equal to the diameter D_{Eyes} of the respective area D^{A}, i.e.:

$$d_{Faces}(g,h) \leq D_{Eyes} \leq 2\alpha \times l_{Eyes}. \quad (3)$$

A merged face based on a group of accurate algorithms may be treated as a correctly detected face, too. If some algorithms in the group have incorrectly detected a face, the merged face based on all results of the group may still be a correctly detected face, depending on the number of incorrectly estimated positions and the errors of estimates.

Definition 3. *A given pair of faces (g, h) is mergeable if and only if the distance between them is at least 2α times smaller than the interocular distance of the corresponding merged face* $l_{Eyes}^{Merge}(g, h)$, *i.e.* $d_{Faces}(g,h) \leq 2\alpha \big(l_{Eyes}^{Merge}(g, h) \big)$.

In other words, the pair of *mergeable* faces will be correctly detected, if position of the corresponding merged face on the image is the true face location. Examples of *mergeable* and *nonmergeable* faces are given in Fig. 2.

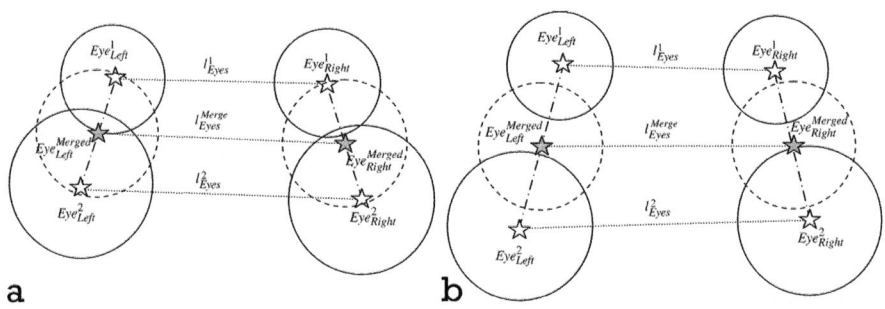

Fig. 2. Examples of mergeable (*a*) and non mergeable (*b*) pairs of faces

In practice, we don't know whether an algorithm has detected a face correctly or incorrectly, nor the true location of the face in the image, and even nor whether a face is in the image at all. Therefore, we suppose that correctly detected faces form clusters around the "true face location", whereas incorrectly detected faces must be scattered in the image. Such clusters may be defined as follows.

Definition 4. *A group of faces forms a cluster if and only if there is at least one face (further called the "center") among the group, that is meargeable with all other faces (in the group).*

Such definition of a cluster of faces is less strict than (3), and allows for intersecting clusters of detected faces, i.e., one face can be member of several different clusters. However, the merged face based on the largest cluster is more likely to be result of correct detection than one based on all other clusters. This consideration is the essence of Algorithm 1. The algorithm consists in repeatedly replacing all faces in the largest cluster by corresponding merged faces, until there exists at least one non-trivial cluster (i.e. with size greater than 1). When at some step the remaining clusters become trivial, namely, each of them contains one face, the algorithm selects the merged face produced by the greatest number of originally detected faces, but not less than the preset *Threshold*, which is the only parameter of the algorithm, otherwise, the algorithm makes the decision that the image contains no face.

It should be emphasized that after replacing clusters by corresponding merged faces, each merged face can become a part of other clusters, etc. A simulated example of combining Face Detectors by this algorithm is shown in Fig. 3. The proposed method is discussed in Section 6.

Algorithm 1. A Geometrical Method of Face Detectors Combining.

Require: α, Threshold;

DetFaces $= \{(Eye_{Left}, Eye_{Right}, merge_{count} = 1), \ldots\}$;

Ensure: $(Eye_{Left}, Eye_{Right})$

  loop

    Determinate the largest cluster of the faces (see Definition 4);

    if the size of the found cluster is > 1 **then**

      Replace all faces in the cluster by the corresponding merged face in DetFaces

    else

      Since there are no clusters consisting of more than one face, select merged face ($Face_m$), that has been originated by the greatest, but not least than the Threshold number of initial faces.

      if such merged face is founded **then**

        return the merged face ($Face_m$);

      else

        return NOT_FACE;

      end if

    end if

  end loop

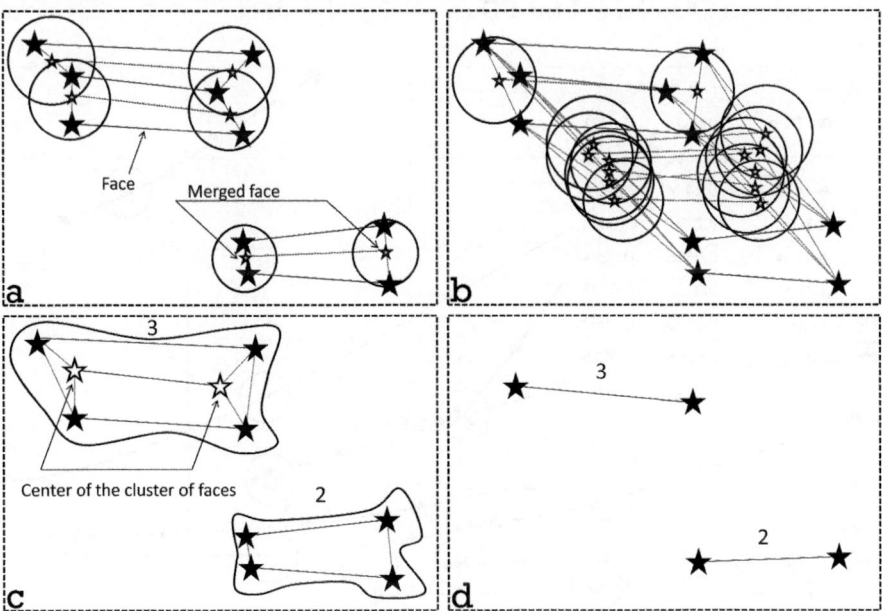

Fig. 3. Successive steps of combining simulated outputs of five Face Detectors by Algorithm 1: mergeable (a) and nonmergeable (b) pairs of faces; clusters of faces (c); substitution of clusters by the corresponding merged faces (d)

4 Experimental Procedure

In this work, we combined the following implementations of different algorithms: Viola-Jones [14] (OpenCV 1.0, OCV); Luxand FaceSDK (FSDK, http://www.luxand.com); Face Detection Lib. (FDLib) [9]; SIFinder (SIF) [10]; Algorithm of the University of Surrey (UniS); FaceOnIt [12] (FoI, http://www.faceonit.ch); Neurotechnology VeriLook (VL, http://www.neurotechnology.com).

The result of each algorithm was evaluated by the following parameters:

- *False Rejection Rate* (FRR) — Ratio of type I errors, which indicates the probability of misclassification of the images containing a face;
- *False Acceptance Rate* (FAR) — Ratio of type II error, which indicates the probability of misclassification of the images not containing a face.

The total size of the test dataset is 59 888 images, namely, 11 677 faces and 48 211 non-faces. More information on experimental data and comparative testing of FD algorithms can be found in [3].

5 Results

The proposed face detector combining algorithm had been routinely tested in accordance with the procedure described in Degtyarev et al. [3]. The main idea of the procedure consists in computing FRR, FAR and vectors of algorithm's errors for each *Threshold* of the FD combining algorithm.

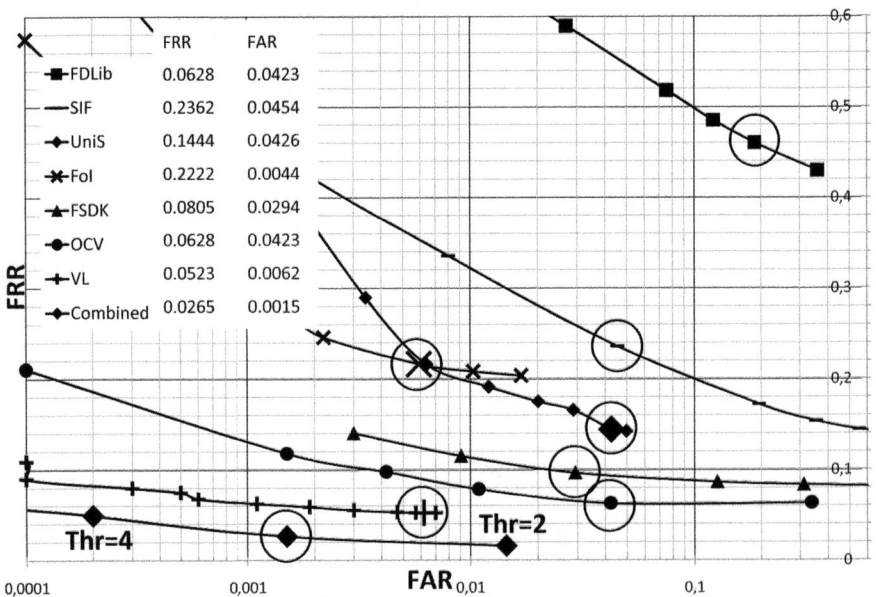

Fig. 4. The ROC plots "FAR in the log scale against FRR" as functions of the tuning parameter *Threshold*. The perfect performance would be the bottom left corner: $FRR = FAR = 0$. Circled points correspond to the tuning parameter value that delivers the minimal detection error for each algorithm.

The receiver operating characteristic (ROC curves) "FAR in the log scale against FRR in the standard biometric sense" [6,15] for all the tested face detectors and their combination by our algorithm are presented in Fig. 4. These curves let us to identify the algorithm with the best overall performance, because the closer the curve to the perfect-performance-point $FRR = FAR = 0$ (the bottom left corner), the better the performance. As we can see, the proposed method of face detector combining does improve the performance of each of the FD algorithms to be combined for 2-3%. Nevertheless, there is still room for future development.

It is obvious that each of the algorithms have unique peculiarities of detection. One way to perform their numerical evaluation is to compare the number of images uniquely classified by each algorithm, as well as the numbers of "challenging" and "easy" images (see Table 1). Here the term *easy images* means the images detected by all algorithms, in the opposite case images are considered as *challenging*.

Table 1. Peculiar images distribution on the datasets

Cases	Number (faces)	% in DB
easy images	38 478 (4 385)	64,25
challenging images	78 (78)	0,13
only OCV	5 (5)	< 0,01
only SIF	5 (5)	< 0,01
only FDL	3 (3)	< 0,01
only FSDK	10 (10)	0,02
only UniS	20 (20)	0,04
only FoI	22 (22)	0,04
only VL	49 (49)	0,08
only Comb	3 (3)	< 0,01

For a better understanding of the potential of the proposed FD combining method, let us take a look to some exemplary cases. As we can see in Fig. 5, all seven algorithms and their combination by the proposed method correctly detected the given faces. In Fig. 6 we can see two examples of non-face images where the two best algorithms (OCV and VL) falsely detected faces, whereas the proposed FD combining method did not find any faces in them, because there were no clusters of faces containing at least 3 elements. In Fig. 7, a much more interesting case is presented – the face correctly detected by the proposed FD combining method, whereas all the original algorithms failed, each of them correctly marked no more than one eye in this face. This case is noteworthy, because our algorithm does not detect any new faces, it only successively finds and merges the largest cluster of faces found by other face detectors. According Table 1, there are also two cases like this one.

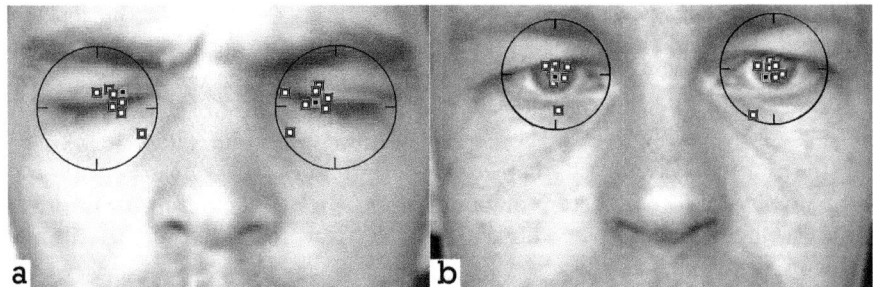

Fig. 5. Results of face detector combining on easy images containing faces; (a) – the pupils of the eyes are not visible; (b) – the pupils are visible

Fig. 6. Face detectors combining results on images not containing faces; two leading FD algorithms (OCV and VL) incorrectly detected faces; Combined algorithm did not find any suitable clusters of faces in given images, thus this images were considered to be *non faces*

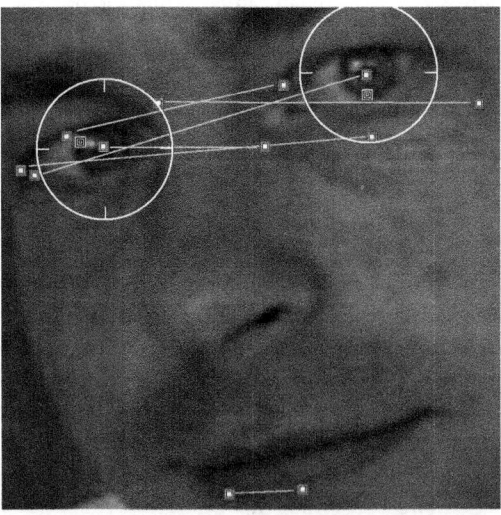

Fig. 7. "Challenging" case of face detectors combining; face correctly detected only by FD combining, because 5 algorithms' outputs (faces) formed a cluster

6   Discussion and Conclusion

We have demonstrated in our experiments that the proposed method of FD combining has better performance (FRR 2.65%, FAR 0.15%) than each of its component algorithms. For comparison, VL and OCV, which are known to merge candidate windows according to some criteria, give, respectively, (FRR 5.23%, FAR 0.62%) and (FRR 6.54%, FAR 2.01%).

The method also has ability to correct some of detection errors made by all "elementary" algorithms (see Fig. 7). Nevertheless, there is a sufficient performance gap of $2-3\%$ FRR that separats it from the desired error-free result. This gap can be eliminated through further development, tuning of the FD combining algorithm and/or adding additional FD algorithms to the collection.

Perhaps, the most significant open question at this stage is the choice of the method of tuning free parameters in single algorithms to be combined. In this work, free parameters in each of the algorithms were chosen to deliver the minimal detection error $\left(\sqrt{FRR^2 + FAR^2} \to \min\right)$, that are not proven to be optimal for the FD combining task. It even may be better to combine a mix of the algorithms with two values of tuning parameters, delivering one the minimal FRR and the other the minimal FAR, because this would prevent forming false clusters and allow to lower the *Threshold* of cluster acceptance.

Exactly the same motivation leads us to another interesting and prospective idea – *self-combining*. It consists in combining results of only one algorithm, but with several different values of tuning parameters. Such an approach would allow us to eliminate misdetected faces, because they must not remain steady as the parameters will be changing, whereas true faces are expected to have fixed intervals of FD tuning parameters outside which they must disappear from the output of the respective algorithm.

Another interesting method of FD combining may consists in selecting a "most likely to be a face" region among originally detected alleged face regions using the decision tree learning approach. As features might be used, for instance distances between the results of algorithms, whether originally detected face regions were detected as faces by other FD algorithms or not, etc. Such method would return only originally detected face positions and not eliminate detection errors made simultaneously by all algorithms (see Fig. 7) in contrast to the proposed geometric method.

As mentioned above, the geometric approach takes into account only face positions in the outputs of detectors. Contrariwise, additional information provided by or linked to some FD algorithms may be helpful for combining face detectors, in particular, for finding the optimal different weights for outputs of different algorithms (FRR and FAR look to be suitable for this role).

It should be emphasized, that the aim of this work is only to show the possibility of face detector combining, and the proposed method is only the first attempt. We believe that disadvantages of this method will lead to much better approaches in solving of the above-mentioned problems.

Acknowledgements

The research leading to these results has received funding from the Russian Foundation for Basic Research, Grant No. 09-07-00394. We also would like to acknowledge many colleagues for discussions about this problem, in particular Glen Ivin, Alexey Kostin and Vadim Mottl.

References

1. Beveridge, J.R., Alvarez, A., Saraf, J., et al.: Face Detection Algorithm and Feature Performance on FRGC 2.0 Imagery. In: Beveridge, J. (ed.) Proceedings of First International Conference on Biometrics, pp. 1–7. IEEE, Los Alamitos (2007)
2. Castrillon, M., Deniz, O., Hernandez, D., Lorenzo, J.: A comparison of face and facial feature detectors based on the Viola Jones general object detection framework. Machine Vision and Applications, pp. 1–14. Springer, Heidelberg (2010), doi:10.1007/s00138-010-0250-7
3. Degtyarev, N., Seredin, O.: Comparative Testing of Face Detection Algorithms. In: Elmoataz, A., Lezoray, O., Nouboud, F., Mammass, D., Meunier, J. (eds.) ICISP 2010. LNCS, vol. 6134, pp. 200–209. Springer, Heidelberg (2010)
4. Degtyarev, N., Seredin, O.: Effect of Eyes Detection and Position Estimation Methods on the Accuracy of Comparative Testing of Face Detection Algorithms. In: 10th International Conference on Pattern Recognition and Image Analysis: New Information Technologies (PRIA-10-2010), St. Petersburg, December 5-12, pp. 261–264. Politechnika (2010)
5. Duin, R.P.W., Tax, D.M.J.: Experiments with classifier combining rules. In: Kittler, J., Roli, F. (eds.) MCS 2000. LNCS, vol. 1857, pp. 16–29. Springer, Heidelberg (2000)
6. Fawcett, T.: An introduction to ROC analysis. Pattern Recognition Letters 27(8), 861–874 (2006)
7. Jesorsky, O., Kirchberg, K.J., Frischholz, R.W.: Robust face detection using the hausdorff distance. LNCS, pp. 90–95 (2001)
8. Jones, M.J.: Face Recognition: Where We Are and Where To Go From Here. IEEJ Trans. on Elect., Information and Systems 129(5), 770–777 (2009)
9. Kienzle, W., Bakir, G., Franz, M., Scholkopf, B.: Face detection – efficient and rank deficient. Advan. in neural inform. process. systems 17, 673–680 (2005)
10. Krestinin, I.A., Seredin, O.S.: Excluding cascading classifier for face detection. In: Proc. of the 19th Int. Conf. on Computer Graphics and Vision, pp. 380–381 (2009)
11. Kittler, J.: Combining classifiers: A theoretical framework. Pattern Analysis & Applications 1(1), 18–27 (1998)
12. Marcel, S., Keomany, J., et al.: Robust-to-illumination face localisation using Active Shape Models and Local Binary Patterns. In: IDIAP-RR, vol. 47 (2006)
13. Mohamed, N.M., Mahdi, H.: A simple evaluation of face detection algorithms using unpublished static images. In: Mohamed, N.M., Mahdi, H. (eds.) Proc. of 10th Intern. Conference on Intelligent Systems Design and Applications, pp. 1–5. IEEE Computer Society Press, Los Alamitos (2010), doi:10.1109/ISDA.2010.5687301
14. Viola, P., Jones, M.J.: Robust Real-Time Face Detection. International Journal of Computer Vision 57, 137–154 (2004)
15. Wechsler, H.: Reliable face recognition methods: system design, implementation and evaluation, 329 p Springer, Heidelberg(2007)

Increase the Security of Multibiometric Systems by Incorporating a Spoofing Detection Algorithm in the Fusion Mechanism

Emanuela Marasco[1,*], Peter Johnson[2], Carlo Sansone[1], and Stephanie Schuckers[2]

[1] Dipartimento di Informatica e Sistemistica,
Università degli Studi di Napoli Federico II
Via Claudio, 21 I-80125 Napoli, Italy
{emanuela.marasco,carlosan}@unina.it
[2] Department of Electrical and Computer Engineering,
Clarkson University
PO Box 5720 Potsdam, NY 13699
{sschucke,johnsopa}@clarkson.edu

Abstract. The use of multimodal biometric systems has been encouraged by the threat of spoofing, where an impostor fakes a biometric trait. The reason lies on the assumption that, an impostor must fake all the fused modalities to be accepted. Recent studies showed that there is a vulnerability of the existing fusion schemes in presence of attacks where only a subset of the fused modalities is spoofed. In this paper, we demonstrated that, by incorporating a liveness detection algorithm in the fusion scheme, the multimodal system results robust in presence of spoof attacks involving only a subset of the fused modalities. The experiments were carried out by analyzing different fusion rules on the Biosecure multimodal database.

1 Introduction

A biological measurement can be qualified as a biometric if it satisfies basic requisite like universality, permanence, distinctiveness, circumvention. The last property concerns the possibility of a non-client being falsely accepted, typically by spoofing the biometric trait of an authorized user [1]. Previous works have shown that it is possible to spoof a variety of fingerprint technologies using spoof fingers made with materials as *Silicon, Play-Doh, Clay* and *Gelatin* (gummy finger) [2].

Multibiometric systems improve the reliability of the biometric authentication by exploiting multiple sources, such as different biometric traits, multiple samples, multiple algorithms. They are able to improve the recognition accuracy, to

---

* Emanuela Marasco is currently a post-doctoral candidate at the Lane Department of Computer Science and Electrical Engineering, West Virginia University, WV (USA).

C. Sansone, J. Kittler, and F. Roli (Eds.): MCS 2011, LNCS 6713, pp. 309–318, 2011.

increase the population coverage, to offer user choice and to make biometric authentication systems more robust to spoofing [3]. Several works in the literature on biometrics demonstrate the efficiency of the multimodal fusion to enhance the recognition accuracy of the unimodal biometric systems [4].

From a security perspective, a multimodal system appears more protected than its unimodal components. The reason is that, one assumes that an impostor must fake all the fused modalities to be accepted and spoofing multiple modalities is harder than spoofing only one [5]. However, a hacker may fake only a subset of the fused biometric traits. Recently, researchers demonstrated that the existing multimodal systems can be deceived also when only a subset of the fused modalities is spoofed [6]. Rodrigues *et al.* proposed an approach to measure the security of a multimodal system, where the contribution provided by each single modality matcher is weighted based on the ease to spoof that biometric trait. For example, the probability of success associated to a spoof attack is high in presence of a sample which gives a low match score. Johnson *et al.* [7] explored the multimodal vulnerability of the score level fusion strategies in a scenario where partial spoofing has occurred.

The goal of this paper is to propose an approach, based on liveness detection techniques, which can improve the security of multimodal biometric systems in presence of spoof attacks involving one fingerprint modality. We have analyzed the performance of different multibiometric systems in presence of partial spoofing when an effective spoofing detection algorithm is incorporated in the fusion mechanism. Our experiments showed that the proposed technique aids to increase the robustness of such systems with respect to the spoofing. In our approach the integration involves match scores, and the spoof attack is detected separately for each modality matcher before fusion. Thus, when a fake sample is detected by the algorithm, the unimodal output does not give any contribution in the fusion which results in a more secure decision.

The current analysis is carried out as a simulation to assess performance of multibiometric systems in presence of spoof attacks. The simulation makes the assumption that live match scores have a similar distribution with respect to spoof match scores. In future work, actual spoof data is needed to assess the performance in a real-world system. However, this simulation can provide a framework for assessing novel algorithms, as well as their relative performance.

The paper is organized as follows. Section 2 presents an overview of our approach, together with the combination rules we considered for our study and the liveness detection algorithm exploited in the fusion. Section 3 describes the adopted dataset and the experiments carried out on it, which show the effectiveness of the proposed technique. Section 4 draws our conclusions.

2   Our Approach

In the current approach, we have analyzed the performance of different multibiometric systems in presence of spoof attacks involving one fingerprint modality, when an effective spoofing detection algorithm is incorporated in the fusion mechanism. The final multimodal decision is made by considering that, when a spoofed sample is detected by the algorithm, the corresponding matcher does not give any contribution in the fusion scheme.

2.1 Score Fusion Rules

When designing a multibiometric system, several factors should be considered. These concern the choice and the number of biometric traits, the level of integration and the mechanism adopted to consolidate the information provided by multiple traits. Fusion at match score level is often chosen since it is easy to access and combine the scores presented by different modalities. The operators which do not contain parameters to be tuned, are known as *fixed* combiners [8]. Based on experimental results, researchers agree that *fixed* rules usually perform well for ensemble of classifiers having similar performance, while *trained* rules handle better matchers having different accuracies. When fusing different modalities, individual matchers often exhibit different performance, thus for this problem *trained* rules should perform better than *fixed* rules [9].

Transformation-based fusion. The match scores provided by different matchers are firstly transformed into a common domain (*score normalization*), then they are combined using a fusion rule. It has been shown that the simple sum rule gives very good accuracy [9]. The technique adopted in our fusion framework is the *min-max*, which retains the original distribution of scores except a scaling factor and transform the scores to a common range from zero to one, based on the minimum and the maximum score values. Given a set of matching scores s_k, $k = 1 \ldots K$, the normalized scores are given by (1).

$$s_k = \frac{s_k - min}{max - min} \tag{1}$$

The operator employed for the current analysis is the simple score sum, defined by (2)

$$s_{sum} = \sum_{k=1}^{N} \frac{1}{N} s_k \tag{2}$$

Density-based fusion. The match scores are considered as random variables, whose class conditional densities are not *a priori* known [10]. So, this approach requires an explicit estimation of density functions from the training data [5]. The model is built by estimating density functions for the genuine and impostor score distributions [11]. A recent method, proposed by Nandakumar et al. in [12], is the framework based on the Likelihood Ratio test, where the scores are modeled as mixture of Gaussians and a statistical test $\Psi(\mathbf{s})$ is performed to discriminate between genuine and impostor classes. The Gaussian Mixture Model (GMM) lets to obtain reliable estimations of the distributions, even if the amount of data needed for it increases as the number of considered biometrics increases. This framework produces high recognition rates at a chosen operating point (in terms of False Acceptance Rate), when it is possible to perform accurate estimations of the genuine and impostor score densities.

Let $\mathbf{s} = [s_1, s_2, ..., s_K]$ denote the scores emitted by multiple matchers, with s_k representing the match score of the k_{th} matcher, $k = 1, ..., K$.

$$\Psi(\mathbf{s}) = \begin{cases} 1, & \text{when } LR(\mathbf{s}) \geq \eta \\ 0, & \text{when } LR(\mathbf{s}) < \eta \end{cases} \qquad (3)$$

where $\mathbf{s} = [s_1, s_2, ...s_K]$ is an observed set of K match scores that is assigned to the genuine class if $LR(\mathbf{s})$ is greater than a fixed threshold η, with $\eta \geq 0$.

2.2   Spoofing Detector

Our multimodal fusion approach is evaluated assuming that *fake-live* match scores are similarly distributed as *live-live* match scores. For each modality, the spoof attack was simulated by substituting a genuine match score in place of an impostor match score. The multimodal system considered in this paper is composed by face and fingerprint traits and it is analyzed under normal operation (i.e., without spoofing), and when only a fingerprint trait is spoofed.

In all the scenarios, a fingerprint liveness detection is integrated in the fusion scheme. In this investigation, we incorporate the performance, known by the literature, of a liveness algorithm which combines perspiration- and morphology-based static features [13]. The classification performance of the adopted algorithm was evaluated by using the parameters of the Liveness Detection Competition 2009 (LivDet09) [14], defined as follows:

- *Ferrlive*: rate of misclassified live fingerprints.
- *Ferrfake*: rate of misclassified fake fingerprints.

In particular, the values of *Ferrlive* and *Ferrfake* were averaged on the three databases (*Biometrika*, *CrossMatch* and *Identix*) that compose the LivDet09 data. On such data, the liveness algorithm exploited in our approach presented an average *Ferrlive* of 12.60% and an average *Ferrfake* of 12.30% [13]. We used the performance obtained on these three different databases taken from LivDet09 as an estimate of the actual performance of the algorithm on the database used in this paper.

When a spoofed modality is detected by the incorporated algorithm, it will not give any contribution to the final decision. In particular, (100%-*Ferrfake*) indicates the percentage of correctly detected live-spoof match scores to be excluded from the combination, and *Ferrlive* indicates the percentage of wrongly detected live-live match scores to be excluded from the combination. In the proposed approach, live genuine match scores are employed for a real live genuine scenario, where FRR can be assessed, and also in place of impostor match scores in order to simulate spoofing.

- In the score sum scheme, this is realized by resetting a percentage of (100%-$Ferrfake$) impostor scores substituted by genuine scores, and a percentage of $Ferrlive$ genuine scores before performing the sum.
- In the likelihood ratio scheme, the detected spoofed modality can be marginalized by employing, for the (100%-$Ferrfake$) of the fake samples and for the $Ferrlive$ of the live samples, the joint density functions involving only the live modalities.

3   Experimental Results

3.1   Dataset

The performance of the proposed strategy was evaluated on a subset of the BioSecure multimodal database. This database contains 51 subjects in the Development Set (training) and 156 different subjects in the Evaluation Set (testing). For each subject, four biometric samples are available over two sessions: session 1 and session 2. The first sample of each subject in the first session was used to compose the gallery database while the second sample of the first session and the two samples of the second session were used as probes (P_1, P_2, P_3). For the purpose of this study, we have employed one face and three fingerprint modalities, denoted as fnf, $fo1$, $fo2$ and $fo3$, respectively [15]. The scores used in our experiments are the output of the matching between the first available sample and the second one for each subject. Our second dataset consists in an unbalanced population composed by 516 genuine and 24,180 (156*155) impostor match scores. The details are reported in Tables 1 and 2.

Table 1. The Biosecure database: Development Set

Biometric	Subjects	Samples	Scores
Face	51	4 per subject	Gen 204×3 Imp $51 \times 50 \times 16$
Fingerprint	51	4 per subject	Gen $(204 \times 3) \times 3$ Imp $(51 \times 50 \times 16) \times 3$

Table 2. The Biosecure database: Evaluation Set

Biometric	Subjects	Samples	Scores
Face	156	4 per subject	Gen 624×3 Imp $156 \times 155 \times 16$
Fingerprint	156	4 per subject	Gen $(624 \times 3) \times 3$ Imp $(156 \times 155 \times 16) \times 3$

3.2   Results

The evaluation of the multibiometric system is carried out by adopting the metric denoted as Spoof False Accept Rate (SFAR) which corresponds to a percentage of times a spoof attack results in success. In this paper, a successful spoof attack is when the sum of match score (in the case of sum rule) is above the threshold when a partial spoof attack has occurred (substitution of genuine score for imposter scores). Such a metric has been introduced in [7] to distinguish from traditional FAR. The complete performance curve which represents the full capabilities of the system at different operating points, is given by the Detection Error Tradeoff (DET) in which FRR is a function of FAR/SFAR obtained using logarithmic scales on both axes.

Table 3 reports our results averaged on 20 iterations where for each iteration the fake samples detected by the algorithm has been randomly varied.

Table 3. Results on Biosecure database in different scenarios where one fingerprint modality is spoofed

Fusion rule	Fused modalities	Spoofed modality	EER No spoof	SFAR 1 spoofed modality	SFAR with algorithm	FRR
sum	3 fg + 1 face	fo1	0.32%	56.36%	28.98%	0.32%
sum	3 fg + 1 face	fo2	0.32%	18.67%	5.67%	0.32%
sum	3 fg + 1 face	fo3	0.32%	9.01%	0.46%	0.32%
avg sum	4 mod	1 fg	0.32%	28.01%	11.70%	0.32%
LR	3 fg + 1 face	fo1	0.004%	91.37%	11.12%	0.004%
LR	3 fg + 1 face	fo2	0.004%	62.47%	0.18%	0.004%
LR	3 fg + 1 face	fo3	0.004%	56.83%	8.81%	0.004%
avg LR	4 mod	1 fg	0.004%	70.22%	6.70%	0.004%

In a multimodal system based on the sum of scores with four modalities, three fingerprints and one face, the EER point fixed on the curve without spoofing corresponds to 0.32%, while for this value of FRR, when the fingerprint $fo3$ is spoofed, SFAR becomes equal to 9.01% (see Fig.1); while incorporating in the fusion the fingerprint liveness detection algorithm, SFAR significantly decreases to a value of 0.46%. See note for Figure 1.

In a multimodal system based on the likelihood ratio involving three fingerprint and one face modalities, the EER point fixed on the curve without spoofing, corresponds to 0.004%, while for this value of FRR, when the fingerprint $fo1$ is spoofed, SFAR becomes equal to 91.37% (see Fig.2 notes); while incorporating in the fusion the fingerprint liveness detection algorithm, SFAR significantly decreases to a value of 11.12%. When $fo2$ is the fingerprint spoofed, SFAR increases to 62.47%, but the error rate can be reduced by introducing the algorithm until a percentage of 0.17%.

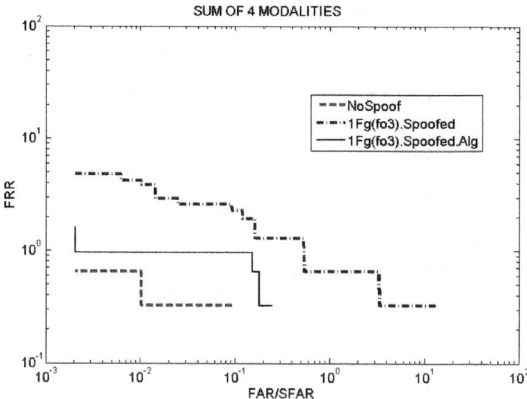

Fig. 1. DET curve of the score sum of three fingerprint and one face modalities taken from Biosecure database over 20 iterations, where one fingerprint is spoofed. Both vertical and horizontal axis of the plot is logarithmically scaled.

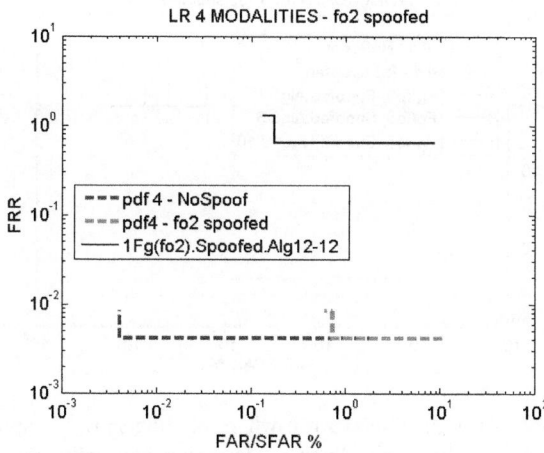

Fig. 2. DET curve of the likelihood ratio involving three fingerprint and one face modalities taken from Biosecure database over 20 iterations, where one fingerprint is spoofed

3.3   Finding the Best Error Rate at Spoof Detection Level

We have extended our investigation by experimentally analyzing how multimodal performance can improve when reducing *Ferrlive* and *Ferrfake*, starting from the values of 12.60% and 12.30% respectively, we used in our previous experiments (see Fig.3 and Fig.4).

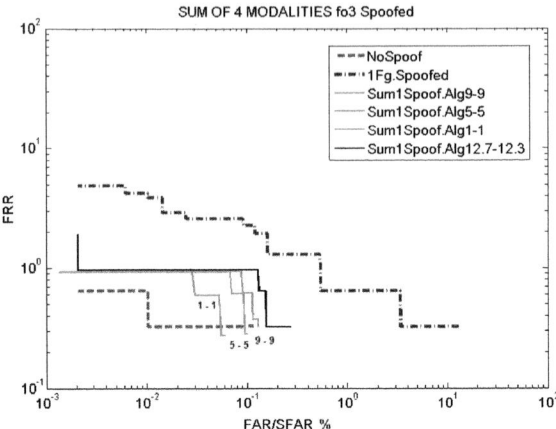

Fig. 3. Performance of the score sum when one fingerprint is spoofed by varying the *Ferrlive* and *Ferrfake* of the liveness detection algorithm incorporated in the fusion

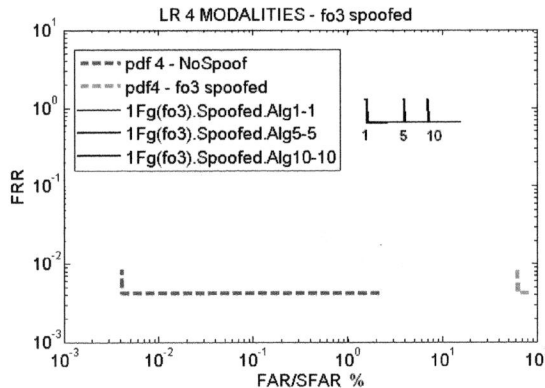

Fig. 4. Performance of the likelihood ratio when one fingerprint is spoofed by varying the *Ferrlive* and *Ferrfake* of the liveness detection algorithm incorporated in the fusion

This step aids to understand which is the best trade-off between the error rate required to a liveness detection algorithm and the fusion performance achieved after incorporating it in the combination scheme. Results are reported in Table 4. The benefits obtained by incorporating the algorithm in the fusion mechanism change by varying the fusion rule. Regarding the LR-based mechanism, the benefits obtained by incorporating the algorithm in the fusion are more significant; in particular, $SFAR$ can be reduced to the value of 0.88% when the spoofing is detected by an algorithm with *Ferrlive* and *Ferrfake* both equal to 1.00%.

Table 4. Results on Biosecure database by varying the error rate of the liveness detection algorithm. In the plot x%-x% indicates the percentage of *Ferrlive* and *Ferrfake*.

Fusion rule	Fused modalities	Modality spoofed	SFAR 1%-1%	SFAR 2%-2%	SFAR 5%-5%	SFAR 7%-7%	SFAR 9%-9%	SFAR 12.60%-12.30%
sum	3 fg + 1 face	fo1	3.87%	5.83%	12.34%	18.93%	26.84%	28.98%
sum	3 fg + 1 face	fo2	0.99%	1.95%	2.34%	3.64%	4.78%	5.67%
sum	3 fg + 1 face	fo3	0.05%	0.08%	0.10%	0.15%	0.38%	0.46%
avg LR	4 mod	1 fg	1.64%	2.62%	4.93%	7.53%	10.67%	11.70%
LR	3 fg + 1 face	fo1	1.02%	1.03%	4.71%	4.81%	8.38%	11.11%
LR	3 fg + 1 face	fo2	0.11%	0.13%	0.13%	0.16%	0.18%	0.18%
LR	3 fg + 1 face	fo3	1.51%	2.28%	3.93%	5.49%	6.79%	8.81%
avg LR	4 mod	1 fg	0.88%	1.15%	2.92%	3.48%	5.12%	6.70%

4 Conclusions and Future Directions

In this paper, we have analyzed the performance of the most efficient fusion approaches at score level under spoof attacks which involve only one fingerprint modality. We have considered a multimodal biometric system in presence of a worst case spoof attack, where the *fake-live* match score distribution is assumed to coincide with the *live-live* match score distribution. Previous works and the results here showed that, when only a subset of the fused modalities is spoofed, multimodal systems can be deceived. Our experiments also demonstrated that a more robust fusion can be realized by incorporating a fingerprint liveness detection algorithm in the combination scheme. Further, we have reduced the error at spoof detection level and found the best trade-off between the optimal *Ferrlive* and *Ferrfake* values and the multimodal performance.

This paper considers the case where spoofing is simulated by substituting with genuine scores. One limitation of the proposed approach lies on the assumption that spoof match scores are distributed as live match scores. Since spoofing is difficult, it may be that the spoof match score distribution has a mean match score which is lower. Therefore, this simulation could be considered as a worst case scenario. Incorporating a spoofing detection, even if it improves FAR under spoof attacks, could have a significant impact on the FRR, as we showed in the case of likelihood ratio-based scheme.

As a future step in this research, the experiments will be extended to additional multimodal databases. A number of fusion algorithms will also be collected and compared using the methods outlined in this paper. Moreover, the performance of the proposed approach will be evaluated by employing real spoofed data.

References

1. Jain, A., Ross, A., Prabhakar, S.: An introduction to biometric recognition. IEEE Transaction on Circuits and Systems for Video 14(1), 4–20 (2004)
2. Yamada, K., Matsumoto, T., Matsumoto, H., Hoshino, S.: Impact of artificial gummy fingers on fingerprint systems. Optical Security and Counterfait Deterrence Techniques IV 4677, 275–289 (2002)

3. Kittler, J., Li, Y.P., Matas, J., Sanchez, M.U.R.: Combining evidence in multimodal personal identity recognition systems. In: International Conference on Audio- and Video-based Biometric Person Authentication (1997)
4. Ross, A., Jain, A.: Information fusion in biometrics. Pattern Recognition Letters 24, 2115–2125 (2003)
5. Ross, A., Jain, A.: Handbook in MultiBiometrics. Springer, Heidelberg (2008)
6. Rodrigues, R.N., Kamat, N., Govindaraju, V.: Evaluation of biometric spoofing in a multimodal system. In: IEEE International Conference on Biometrics, BTAS (2010)
7. Johnson, P.A., Tan, B., Schuckers, S.: Multimodal fusion vulnerability to non-zero effort (spoof) imposters. In: IEEE International Workshop on Information Forensics and Security, WIFS (2010)
8. Poh, N.: École Polytechnique Fédéral de Lausanne. In: Multi-system biometric authentication: optimal fusion and user-specific information (2006)
9. Roli, F., Kittler, J., Fumera, G., Muntoni, D.: An experimental comparison of classifier fusion rules for multimodal personal identity verification systems. In: Roli, F., Kittler, J. (eds.) MCS 2002. LNCS, vol. 2364, pp. 325–336. Springer, Heidelberg (2002)
10. Dass, S., Nandakumar, K., Jain, A.: A principled approach to score level fusion in multimodal biometric systems. In: Fifth AVBPA, July 2005, pp. 1049–1058 (2005)
11. Vatsa, M., Singh, R., Noore, A., Ross, A.: On the dynamic selection of biometric fusion algorithms. IEEE Transaction on Information Forensics and Security 5(3), 470–479 (2010)
12. Nandakumar, K., Chen, Y., Dass, S., Jain, A.: Likelihood ratio-based biometric score fusion. IEEE Transaction on Pattern Analysis and Machine Intelligence 30(2), 342–347 (2008)
13. Marasco, E., Sansone, C.: An anti-spoofing technique using multiple textural features in fingerprint scanners. In: IEEE Workshop on Biometric Measurements and Systems for Security and Medical Applications (BioMs), pp. 8–14 (2010)
14. Marcialis, G.L., Lewicke, A., Tan, B., Coli, P., Grimberg, D., Congiu, A., Tidu, A., Roli, F., Schuckers, S.: First international fingerprint liveness detection competition—livDet 2009. In: Foggia, P., Sansone, C., Vento, M. (eds.) ICIAP 2009. LNCS, vol. 5716, pp. 12–23. Springer, Heidelberg (2009)
15. Poh, N., Bourlai, T., Kittler, J.: A multimodal biometric test bed for quality-dependent, cost-sensitive and client-specific score-level fusion algorithms. Pattern Recognition 43, 1094–1105 (2010)

Cohort Based Approach to Multiexpert Class Verification

Josef Kittler, Norman Poh, and Amin Merati

Centre for Vision, Speech and Signal Processing,
University of Surrey, Guildford GU2 7XH, UK
{J.Kittler,N.Poh,A.Merati}@surrey.ac.uk
http://www.ee.surrey.ac.uk/CVSSP/

Abstract. We address the problem of cohort based normalisation in multiexpert class verification. We show that there is a relationship between decision templates and cohort based normalisation methods. Thanks to this relationship, some of the recent features of cohort score normalisation techniques can be adopted by decision templates, with the benefit of noise reduction and the ability to compensate for any distribution drift.

1 Introduction

Although class identity verification is a two class hypothesis testing problem, the underlying recognition task can be two or multiclass pattern recognition problem. For instance, detection problems inherently involve just two classes. The hypothesis tested is that an observation is consistent with a particular class identity. A typical example is face detection where for every tested locality it is assumed that a face is present. In some verification problems it is relatively easy to collect sufficient number of samples representative of the hypothesised identity and its negation. In such cases one can use conventional machine learning methods to design a verification (detection) system. This situation is typified for instance by face detection where tens of thousands of exemplars of each category are available for training. In anomaly detection, on the other hand, it is relatively easy to collect a huge number of samples representing the normal class but more difficult to capture enough samples of abnormalities. The verification system design then has to be approached as a one class pattern recognition problem.

In this study we address the problem of class identity verification where the underlying task involves multiple classes. For instance, in biometrics, a sample may belong to one of many individuals. The underlying recognition task is a multiclass problem. Then, in a cooperative scenario, where the subject claims a certain identity, the problem becomes one of two-class verification whereby the claimed identity is either accepted or rejected. In principle one can attempt to solve the verification problem as a two class learning problem, by measuring the score for the hypothesised identity. If the score is not high enough, the claim is rejected. Thus for verification, in principle, we need only a model for the

C. Sansone, J. Kittler, and F. Roli (Eds.): MCS 2011, LNCS 6713, pp. 319–329, 2011.

hypothesised class. However, it has been demonstrated that it is beneficial to make use of multiple class models even for verification [2], [8]. Multiple class models jointly model the reject class. By computing the score for each model on line (during testing), we can detect any drift of the reject class distribution. This in turn allows on-line adaptation of the decision rule, which is achieved by on-line modelling of the distribution of the reject class scores as a Gaussian and using it to normalise the hypothesised class identity score. This on-line adaptation is meritorious both for enhancing the performance of a single expert [6] as well as a prerequisite for multiple expert fusion [9].

It has recently been shown that instead of modelling the reject class score distribution using a Gaussian, it is more powerful to model the profile of class conditional scores. The idea behind this approach is that for true claims the profile of class conditional scores is very different from the profile of scores for an untrue claim. The method proposed in [11] involved ordering the class conditional scores in the descending order of their magnitude. Each verification claim is then tested by measuring the average squared deviations between the observed profile and a profile template. We shall show that this method has a close relationship with the decision template method of Kuncheva et al [5]. By virtue of this relationship, one does not need to rank order the class conditional responses to create a profile template. Retaining class identities is sufficient. However, rank ordering opens a new way of modelling the class conditional score profile. The method presented in [7] uses profile models involving a very small number of parameters which has noise reducing effect. More over, in contrast to the decision template method, the rank order profile modelling can also cope with any score distribution drift.

The aim of this paper is to establish the relationship between score normalisation methods and decision templates in the context of class identity verification. We shall show that decision templates correspond to cohort normalisation methods. Thanks to this relationship, some of the recent features of cohort score normalisation techniques can be adopted by decision templates, with the benefit of noise reduction and the capacity for a distribution drift compensation. The effectiveness of modelling rank ordered class conditional scores is illustrated on a problem of fusing multimodal biometric experts.

The paper is organised as follows. In the next section we discuss the problem of cohort score normalisation and establish the relationship between this technique and decision templates. In Section 3 we extend the normalisation method to the problem of fusing the outputs of multiple experts. The methodology is applied to the problem of multimodal biometrics in Section 4. The paper is drawn to conclusion in Section 5.

2   Cohort Based Reasoning

Let us consider an m-class pattern recognition problem. Given a test pattern $\mathbf{x}$, for each class ω_i, a classifier delivers a score s_i representing the degree of membership of $\mathbf{x}$ in class ω_i. The degree of membership could be defined in terms of aposteriori class probability, $P(\omega_i|\mathbf{x})$. However, we shall not impose any such constraint on its essence. Suffice it to say that scores s_i, $i = 1, ..., m$ satisfy $s_i \geq 0$ (their values being non negative). In general, the higher the score value s_i, the more consistent the observation $\mathbf{x}$ is with class ω_i.

In a verification scenario, when a hypothesis regarding the class identity of $\mathbf{x}$ can be generated by an independent process, the decision making task becomes one of accepting or rejecting the hypothesis. In principle, this task can be accomplished by computing the score s_h for the hypothesised model. If the score is high enough (above a decision threshold) the hypothesis is accepted. Otherwise it is rejected. The optimal threshold can be determined using off-line training data.

While it may be advantageous to perform verification using just the model of the hypothesised class, it should be noted, that all the other classes represent samples from the reject class. Thus in principle, by computing the score for each and every class model, given $\mathbf{x}$, we get an on-line empirical distribution of scores representing the reject class. This distribution conveys valuable information, as it can be compared with the off-line distribution to determine any drift in score values. This may be caused, for instance, by changing environmental conditions in which measurement $\mathbf{x}$ is acquired. A more practical proposition is to use the on-line distribution of scores for all the alternative hypotheses for the normalisation of the score for the class identity being verified. The well known *t-norm* [8] is designed to achieve such normalisation.

Let μ_h be the mean of the empirical distribution of scores delivered by our classifier for all classes other than class ω_h to be verified, i.e. $\mu_h = \frac{1}{m-1} \sum_{i \neq h} s_i$ and denote the standard deviation of these scores by σ_h. Then t-normalised score $\hat{s}_h$ is defined as

$$\hat{s}_h = \frac{s_h - \mu_h}{\sigma_h} \tag{1}$$

From (1) it is evident that the underlying model for the distribution of cohort scores is a Gaussian, with mean μ_h and standard deviation σ_h. Since the score values are non negative, the assumption of normality of the cohort distribution is somewhat unrealistic. For this reason other normalisation schemes have been proposed [6]. Most methods rely on some rank order statistics [11,1]. The method proposed in [7] establishes a complete ordering of the cohort scores, for some off-line cohort set, in the descending order of their magnitudes. Let $r(i)$ be an index mapping function which relates class index i to its rank position in the list of ordered scores, and let $i(r)$ be the inverse index mapping function. Then function $s(r) = s_{i(r)}$ is monotonic. Clearly, for every hypothesis h to be verified, the index mapping function, as well as function $s(r)$ would be different. Now if a query sample belongs to the hypothesised class ω_h, then the class conditional scores $S(r) = S_{i(r)}$ computed for the input test pattern would adhere to the profile $s(r)$. On the other hand, when the observation is inconsistent with the hypothesis, then the observed function $S(r)$ would be random. Thus, the mean squared error between $S(r)$ and $s(r)$, can be used as a basis for accepting or rejecting the hypothesis, i.e.

$$\begin{aligned} \mathbf{x} &\rightarrow \omega_h & if\ \theta \leq \rho \\ reject & & \omega_h\ if\ \theta > \rho \end{aligned} \tag{2}$$

where the test statistics θ is defined as

$$\theta = \frac{1}{m} \sum_{r=1}^{m} [S(r) - s(r)]^2 \tag{3}$$

and ρ is a suitable threshold.

It is interesting to note that matching class conditional scores produced for a test pattern to a score profile is the basis of a multiple classifier fusion method known as *decision templates*. A single classifier version of the decision template method then compares the decision template entries $s_i,\ i = 1, ..., m$ to the class conditional scores S_i obtained for a given test pattern $\mathbf{x}$. There are a number of norms that can be used to measure the similarity of s_i and $S_i,\ \forall i$ [5] but the quadratic norm in (3) is among the recommended possibilities. Clearly the reordering of scores will have no effect on the value of the test statistics and therefore these two methods are equivalent.

The decision template method was devised for multiple classifier fusion and as such it is pertinent to ask what relevance it has for a hypothesis testing involving a single classifier. Clearly the answer is not much, but looking at a single classifier decision template (one column of the decision template matrix) can help to understand the properties of this post-processing method.

- In principle, one can look at the class conditional scores as features and the hypothesis testing is then a process of decision making in this feature space. When the class conditional scores are normalised so that they sum up to one (i.e. they represent aposteriori class probabilities), then these features have been shown to be optimal [3]. Thus decision making in this new feature space should in theory be as good as decision making in the original feature space. The benefit of the decision template method is that it is readily extensible to a multiple expert scenario.
- When the decision making problem involves a large number of classes, most of the dimensions of this new feature space do not convey discriminative information and will only inject noise into the decision making process.
- Where as a simple modelling of the cohort scores using a Gaussian is far from perfect, it had the advantage of enabling the hypothesis testing process to compensate for any drift in the score distribution.

Thus our attempt to introduce a better distribution models resulted in the loss of some of the attractive features of the simple t-norm. In the following we shall discuss how these features can be restored while retaining the benefit of working with more flexible score distribution models. We shall see that there are two aspects to the problem

1. cohort score distribution modelling
2. the use of cohort score distribution model in decision making and fusion.

2.1   Cohort Score Distribution Modelling

As pointed out earlier, the direct use of decision templates for hypothesis testing according to (3), whether the scores are ranked or not, suffers from several disadvantages. However, the benefit of score ranking is that the off-line cohort score function $s(r)$ is monotonic. This opens the possibility for modelling $s(r)$ in terms of a function of few parameters. In particular, we can model $s(r)$ as

$$s(r) = f(r, \mathbf{a}) \tag{4}$$

The function, $f(r, \mathbf{a})$ would typically be a low order polynomial defined by a set of parameters $\mathbf{a}$. Given a set of rank order cohort scores, the function can easily be fitted to the data. Note, that such a function fitting would be very difficult for decision templates, as the evolution of class conditional scores s_i as a function of i is potentially much more complex. The fitting process allows us to represent a multidimensional feature space in terms of just a few parameters. For large cohorts, the information compression achieved through this process is enormous. This has a number of benefits. First of all, it helps to minimise overfitting. Second, it helps to reduce the amount of noise injected into the decision making process. There is a chance that some of the parameters of the rank ordered cohort score distribution fitting computed for a test pattern will be invariant to a distribution drift. For instance, if the test sample quality changes and the score values for all the classes are lower, this would be reflected only in an offset parameter for the function $f(r, \mathbf{a})$, with the rest of the model being unaffected.

2.2   Cohort Score Models in Decision Making

In principle, the decision making can be conducted in the cohort score feature space, as in the case of the decision template method. When the dimensionality of the cohort model is reduced, the benefit of noise reducing property of the fitting process is manifest in improved performance. However, the cohort score model can alternatively be used in conjunction with the raw score for the hypothesis being tested. We shall demonstrate that this latest option is the most effective.

3   Cohort Based Fusion of Multiple Experts

The discussion so far assumes a single classifier involving a set of cohort scores. In order to consider the multiple classifiers setting, we shall further augment the score variable with the subscript $p \in \{1, \ldots, P\}$ to denote the output of the p-th classifier. Thus, s_h^p denotes the reference score; $\hat{s}_h^p$, the T-normalised score; $S^p(r)$, the cohort score profile; and $f(r, \mathbf{a}^p)$, the fitted score profile.

We shall discuss the baseline methods followed by three other competing methods that take the cohort information into account.

Baseline: In order to combine the cohort-based scores, one takes a weighted sum of the $\{s_h^p\}$ scores. To this end, we use logistic regression. It has a number of advantages. First, it is a linear classifier, and by being linear, the risk of overfitting is significantly reduced. Second, its optimisation procedure, known as Gradient Ascent [4], has a unique global solution. Logistic regression computes $P(\omega_h|s_h^1, \ldots, s_h^P)$ in support of the hypothesis that a query pattern belongs to ω_h versus its alternative that it belongs to the remaining classes $\{\omega_i|i \neq h\}$.

T-norm: Since T-norm is a function that takes s_h^p as input and produces a normalised output $\hat{s}_h^p$, the most straightforward way combine the classifier outputs is to take a weighted sum of $\{\hat{s}_h^p\}$. To this end, we use logistic regression that computes $P(\omega_h|\hat{s}_h^1, \ldots, \hat{s}_h^P)$.

Decision Template: The above weighted-sum approach cannot be used directly to combine the cohort score profile. The Decision Template approach does so by considering the set of profile vectors $\{S^p(r)|\forall_p\}$ as a single large vector S. This vector forms the decision template. A distance is then defined in order to compare the decision template with the set of profile vectors computed from a query sample.

Our proposal – cohort distribution modelling: Rather than using the entire profiles $\{S^p(r)|\forall_p\}$ that are noisy for the alternative classes $i \neq h$, we shall use the fitted profile instead. Combining the multiple the classifier outputs along with their respective cohort score profile amounts to combining the vectors: $\{s_h^p, \mathbf{a}^p|\forall_p\}$. Let the concatenation of these vectors be $\mathbf{A}$. One can then use logistic regression to classify $\mathbf{A}$. In this case, logistic regression computes $P(\omega_h|\mathbf{A})$ in support of the hypothesis that a query pattern belongs to ω_h versus its alternative that it belongs to the remaining classes.

It is worth noting that $\mathbf{A}$ is much smaller in dimension compared to the decision template S. This compression is achieved by summarising each of the p-th score profile $S^p(r)$ (a constituent element in S) by its fitted parameters $\mathbf{a}^p$ (a constituent element in $\mathbf{A}$). This compression has two implications. First, a classifier such as logistic regression can be used to classify $\mathbf{A}$ which is relatively low in dimension. Second, the representation $\mathbf{A}$ is much less noisier yet retains all the discriminative power subsumed in S.

4   Experimental Support

In order to illustrate the effectiveness of our proposal, we will compare the three competing methods along with the baseline approach according to one of the two roles they play: as a score normalization scheme and as a fusion mechanism. The main difference between the two is that the former involves only a single classifier/system whereas the latter involves $P > 1$ systems.

Although Decision Template is commonly used for fusion only, it is recalled here that it can also be used as a score normalization scheme by simply computing a distance metric, θ, between two cohort score profiles (in which one is a decision template and another is a query pattern). We have experimented with several distance metrics and found that normalized correlation is the most effective distance metric. Therefore, only this metric is used when reporting the performance of Decision Template.

Another point to note is that, as a score normalization procedure, logistic regression produces no effect, since it is a non-decreasing monotonic function. As a result, the baseline method is represented by the reference score s_h, whereas for the T-norm method, $\hat{s}_h$ is used as the system output that is used for subsequent performance evaluation.

In order to evaluate the performance in the context of verification (as opposed to identification), we shall plot the Receiver's Operating Characteristic (ROC) curve. It is a plot of False Acceptance Rate (FAR) versus False Rejection Rate (FRR). FAR corresponds to the probability of falsely accepting a negative class (the altnerative class labels where $i \neq h$) whereas FRR is the probability of falsely rejecting a positive class (the reference class h). The unique point in which FAR equal FRR is called Equal Error Rate (EER). This operating point is used to characterise the performance of the abovementioned methods. The lower EER, the better the system performance.

The Biosecure DS2 dataset is used for this purpose [10]. This data set contains the impressions of six fingers of some 400 subjects. For each finger, four impressions were acquired over two sessions. Two devices were used: thermal and optical device. In order to report the performance, we shall divide the experiments into 12 sets spanned by 6 fingers and the 2 acqusition devices.

Five *disjoint* groups of subjects were identified, with the first four groups (respectively referred to as g1–g4), constituting enrollees, and the final group forming a separate set of cohort users to provide a pool of cohort models. Subjects in g1 and g2 were used as enrollees in the *development* (*dev*) set; and, g3 and g4 as enrollees in the *evaluation* (*eva*) set. Subjects in group g5 were used as cohort models. The total number of subjects in g1–g4 are $\{84, 83, 83, 81\}$ respectively. The total number of cohort users is 84 for both modalities. For the purpose of obtaining a cohort scores, only the first of the four samples of the cohort was used.

We require that each of the *dev* and *eva* sets to have its own enrollment and query data sets, i.e.,$\mathcal{D}_{d,enrol}$, $\mathcal{D}_{d,query}$ for $d \in \{dev, eva\}$. Recall that there are four impressions (images) per finger, per subject and per device. The first fingerprint impression was used as the enrollment template for the target user. In order to generate match (genuine) scores, the second impression was used to produce scores for $\mathcal{D}_{dev,query}$ whereas the remaining two query samples were used to produce scores for $\mathcal{D}_{eva,query}$.

To generate the non-match scores, for $\mathcal{D}_{dev,enrol}$ we used query samples of $g3$; for $\mathcal{D}_{dev,query}$, $g4$; for $\mathcal{D}_{eva,enrol}$, $g1$; and, for $\mathcal{D}_{dev,query}$, $g2$. In this way, the non-match scores in *all* four data sets are completely disjoint.

In the empirical evaluation to be reported in the next section, we use $\mathcal{D}_{dev,query}$ as our training set and $\mathcal{D}_{eva,query}$ as our test set. Note that the enrollees and non-match subjects in these two match scores are completely *disjoint*. This simulates a scenario where the development and operational data have disjoint subjects, a very realistic condition in practice.

Using the conventional machine learning terms, we shall treat $\mathcal{D}_{eva,query}$ as *the test* set, whereas the remaining three data sets as *the training* set.

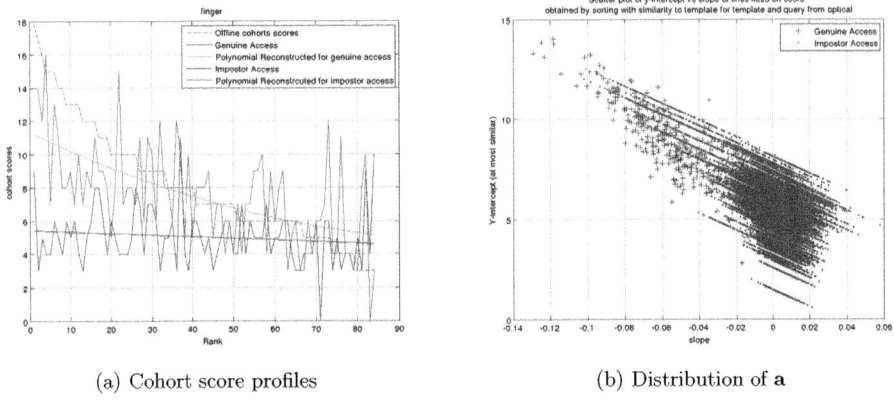

(a) Cohort score profiles   (b) Distribution of **a**

Fig. 1. (a) Cohort score profiles as well as their respective reconstructed versions for the reference class (genuine) and the remaining classes (impostor in this case). The *offline* cohort score profile is, by defition, a decreasing function as it was used to determine the rank order of the cohorts. (b) The distribution of the fitted parameters when the cohort score profiles are fitted with a line. "+" denotes the parameters of the reference class (genuine matching); and "·", the remaining classes (impostor matching).

Figure 1(a) shows the various cohort score profiles sorted by the rank order $S(r)$ determined from the training data set. The "offline cohort score profile" corresponds to the decision template. By fitting this profile using a polynomial function, one obtains a smoothed curve with a noticeable positive gradient under matching with the reference class (genuine or client access in biometric application). In comparison, the score profile subjecting to matching with the alternative class (impostor in biometric application) does not exhibit trend. This shows that cohort score profile contains *discriminatory* information.

Figure 1(b) shows the scatter plot of the fitted regression parameters **a** in terms of slope versus bias (intercept), hence, representing the cohort score profiles with a line. This figure shows that the cohort information along, without the reference score, s_h, contains highly discriminative information.

Table 1 compares the effectiveness of the four methods as a score normalization scheme in terms of EER in percentage. As can be seen, our proposed method which fits the cohort score profile attains the best generalization performance in 11 out of the 12 data sets. We then analysed the *relative merit* of these methods by comparing the performance with the baseline method. This was done by computing the relative change of EER, defined as

$$\text{rel. change of EER} = \frac{\text{EER}_{algo} - \text{EER}_{baseline}}{\text{EER}_{baseline}},$$

where EER_{algo} is the EER of a given score normalization procedure whereas $\text{EER}_{baseline}$ is the EER of the original system without any score normalization. A boxplot summarizing the relative gain of each method is shown in Figure 2.

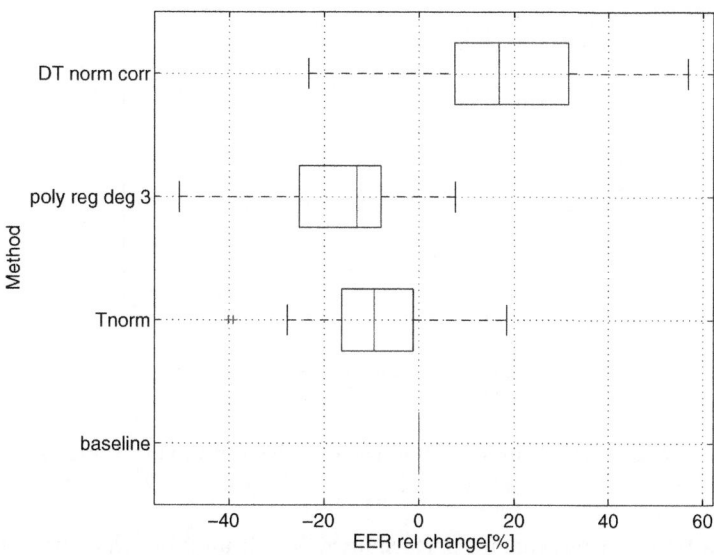

Fig. 2. Relative change of EER profiles

Table 1. Comparison of different cohort-based normalization schemes

Dataset	baseline	T-norm	poly	DT
fo1	2.79	2.42	**2.13**	2.59
fo2	1.80	1.39	**1.14**	2.01
fo3	3.11	2.45	**2.21**	3.52
fo4	3.69	2.84	**2.70**	3.84
fo5	3.41	**2.95**	3.22	3.45
fo6	3.05	2.76	**2.69**	3.61
ft1	9.61	9.61	**9.45**	12.08
ft2	5.41	4.71	**4.37**	6.18
ft3	8.78	8.37	**8.05**	11.08
ft4	12.61	12.09	**12.03**	16.78
ft5	6.89	7.28	**6.30**	9.02
ft6	8.40	7.84	**7.43**	12.12

A negative change implies improvement over the baseline. As can be observed, across the 12 data sets, one can expect a relative reduction of error between 5% and 25% using the proposed method.

The final experiment reports the performance of fusion of all the fingers. For illustration, we show the result of one of the fusion systems comparing all the four methods in Figure 3. In this figure, the Decision Template (DT) approach was implemented with two distance metrics, namely, Euclidean distance and normalized correlation. As can be observed, DT with normalized correlation

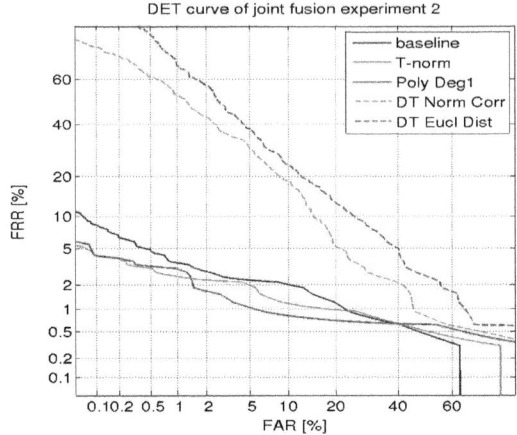

Fig. 3. DET curves of the fusion problem involving 3 fingers

performs better. However, its performance is still far from that attained by T-norm or our proposed method. This shows that the importance of correctly exploiting the cohort information.

5 Conclusion

We discussed the problem of cohort based normalisation in multiexpert class verification. We exposed the existence of a close relationship between decision templates and cohort based normalisation methods and showed that thanks to this relationship, some of the recent features of cohort score normalisation techniques can be adopted by the decision templates approach. The benefit of this includes noise reduction and a distribution drift compensation. This has been demonstrated on an application in multimodal biometrics.

Acknowledgements. This work was supported by the Engineering and Physical Sciences Research Council (EPSRC) Research Grant EP/F069421/1 and by the prospective researcher fellowship PBEL2-114330 of the Swiss National Science Foundation.

References

1. Aggarwal, G., Ratha, N.K., Bolle, R.M.,Chellappa, R.: Multi-biometric cohort analysis for biometric fusion. In: IEEE Int. Conf. on Acoustics, Speech and Signal Processing (2008)
2. Auckenthaler, R., Carey, M., Lloyd-Thomas, H.: Score normalization for text-independant speaker verification systems. Journal of Digital Signal Processing (DSP) 10, 42–54 (2000)

3. Fukunaga, K., Ando, S.: The optimum non-linear features for a scatter criterion in discriminant analysis. IEEE Tans. Information Theory 23(4), 453–459 (1977)
4. Hastie, T., Tibshirani, R., Friedman, J.: The Elements of Statistical Learning. Springer, Heidelberg (2001)
5. Kuncheva., L., Bezdek, J.C., Duin, R.P.W.: Decision Template for Multiple Classifer Fusion: An Experimental Comparison. Pattern Recognition Letters 34, 228–237 (2001)
6. Mariethoz, J., Bengio, S.: A unified framework for score normalization techniques applied to text independent speaker verification. IEEE Signal Processing Letters 12 (2005)
7. Merati, A., Poh, N., Kittler, J.: Extracting discriminative information from cohort models. In: Fourth IEEE International Conference on Biometrics: Theory Applications and Systems (BTAS) pp.1–6, (September 2010)
8. Navratil, J., Ramaswamy, G.N.: The awe and mystery of t-norm. In: EUROSPEECH 2003, pp. 2009–2012 (2003)
9. Poh, N., Merati, A., Kitter, J.: Making better biometric decisions with quality and cohort information: A case study in fingerprint verification. In: Proc. 17th European Signal Processing Conf., Eusipco (2009)
10. Poh, N., Bourlai, T., Kittler, J.: A multimodal biometric test bed for quality-dependent, cost-sensitive and client-specific score-level fusion algorithms. Pattern Recogn. 43(3), 1094–1105 (2010)
11. Tulyakov, S., Zhang, Z., Govindaraju, V.: Comparison of combination methods utilizing t-normalization and second best score model. In: IEEE Conf. on Computer Vision and Pattern Recognition Workshop (2008)

A Modular Architecture for the Analysis of HTTP Payloads Based on Multiple Classifiers*

Davide Ariu[1] and Giorgio Giacinto[1]

Department of Electrical and Electronic Engineering, University of Cagliari, Italy
{davide.ariu,giacinto}@diee.unica.it
http://prag.diee.unica.it

Abstract. In this paper we propose an Intrusion Detection System (IDS) for the detection of attacks against a web server. The system analyzes the requests received by a web server, and is based on a two-stages classification algorithm that heavily relies on the MCS paradigm. In the first stage the structure of the HTTP requests is modeled using several ensembles of Hidden Markov Models. Then, the outputs of these ensembles are combined using a one-class classification algorithm. We evaluated the system on several datasets of real traffic and real attacks. Experimental results, and comparisons with state-of.the.art detection systems show the effectiveness of the proposed approach.

Keywords: Anomaly Detection, IDS, HMM, Payload Analysis.

1   Introduction

The always increasing number of Web-based applications that are deployed worldwide, makes their protection a key topic in computer security. The traditional defense systems (e.g. Intrusion Detection/Prevention Systems) are based on a database of signatures that describe known attacks. Unfortunately, the large number of new attacks that appears everyday, and the wide use of custom applications on web servers, make almost impossible to have signature-based systems always updated to the most recent and effective attacks. A possible solution to this problem is offered by the "anomaly based" approach to intrusion detection.

An anomaly based system builds a model of the "normal" behavior of the resource to be protected. An attack pattern is detected if it appears "anomalous" with respect to the normal behavior, that is if it significantly deviates from the statistical model of the normal activity. The normal behavior is defined as a set of characteristics that are observed during normal operation of the resource to be protected, e.g., the distribution of the characters in a string parameter, the mean and standard deviation of the values of integer parameters [[6], [7]]. One of

---

* This research was sponsored by the RAS (Autonomous Region of Sardinia) through a grant financed with the "Sardinia PO FSE 2007-2013" funds and provided according to the L.R. 7/2007. Any opinions, findings and conclusions expressed in this material are those of the authors and do not necessarily reflect the views of the RAS.

C. Sansone, J. Kittler, and F. Roli (Eds.): MCS 2011, LNCS 6713, pp. 330–339, 2011.

Fig. 1. An example of legitimate HTTP payload

GET /pra/ita/home.php HTTP/1.1	$\Longrightarrow$	Request-Line
Host: prag.diee.unica.it Accept: text/*, text/html User-Agent: Mozilla/4.0	$\Longrightarrow$	Request-Headers

the reasons that initially prevented anomaly-based IDS from becoming popular is the fact that they tend to generate too high rates of false alarms. In fact the false alarm rate is a crucial parameter in the evaluation of an IDS since an IDS is generally required to manage large amounts of patterns (hundreds of thousands) every day. A strategy which is usually employed to mitigate this problem is that of realizing IDSs based on multiple classifiers in order to increase the overall classification accuracy [[2],[6],[7],[9],[11]].

The anomaly based IDS recently proposed in the literature for the protection of web servers and web applications basically analyze the requests received by the web server. HTTP requests are carried in the data portion of the network packet that is generally called "HTTP payload". An example of HTTP payload is presented in Figure 1. The HTTP protocol is defined by RFC 2616 [1]. According to this RFC, a HTTP payload contains a Request-Line plus a certain number of Request-Header fields. More in detail:

- The Request-Line begins with a method token (e.g. POST, GET), followed by the Request-URI and the protocol version, and ending with CRLF. The Request-URI contains the name of the resource requested on the web server. In Figure 1 the resource requested is the page /pra/ita/home.php.
- The Request-Header fields are used by the client to provide additional informations to the web server. For example, with the User-Agent header, the client host notifies to the web server the type and the version of the web browser. This information can be used by the web server to optimize the response sent back to the client according to the version of the browser. In Figure 1 the value of the User-Agent header is Mozilla/4.0.

Anomaly based IDS use statistical models to represent and analyze HTTP requests. Basically they create a statistical model of the bytes' distribution within the payload. Some of them, such as *HMM-Web* [5] or *Spectrogram* [11], focus on the Request-Line only and perform an analysis based on Hidden Markov Models (HMM), and on Mixture of Markov-chains respectively. Other approaches, such as *PAYL* [13] and *HMMPayl* [2] analyze the bytes' distribution of the whole payload using $n - gram - analysis$ or HMM. These IDSs are based on the assumption that the bytes' statistics of HTTP payloads containing attacks are different from the bytes' statistics of the legitimate traffic. Nevertheless, at the best of our knowledge, none of the IDSs proposed in the literature, exploits the a-priori knowledge of the structure of the HTTP payload.

In this paper, we propose an IDS based on HMM that effectively exploit the analysis of the different portions of the HTTP payload structure. In particular, for each header of the payload, we use a different ensemble of HMM to analyze the related values. Another ensemble of HMM is used to analyze the Request-Line. Attack detection is performed by stacking the outputs of the HMM ensembles, and using this vector as input for a one-class classifier. The experimental results achieved on several datasets of legitimate traffic and attacks confirm the effectiveness of the proposed approach, and its superiority with respect to similar IDS proposed in the literature.

The rest of the paper is organized as follows. In section 2 a review of the State of Art is provided. In Section 3 a detailed description of the IDS is provided. In sections 4 and 5 we describe respectively the experimental setup and results. We then draw the conclusions in section 6.

2   State of the Art

In the recent years, several anomaly based IDS have been proposed for the protection of web servers and web applications. They usually rely on multiple classifiers or models for two main reasons. First, multiple classifiers generally lead to better classification accuracy. In the case of IDSs, this means an higher percentage of detected attacks and a smaller percentage of false alarms. Examples of applications of Multiple Classifiers are [[2],[6],[7],[9],[11]]. The second reason for using multiple classifiers is that they usually increase the robustness of the system against attempts of evasion. This topic is gaining an increasing attention in the last years not only in the Intrusion Detection area but also in related fields such as spam detection and biometric authentication [4,8].

Intrusion detection techniques such as those proposed in [[5],[7],[11]] limit their analysis to the structure of the Request-Line, and in particular they focus on the value of the input parameters received by the web applications. These approaches are tailored for the detection of the most frequent attacks against web applications, e.g., SQL-Injection, Cross-Site Scripting, that basically exploit the flaws of the web applications in the validation of the received input. Nevertheless, these IDS are completely ineffective against attacks that exploit other vulnerabilities of web applications.

IDS such as [[2],[9],[13]] cover a broader range of attacks since they model the bytes' distribution of the whole payload. As a consequence, they are theoretically able to detect any kind of attack that makes the payload statistics deviating from those of the legitimate traffic. The bytes' distribution of the payload can be modeled in several ways. $PAYL$ [13] performs an $n - gram$ analysis using a very small value for n, since the size of the features space exponentially increases with n. This represent a severe limitation for $PAYL$ since the IDS can be easily evaded if the attacker is able to mimic the statistics of the legitimate traffic. $HMMPayl$ performs an analysis of the payload based on HMM. This analysis is equivalent to the $n - gram$ analysis, but it is able to circumvent the limitation on the value of n from which $PAYL$ suffered. This lead to an increased classification accuracy of $HMMPayl$ with respect to $PAYL$. Nevertheless, the analysis performed by $HMMPayl$ is quite complex. This might be an issue since an IDS such as $HMMPayl$ must be able to keep up with the network speed. For this reason, in this paper we propose to exploit the a-priory knowledge of the payload structure in order to significantly reduce the complexity of the classification algorithm without affecting the classification accuracy.

The largest part of the IDSs proposed in the literature are based on outliers detection techniques, and one-class classifiers. This means that the classifiers, or, more in general, the statistical models on which they are based, are built using samples of legitimate patterns only. There are various reasons for this choice. First of all, the main aim of anomaly based IDS is to recognize those patterns that are anomalous with respect to those assumed to be legitimate. In addition, a two-class model (normal vs. attack) would not be probably the one that best fits the problem. In fact the attack class would contain patterns that are completely different each other, as they exploit vulnerabilities of different type and, as a consequence, they exhibit statistical properties that are completely different [2]. Another (and more "practical") reason for using one-class classifiers is that collecting representative samples of the attack class is usually quite difficult. In fact, the attacks a web server might be susceptible to, depend on several elements such as the platform (e.g. the operating system), the hosted applications, the network topology, and so on. One could certainly scan the web server and analyze the web applications looking for possible vulnerabilities, and then create samples of attacks that exploit them. Nevertheless, the assessment of all the possible attacks a web server might be subject to, remains a task quite difficult and time consuming. If such a knowledge should be available, then there would be better ways to protect against known attacks than training a classifier, that is patching the vulnerabilities. Another possibility is to extract the signatures for that attacks, and deploy them in a signature-based IDS.

3   A Modular Architecture for the Analysis of HTTP Payloads

This section provides the details of the IDS proposed in this paper. A simplified scheme of the system is presented in Figure 2. Several solutions proposed in the literature (e.g. *HMM-Web*[5], *Spectrogram* [11]) focus their analysis only on the Request-Line (which is in red in the figure). Otherwise, solutions such as *PAYL* [13], or *HMMPayl* [2] analyze the bytes' distribution of the whole payload but do not take into account its structure. This paper aim to investigate the use of a model of the HTTP payload which reflects its structure as it has been defined by the RFC 2616 [1]. This analysis is performed in two steps. First, the payload is split in several "fields", that are analyzed by several HMM ensembles. Second, the outputs of the HMM are combined using a one-class classifier that finally assigns a class label to the payload. A more detailed description of these steps follows.

HMM Ensembles. We briefly reminded the HTTP payload structure in section 1, recalling that this structure consists of a Request-Line plus (eventually) one or more Request-Headers. We also remind here that the Request-Line

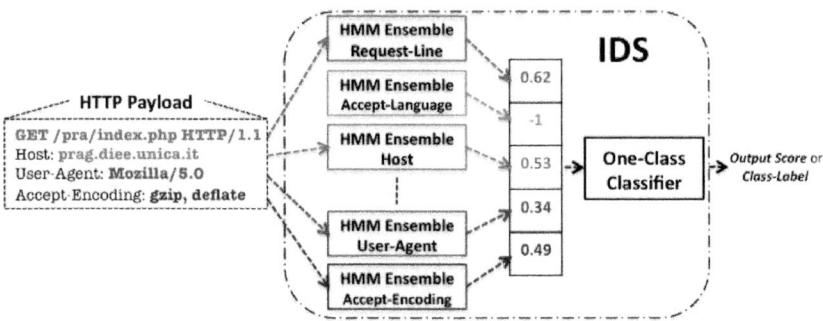

Fig. 2. A simplified representation of the IDS architecture. An exhaustive list of all the Request-Headers analyzed by the IDS is reported in Table 1. The analysis of each header can be carried out by a single HMM or by an ensemble of HMM.

specifies the resource requested by the browser to the web server (e.g. the index.php page), whereas the Request-Headers are used by the web browser to provide additional information to the web server (thus their presence is optional). The RFC 2616 defines for the HTTP protocol a large number of possible Request-Headers, the most part of which is generally unused. In order to simplify the analysis, a list of the headers that a web server is actually using can be easily produced through a simple inspection of the incoming network traffic. Table 1 reports the list of the 18 headers that we observed within our datasets during the experimental evaluation of the proposed technique.

Once the set of the Request-Headers to be analyzed has been defined, the payload is processed as follows. First, a probability is assigned to each Request-Header and to the Request-Line by a different HMM ensemble through the *Forward-Backward* procedure [10]. With respect to the example shown in figure 2, the strings analyzed by the different HMM ensembles are the following: the string from "GET" to "HTTP/1.1" is analyzed by the the Request-Line ensemble (red); the string "prag.diee.unica.it" is analyzed by the Host ensemble (green); the string "Mozilla/5.0" is analyzed by the User-Agent ensemble (blue) and so on. Details about the setting of the HMM parameters will be provided in the following section. It is just worth noting that the use of HMM ensembles instead of single HMM allows mitigating the risk of having a single HMM that performs poorly, due to the random initialization of the parameters. The output of the HMM ensemble is thus computed by averaging the outputs of the individual HMMs, as they differ for the parameter initialization only.

One Class classifier. The analysis performed by the HMM ensembles produces as output a set of probabilities assigned to the Request-Line and Headers by the ensembles (see figure 2). Obviously a fusion stage is required in order to combine the outputs of the the different ensembles. We considered static rules (e.g. the mean or the product rule) to perform the combination. Unfortunately, as we will show in section 5, they do not result suitable for this purpose. A further possibility is to concatenate the outputs of the ensembles within an array

Table 1. List of the Request-Headers analyzed by the IDS. A detailed description of the role played by each header within the HTTP protocol can be found in [1].

Accept	Connection	Cache-Control	Via	User-Agent
Accept-Charset	From	If-None-Match	UA-CPU	Transfer-Encoding
Accept-Encoding	Host	If-Modified-Since	UA-OS	X-Forwarded-For
Accept-Language	Referer	Keep-Alive		

that will be provided as input to a one-class classifier. In this case, the outputs of the ensemble are used as features and a label (attack or legitimate) is assigned to the payload as the result of a classification in this features space.

It can be observed that a legitimate payload typically contains a number of five or six headers. Which headers are included in the payload depends on the settings of the HTTP client. On the other hand, the IDS must be able to analyze all the headers that occur in the network traffic (we observed the presence of 18 different headers in the network traces used in our experiments). As a consequence, the one class classifier will be designed to work in a features space of size equal to the number of observed headers (18 in our case) plus one, since the Request-Line must be also analyzed. From the perspective of the one class classifiers, the absent headers represent "missing features", since a probability will be associated by the ensembles only to the headers within the payload. The problem of managing these missing features is approached differently in the Training and in the Detection phases. During the training of the one class classifier, the missing features are replaced by their average value (computed over the payloads in which they are present). This is a practice well known in the literature [12]. This choice does not affect the results of classifier training since the most important features (e.g. the Request-Line) are present in the largest part of the traffic. In the detection phase, the missing components are set to the value of -1, that is a value outside the output range of the HMM (the output range is in [0,1], as they are probabilities).

Complexity Evaluation Since the training of the IDS is performed off-line, the complexity is estimated for the detection phase only. Let K the number of Request-Headers analyzed by the IDS (K typically assumes a value in the range between 15 and 20). If all of these headers appears in the payload, the IDS has to analyze $K+1$ sequences (the headers plus the request line). In addition a further classification step has to be performed in a features space of size $K+1$. Just to provide a brief comparison with a similar approach let us consider the *HMMPayl* algorithm [2]. In the case of *HMMPayl* the number of sequences analyzed by the IDS is approximately as high as the length (in terms of number of bytes) of the payload. A typical legitimate payload has a length of several hundreds of bytes. Thus, the solution proposed here offers two main advantages: the first is that the number of sequences analyzed is significantly smaller; the second is that this number is known and depends only on the setup of the IDS.

4   Experimental Setup

In this Section we describe the experimental setup on which we performed the experiments.

HMM Parameters. The parameter that influences the most the performance of a discrete HMM is the number of (hidden) states. A rule does not exist to estimate the optimum value for the number of states for a bunch of data. Here, we used the "effective-length" of the training sequences, which is an heuristic that has been successfully used also in [5]. The effective length basically counts the number of different characters in a string. For instance the effective length of the string "abc" is 3, that of "abcd" is 4, and that of "aabcdd" is still 4. Thus, for each ensemble of HMM, we set the number of states of every single HMM equal to the average effective-length calculated on the corresponding training set. In addition, the transition and emission matrices are randomly initialized for each HMM. Then, the estimate of the model parameters that maximize the probability assigned by the model to the sequences within the training set is calculated by resorting to the *Baum-Welch* algorithm [3].

Datasets. The intrusion detection algorithm proposed has been deeply tested on two different datasets of normal traffic, and on three datasets containing different kinds of attacks. For what concerns datasets of normal traffic both of them consists of real traffic traces collected at academic institutions. One dataset is made up of HTTP requests towards the website of the College of Computing at the Georgia Tech (GT), USA. The other one consists of HTTP request toward the website of our department (DIEE) at the University of Cagliari, Italy.

They consist respectively of seven and six days of traffic. It is worth to remark that both the GT and the DIEE datasets are completely unlabeled. We considered the GT and DIEE datasets as "clean" from known attacks for the purpose of measuring the false positive rate since any evidence of occurring attacks has not been reported in the period in which we collected the traffic.

The experiments have been carried out in the same way on both datasets, for what concerns both training and testing. A *k*-fold cross validation has been realized, using in rotation one day of traffic for training and all the remaining days for testing purposes. Details about the number of packets and the size (in MB) of each trace are provided in table 2.

We evaluated the detection rate of the IDS on several datasets consisting of attacks frequently observed against web applications. Attack datasets are briefly

Table 2. Details of the legitimate traffic datasets used for the training and to evaluate the false alarms rate

Dataset	Day	1	2	3	4	5	6	7
DIEE	Packets	10,200	10,200	10,200	10,200	10,200	10,200	—
	Size (MB)	7.2	7.4	6.6	6	6.4	6.7	—
GT	Packets	307,929	171,750	289,649	263,498	195,192	184,572	296,425
	Size (MB)	131	72	124	110	79	78	127

Table 3. Details of the attacks datasets used to evaluate the detection rate

Dataset Name	# of Attacks	Description
Generic Attacks	66	Shell-code, Denial of Service or Information Leakage
Shell-code Attacks	11	Shell-code attacks from the Generic Attack dataset
XSS-SQL Attacks	38	Cross-site Scripting and SQL-Injection attacks

described in Table 3. Generic and Shell-code attacks are the same used in [[2,9]]. Attacks into the XSS-SQL dataset are the same used in [[2,5]].

Performance Evaluation. In order to validate the classification performance of our detector, we use the ROC curve analysis, and the related Area Under the ROC Curve (AUC). Since we are interested in evaluating the IDS for small values of the false positive rate, we computed the area under the ROC curve in the range $[0, 0.1]$ of the false positive rate. In order to obtain a performance value in the range $[0, 1]$, we normalized the "partial" AUC (AUC_p) computed in $[0, 0.1]$ by dividing it by 0.1.

5   Experimental Results

This section provides a discussion of the experimental results achieved. The performance, evaluated in terms of AUC_p, has been calculated considering several one-class classification algorithms. In addition, we also varied the number of HMM within each ensemble. A number of HMM from 1 to 3 has been considered.

We first considered the static rules as a possible choice for the one-class classifier. We considered two rules, respectively the average and the product rules. The missing features have been excluded from the computation. For the sake of brevity we report just some examples of the results achieved. The AUC_p was equal to 0.440 on the **Generic** attacks and equal to 0.464 on the **Shell-code** attacks (DIEE legitimate traffic) for the average rule. The results achieved using the product rule (and on the GT dataset) were equivalent. These results clearly show that static rules are not suitable in this scenario, thus confirming our choice of one-class classifiers as trainable fusion rules.

We experimented using several classifiers to combine the outputs produced by the HMM ensembles. In particular, we considered the Gaussian (**Gauss**) distribution, the Mixture of Gaussians (**MoG**), the **Parzen** density estimators, and the **SVM**. For the first three classifiers we used the implementation provided within the dd_tools[1]. For the SVM, we used the implementation provided by LibSVM[2]. We used a Radial Basis function for the SVM kernel. We left the setting of the other parameters to the default values.

Table 4(a) and 4(b) report the results achieved on the DIEE and GT dataset respectively. The same tables also reports the average values of AUC_p achieved by *HMMPayl* under the same conditions. From a deep comparison between

---

[1] Dd_tools - http://prlab.tudelft.nl/david-tax/dd_tools.html

[2] LibSVM - http://www.csie.ntu.edu.tw/~cjlin/libsvm

Table 4. Average and Standard Deviation values of AUC_p. The rightmost column reports the performance achieved by *HMMPayl* [2].

(a) DIEE Dataset.

Attack Dataset	HMM	Gauss	Parzen	MoG	SVM	HMMPayl
Generic	1	0.656 (0.216)	0.931 (0.022)	0.874 (0.097)	0.843 (0.036)	
	2	0.659 (0.207)	0.931 (0.023)	0.857 (0.125)	0.848 (0.029)	0.922 (0.058)
	3	0.659 (0.226)	0.933 (0.021)	0.865 (0.129)	0.851 (0.031)	
XSS-SQL	1	0.937 (0.030)	0.941 (0.031)	0.923 (0.046)	0.838 (0.203)	
	2	0.936 (0.030)	0.940 (0.033)	0.915 (0.059)	0.863 (0.175)	0.847 (0.032)
	3	0.935 (0.030)	0.939 (0.034)	0.924 (0.046)	0.871 (0.161)	
Shell-code	1	0.935 (0.030)	0.946 (0.022)	0.923 (0.032)	0.889 (0.061)	
	2	0.942 (0.033)	0.946 (0.022)	0.916 (0.028)	0.899 (0.055)	0.996 (0.002)
	3	0.944 (0.035)	0.945 (0.023)	0.924 (0.028)	0.908 (0.056)	

(b) GT Dataset.

Attack Dataset	HMM	Gauss	Parzen	MoG	SVM	HMMPayl
Generic	1	0.686 (0.107)	0.920 (0.082)	0.915 (0.035)	0.801 (0.102)	
	2	0.695 (0.087)	0.922 (0.087)	0.917 (0.037)	0.809 (0.095)	0.866 (0.071)
	3	0.709 (0.024)	0.923 (0.093)	0.919 (0.028)	0.816 (0.093)	
XSS-SQL	1	0.718 (0.107)	0.972 (0.018)	0.870 (0.055)	0.806 (0.043)	
	2	0.725 (0.095)	0.972 (0.018)	0.896 (0.052)	0.813 (0.037)	0.827 (0.056)
	3	0.737 (0.083)	0.973 (0.018)	0.904 (0.037)	0.816 (0.030)	
Shell-code	1	0.848 (0.060)	0.928 (0.079)	0.930 (0.044)	0.909 (0.073)	
	2	0.837 (0.041)	0.926 (0.084)	0.910 (0.043)	0.917 (0.075)	0.988 (0.003)
	3	0.837 (0.036)	0.925 (0.088)	0.909 (0.043)	0.917 (0.072)	

HMMPayl and other similar algorithms (e.g [5,9,11]) *HMMPayl* resulted as the most effective IDS on the same datasets we used here [2]. Thus, in this paper we consider only *HMMPayl* for the sake of comparison.

It can be easily observed that if we exclude the case on which the **Gauss** classifier is used, the proposed solution performs generally well with respect to *HMMPayl*. A result which is worth to notice is that achieved using the **Parzen** classifier. In fact in this case the IDS works significantly better than *HMMPayl* against the **XSS-SQL** attacks (especially on the GT dataset) and it works better also against the **Generic** attacks. On the contrary, *HMMPayl* performs better when **Shell-code** attacks are considered. This is not surprising since *HMMPayl* basically creates a detailed model of the bytes' distribution of the payload, that in the case of the **Shell-code** attacks significantly deviates from that of the legitimate traffic. Nevertheless, we are quite convinced that the effectiveness of our IDS can be easily improved also against attacks of this type by designing more carefully the HMM ensemble. In fact, we observed that for certain headers the length of the sequences can variate heavily from payload to payload. Since the probability assigned by a HMM to a sequence significantly depends on the sequence length, a model that takes into account also the length of the header would be probably preferable in the case of those headers. It must be also considered that the one-class classifier can be further optimized since we left the setting of the parameters to the default values.

We can also notice that increasing the number of classifiers within the HMM ensembles does not provide remarkable benefits. It can be observed that in some cases the AUC_p increases with the number of HMM (e.g. in the **SVM** column) whereas in other cases the AUC_p slightly reduces (e.g. in the **Parzen** column). Notwithstanding, the observed variations are very low for both the average and the standard deviation of the AUC_p.

6   Conclusions

This paper we proposes an IDS that models the HTTP payload structure for the purpose of detecting the attacks against a web server. The IDS heavily relies on the MCS paradigm, since the outputs provided by a set of HMM ensembles are combined using a one-class classifier. The experimental results achieved confirm the effectiveness of the proposed solution and also show that the IDS works generally better than analogous algorithms. In addition, as a consequence of its small complexity, this IDS would be easily implemented in a real system.

References

1. RFC 2616 - Hypertext Transfer Protocol – HTTP/1.1(1999)
2. Ariu, D., Tronci, R., Giacinto HMMPayl, G.:HMMPayl: An intrusion detection system based on Hidden Markov Models. In: Computers & Security (in Press, 2011)
3. Baum, L.E., Petrie, T., Soules, G., Weiss, N.: A maximization technique occurring in the statistical analysis of probabilistic functions of markov chains. The Annals of Mathematical Statistics 41(1), 164–171 (1970)
4. Biggio, B., Fumera, G., Roli, F.: Multiple classifier systems for adversarial classification tasks. In: Benediktsson, J.A., Kittler, J., Roli, F. (eds.) MCS 2009. LNCS, vol. 5519, pp. 132–141. Springer, Heidelberg (2009)
5. Corona, I., Ariu, D., Giacinto, G.: HMM-Web: A framework for the detection of attacks against web applications. In: IEEE International Conference on Communications, Dresden, Germany (2009)
6. Kruegel, C., Vigna, G.: Anomaly detection of web-based attacks. In: ACM conference on Computer and Communications Security, New York, USA (2003)
7. Kruegel, C., Vigna, G., Robertson, W.: A multi-model approach to the detection of web-based attacks. Computer Networks 48(5), 717–738 (2005)
8. Marcialis, G.L., Roli, F., Didaci, L.: Personal identity verification by serial fusion of fingerprint and face matchers. Pattern Recognition 42(11), 2807–2817 (2009)
9. Perdisci, R., Ariu, D., Fogla, P., Giacinto, G., Lee, W.: McPAD: A multiple classifier system for accurate payload-based anomaly detection. Computer Networks 53(6), 864–881 (2009)
10. Rabiner, L.R.: A tutorial on hidden markov models and selected applications in speech recognition. Proceedings of the IEEE 77(2), 257–286 (1989)
11. Song, Y., Keromytis, A.D., Stolfo, S.J.: Spectrogram: A mixture-of-markov-chains model for anomaly detection in web traffic. In: NDSS, The Internet Society (2009)
12. Friedman, J., Hastie, T., Tibshirani, R.: The Elements of Statistical Learning: Data Mining, Inference, and Prediction, 2nd edn. Springer, Heidelberg (2009)
13. Wang, K., Stolfo, S.J.: Anomalous payload-based network intrusion detection. In: Jonsson, E., Valdes, A., Almgren, M. (eds.) RAID 2004. LNCS, vol. 3224, pp. 203–222. Springer, Heidelberg (2004)

Incremental Boolean Combination of Classifiers[*]

Wael Khreich[1], Eric Granger[1], Ali Miri[2], and Robert Sabourin[1]

[1] Laboratoire d'imagerie, de vision et d'intelligence artificielle
École de technologie supérieure, Montreal, QC, Canada
wael.khreich@livia.etsmtl.ca,
{eric.granger,robert.sabourin}@etsmtl.ca
[2] School of Computer Science, Ryerson University, Toronto, Canada
ali.miri@ryerson.ca

Abstract. The incremental Boolean combination (*incrBC*) technique is a new learn-and-combine approach that is proposed to adapt ensemble-based pattern classification systems over time, in response to new data acquired during operations. When a new block of training data becomes available, this technique generates a diversified pool of base classifiers from the data by varying training hyperparameters and random initializations. The responses of these classifiers are then combined with those of previously-trained classifiers through Boolean combination in the ROC space. Through this process, an ensemble is selected from the pool, where Boolean fusion functions and thresholds are adapted for improved accuracy, while redundant base classifiers are pruned. Results of computer simulations conducted using Hidden Markov Models (HMMs) on synthetic and real-world host-based intrusion detection data indicate that *incrBC* can sustain a significantly higher level of accuracy than when the parameters of a single best HMM are re-estimated for each new block of data, using reference batch and incremental learning techniques. It also outperforms static fusion techniques such as majority voting for combining the responses of new and previously-generated pools of HMMs. Pruning prevents pool sizes from increasing indefinitely over time, without adversely affecting the overall ensemble performance.

1 Introduction

In practice, pattern recognition systems are typically designed a priori using limited and imbalanced data acquired from complex changing environments. Various one- and two-class neural and statistical classifiers have been applied to detection tasks, for instance, to learn and detect normal or abnormal system behavior. Since the collection and analysis of representative training data for design an validation is costly, the classifier may represent an incomplete view of system behavior. Since new training data may become available after a classifier has originally been deployed for operations, it could be adapted to maintain or improve performance over time.

---

[*] This research has been supported by the Natural Sciences and Engineering Research Council of Canada.

C. Sansone, J. Kittler, and F. Roli (Eds.): MCS 2011, LNCS 6713, pp. 340–349, 2011.

Given a new block of training data, incremental re-estimation of classifier parameters raises several challenges. Parameters should be updated from new data without requiring access to the previously-learned data, and without corrupting previously-acquired knowledge [6]. State-of-the-art batch learning classifiers must accumulate new training data in memory, and retrain from the start using all (new and previously-accumulated) data. A number of classifiers in literature have been designed with the inherent ability to perform supervised incremental learning. However, the decline in performance caused by knowledge corruption remains an issue. Indeed, single classifier systems for incremental learning may not adequately approximate the underlying data distribution when there are multiple local maxima in the solution space [1].

Ensemble methods have been employed to overcome such limitations [6]. Theoretical and empirical evidence suggests that combining the responses of several accurate and diverse classifiers can enhance the overall accuracy and reliability of a pattern classification system [4,10]. Despite reducing information to binary decisions, combining responses at the decision level, in the Receiver Operating Characteristic (ROC) space, allows to combine across a variety of classifiers trained with different hyperparameters, feature subsets and initializations.

In this paper, a new ensemble-based technique called incremental Boolean combination (*incrBC*) is proposed for incremental learning of new training data according to a learn-and-combine approach. When a new block of training data becomes available, it is used to generate a new pool of classifiers by varying training hyperparameters and random initializations. The responses from the newly-trained classifiers are then combined to those of the previously-trained classifiers by applying Boolean combination in the ROC space. The proposed system allows to improve overall accuracy by evolving an ensemble of classifiers (EoCs) in which Boolean fusion functions and decision thresholds are adapted. Since the pool size grows indefinitely over time, *incrBC* integrates model management strategies to limit the pool size without significantly degrading performance.

For proof-of-concept, *incrBC* is applied to adaptive anomaly detection from system call sequences with Hidden Markov Models (HMMs). The experiments are conducted on both synthetically generated and sendmail data from the University of New Mexico [11]. Learning new data allows to account for rare events, and hence improve detection accuracy and reduce false alarms. The performance of the proposed system is compared to that of the reference algorithms for batch and incremental learning of HMM parameters. In addition, the performance achieved with Boolean fusion functions is compared to that of median (MED) and majority vote (VOTE) functions combining the outputs from pool of HMMs.

2   Learn-and-Combine Approach Using Incremental Boolean Combination

Boolean combination (BC) has recently been investigated to combine the decision of multiple crisp or soft one- or two-class classifiers in the ROC space [8]. These threshold-optimized decision-level combination techniques can outperform several techniques in the Neyman-Pearson sense, yet they assume that the classifiers are conditionally-independent, and that their corresponding ROC curves are convex and proper. These assumptions are rarely valid in practice, where classifiers are designed using limited and imbalanced data. In previous research, the authors proposed BC techniques [3] for efficient fusion of multiple ROC curves using all Boolean functions, without any prior assumptions. These technique apply to batch learning of a fixed-size data set.

In this paper, an extension to the batch BC techniques (proposed in [3]) – called incremental BC (*incrBC*) technique – is proposed for incremental learning of new training data during operations. In response to a new block of data, this learn-and-combine approach consists in generating a new pool of classifiers, and then applying *incrBC* to combine ROC curves of the new pool with the ROCCH obtained with previously-obtained data.

As described in Algorithm 1, *incrBC* uses each Boolean function to combine the responses corresponding to each decision threshold from the first classifier to those from the second classifier. Fused responses are then mapped to vertices

---

Algorithm 1. $incrBC(\lambda_1, \lambda_2, \ldots, \lambda_K, \mathcal{V})$: Incremental Boolean combination of classifiers

> **input** : K classifiers $(\lambda_1, \lambda_2, \ldots, \lambda_K)$ and a validation set $\mathcal{V}$ of size $|\mathcal{V}|$
> **output**: ROCCH of combined classifiers where each vertex is the result of 2 to K combination of crisp classifiers. Each combination selects the best decision thresholds (λ_i, t_j) and Boolean function, which are stored in the set $(\mathcal{S})$

1   $n_k \leftarrow$ no. decision thresholds of λ_k using $\mathcal{V}$ // no. vertices on ROC(λ_k)
2   $BooleanFunctions \leftarrow \{a \wedge b, \neg a \wedge b, a \wedge \neg b, \neg(a \wedge b), a \vee b, \neg a \vee b, a \vee \neg b, \neg(a \vee b), a \oplus b, a \equiv b\}$
     compute $ROCCH_1$ of the first two classifiers (λ_1 and λ_2)
3   **allocate** $\boldsymbol{F}$ an array of size: $[2, n_1 \times n_2]$ // temporary storage of combination results
4   **foreach** $bf \in BooleanFunctions$ **do**
5   **for** $i \leftarrow 1$ **to** n_1 **do**
6           $\boldsymbol{R_1} \leftarrow (\lambda_1, t_i)$ // responses of λ_1 at decision threshold t_i using $\mathcal{V}$
7   **for** $j \leftarrow 1$ **to** n_2 **do**
8               $R_2 \leftarrow (\lambda_2, t_j)$ // responses of λ_2 at decision threshold t_j using $\mathcal{V}$
9               $\boldsymbol{R_c} \leftarrow bf(\boldsymbol{R_1}, \boldsymbol{R_2})$ // combine responses using current Boolean function
10   compute (tpr, fpr) of $\boldsymbol{R_c}$ using $\mathcal{V}$ // map combined responses to ROC space
11   push (tpr, fpr) onto $\boldsymbol{F}$

12   compute $ROCCH_2$ of all ROC points in $\boldsymbol{F}$
13      $n_{ev} \leftarrow$ number of emerging vertices
14      $\mathcal{S}_2 \leftarrow \{(\lambda_1, t_i), (\lambda_2, t_j), bf\}$ // set of selected decision thresholds from each classifier and Boolean functions for emerging vertices

15 **for** $k \leftarrow 3$ **to** K **do**
16   **allocate** $\boldsymbol{F}$ of size: $[2, n_k \times n_{ev}]$
17   **foreach** $bf \in BooleanFunctions$ **do**
18   **for** $i \leftarrow 1$ **to** n_{ev} **do**
19              $\boldsymbol{R_i} \leftarrow \mathcal{S}_{k-1}(i)$ // responses from previous combinations
20   **for** $j \leftarrow 1$ **to** n_k **do**
21                  $\boldsymbol{R_k} \leftarrow (\lambda_k, t_j)$
22                  $\boldsymbol{R_c} \leftarrow bf(\boldsymbol{R_i}, \boldsymbol{R_k})$
23   compute (tpr, fpr) of $\boldsymbol{R_c}$ using $\mathcal{V}$
24   push (tpr, fpr) onto $\boldsymbol{F}$

25   compute $ROCCH_k$ of all ROC points in $\boldsymbol{F}$
26      $n_{ev} \leftarrow$ number of emerging vertices
27      $\mathcal{S}_k \leftarrow \{\mathcal{S}_{k-1}(i), (\lambda_k, t_j), bf\}$ // set of selected subset from previous combinations, decision thresholds from the newly-selected classifier, and Boolean functions for emerging vertices

28 **store** $\mathcal{S}_k : 2 \leq k \leq K$
29 **return** $ROCCH_K$

---

in the ROC space, and their ROC convex hull (ROCCH) is computed. Vertices that are superior to the ROCCH of original classifiers are then selected. The set (S) of decision thresholds from each classifier and Boolean functions corresponding to these vertices is stored, and the ROCCH is updated to include emerging vertices. The responses corresponding to each decision threshold from the third classifier are then combined with the responses of each emerging vertex, and so on, until the last classifier in the pool is combined. The BC technique yields a final ROCCH for visualization and selection of operating points, and the set of selected thresholds and Boolean functions, S, for each vertex on the composite ROCCH to be applied during operations. For a pool of K soft classifiers each comprising n crisp classifiers, the worst-case time complexity required for combinations (during the design phase) is $\mathcal{O}(Kn^2)$ Boolean operations.

Selecting crisp detectors from all HMMs in the pool leads to unnecessarily high computational and memory complexity since the pool size grows as new blocks of data become available. In addition, HMMs are combined according to the order in which they are stored in the pool. An HMM trained on new data may capture different underlying data structure and could replace several previously-selected HMMs. Model management mechanisms are therefore required for ensemble selection and pruning less relevant members of the pool.

Model Selection. Ensemble selection is performed at each new block of data, and the best ensemble of classifiers is selected from the pool based on different optimization criteria and selection strategies, each one seeking to increase accuracy and reduce the computational and memory complexity [9]. Optimization criteria include ensemble accuracy, entropy and diversity. Ensemble selection strategies include ranking- and search-based techniques. In ranking-based techniques, the pool members are ordered according to some performance measure on a validation set, and the top classifiers are then selected to form an ensemble. Search-based techniques first combine the outputs of classifiers and then select the best performing ensemble evaluated on a validation set.

The proposed ensemble selection algorithm, BC_{search}, (see Algorithm 2) employs both rank- and search-based selection strategies to optimize the area under the ROCCH (AUCH)[1]. In contrast with existing techniques, BC_{search} exploits the monotonicity of the $incrBC$ algorithm for an early stopping criterion. Furthermore, the final composite ROCCH is always stored which allows for visualization of the whole range of performance and adaptation to changes in prior probabilities and costs of errors. This can be achieved by adjusting the desired operational point, which activates different classifiers, decision thresholds and Boolean functions.

As described in Algorithm 2, the BC_{search} algorithm employs an incremental search strategy. It starts by ranking all classifiers in decreasing order of their AUCH accuracy on a validation set and then selects the classifier with the largest AUCH value. Next, it applies the $incrBC$ algorithm to combine the selected

---

[1] Other ROC-based measures, such as the partial AUCH or the true positive rate at a required false positive rate, can also be measured for a region-specific accuracy.

Algorithm 2. $BC_{search}(\mathcal{P}, \mathcal{V})$: Boolean combination – incremental search

---

    input : Pool of classifiers $\mathcal{P} = \{\lambda_1, \lambda_2, \ldots, \lambda_K\}$ and a validation set $\mathcal{V}$
    output: Ensemble of base classifiers (E) selected from the pool $\mathcal{P}$

1 set tol // set the tolerance value for AUCH
2 **foreach** $\lambda_k \in \mathcal{P}$ **do**
3   ⌊ compute ROC curves and their $ROCCH_k$, using $\mathcal{V}$

4 **sort** classifiers $(\lambda_1, \ldots, \lambda_K)$ in descending order of $AUCH(ROCCH_k)$ values
5 $\lambda_1 = \arg\max_k\{AUCH(ROCCH_k)\} : \lambda_k \in \mathcal{P}\}$
6 $E \leftarrow \lambda_1$ // select classifier with the largest AUCH value
7 **foreach** $\lambda_k \in \mathcal{P} \backslash E$ **do**   // remaining classifiers in $\mathcal{P}$
8   ⌊ $ROCCH_k = incrBC\,((\lambda_1, \lambda_k), \mathcal{V})$

9 $\lambda_2 = \arg\max_k\{AUCH(ROCCH_k)\} : \lambda_k \in \mathcal{P} \backslash E\}$
10 $E \leftarrow E \cup \lambda_2$ // select classifier with largest AUCH improvement to E
11 $ROCCH_1 \leftarrow ROCCH_2$ // update the convex hull
12 $\mathcal{S}_2 \leftarrow \{(\lambda_1, t_i), (\lambda_2, t_{i'}), bf\}$ // Set of selected decision thresholds from each
    classifier and selected Boolean functions (from *incrBC* algorithm)
13 $j \leftarrow 3$
14 **repeat**
15   **foreach** $\lambda_k \in \mathcal{P} \backslash E$ **do**
16   ⌊ $ROCCH_k = incrBC\,((\mathcal{S}_{j-1}, \lambda_k), \mathcal{V})$

17     $\lambda_j = \arg\max_k\{AUCH(ROCCH_k)\} : \lambda_k \in \mathcal{P} \backslash E\}$
18     $E \leftarrow E \cup \lambda_j$
19     $ROCCH_{j-1} \leftarrow ROCCH_j$
20     $\mathcal{S}_j \leftarrow \{\mathcal{S}_{j-1}(i), (\lambda_j, t_{i'}), bf\}$ // Set of selected subset from previous combinations,
    decision thresholds from the newly-selected classifiers, and Boolean functions
    (derived from *incrBC* algorithm)
21     $j \leftarrow j + 1$
22 **until** $AUCH(ROCCH_j) \leq AUCH(ROCCH_{j-1}) + tol$ // no improvement
23 **store** $\mathcal{S}_j, \ 2 \leq j \leq K$
24 **return** E

---

classifier with each of the remaining classifiers in the pool. The classifier that most improves the AUCH accuracy of the EoCs is then selected. The cumulative EoCs are then combined with each of the remaining classifiers in the pool (using *incrBC*), and the classifier that provides the largest AUCH improvement to the EoCs is selected, and so on. The algorithm stops when the AUCH improvement of the remaining classifiers is lower than a user-defined tolerance value, or when all classifiers in the original ensemble are selected. Given a pool of K classifiers, the worst-case time complexity of this selection algorithm is $O(K^2)$ w.r.t. the number of Boolean combination of *incrBC*. However, this complexity is only attained when the algorithm selects all classifiers (zero tolerance). In practice, the computational time is typically lower depending on the tolerance value.

Model Pruning. As new blocks of data are learned incrementally, pruning less relevant models is essential to limit the pool size $|\mathcal{P}|$ (and memory resources) from growing indefinitely. With *incrBC*, classifiers that go unselected over time are discarded. A counter is therefore assigned to each classifier in the pool indicating the number of blocks for which an classifier was not selected as an ensemble member. A classifier is then pruned from the pool, according to a user-defined life time (LT) expectancy value of unselected models. For instance, with an $LT = 3$ all classifiers that have not been selected after receiving three blocks of data, as indicated by their counters, are discarded from the pool.

3   Simulation Results

Host-based intrusion detection systems applied to anomaly detection (AD) typically monitor for significant deviations in system call sequences, since system calls are the gateway between user and operating system's kernel mode. Among various neural and statistical classifiers, techniques based on HMMs have been shown to produce a high level of performance [11], although standard techniques for re-estimating HMM parameters involve batch learning [2]. Designing an HMM for AD involves estimating HMM parameters and the number of hidden states (N) from the training data. *incrBC* allows to adapt AD to newly-acquired data based on a learn and combine approach.

Proof-of-concept simulations are conducted on both synthetically generated data and sendmail process data collected at the University of New Mexico (UNM)[2] [11]. The synthetic data generator is based on the conditional relative entropy (CRE) [3,7], which controls the irregularity of the generated data ($CRE = 0$, perfect regularity and $CRE = 1$, complete irregularity). The synthetically generated data simulate a complex process, with an alphabet $\Sigma = 50$ symbols and $CRE = 0.4$. The sizes of injected anomalies (AS) are assumed equal to the detector window (DW) sizes. The training is conducted on ten successive blocks D_k, for $k = 1, \ldots, 10$, of normal system call sequences, each of length $DW = 4$ symbols. For the synthetic data, each block comprises 500 sub-sequences, whereas the validation ($\mathcal{V}$) and test ($\mathcal{T}$) sets are comprised of $2,000$ and $5,000$ labeled sub-subsequences (normal or anomalous). For sendmail data, each block comprises 100 sub-sequences, and each $\mathcal{V}$ and $\mathcal{T}$ comprise 450 sub-sequences. In both cases, the anomaly size is $AS = 4$ symbols and the ratio of normal to anomalous sequences is $4 : 1$.

For each D_k, 20 new base HMMs are generated and appended to the pool ($\mathcal{P}$). These ergodic HMMs are trained with 20 different number of states ($N = 5, 10, \ldots, 100$) according to the Baum-Welch (BW) algorithm. For each N value, 10-fold cross-validation and ten different random initializations are employed to select the HMM (λ_N^k) that gives the highest AUCH on $\mathcal{V}$. The HMMs from the pool (of increasing size $|\mathcal{P}| = 20, 40, \ldots, 200$ HMMs) are then provided for incremental combination according to the *incrBC* technique. The same training, validation and selection procedures are applied to the other techniques. However, for the reference batch Baum-Welch (BBW) the training is conducted on cumulative data blocks ($D_1 \cup D_2 \ldots \cup D_k$), and both online Baum-Welch (OBW) [5] and IBW [2] algorithms resume the training from the previously-learned HMMs using only the current block of data (D_k).

Incremental BC without model management. The performance of the proposed *incrBC* technique is first presented without employing the model management strategy. For a given block D_k, all available HMMs in $\mathcal{P}$ were selected and combined according to *incrBC*. As illustrated in Figure 1, the average AUCH accuracy achieved by applying this technique is highest overall. It is significantly

---

[2]   http://www.cs.unm.edu/~immsec/systemcalls.htm

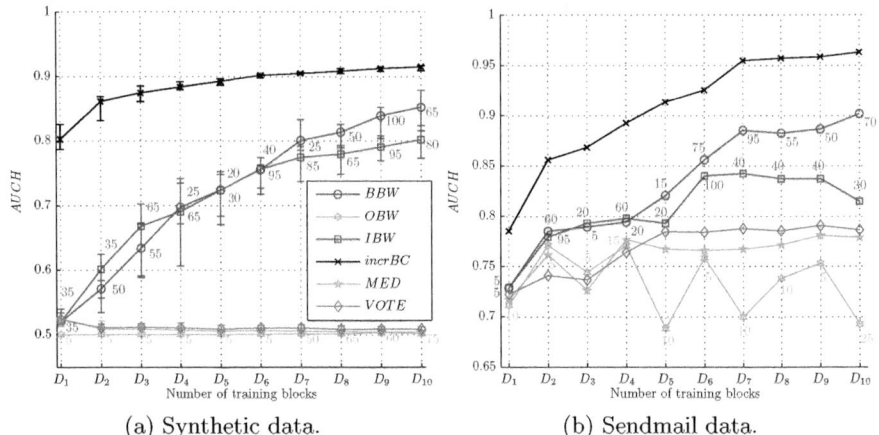

(a) Synthetic data.   (b) Sendmail data.

Fig. 1. Average AUCH of techniques used for incremental learning over 10 successive blocks of data D_k. For each block, the HMMs are trained according to each
technique with 20 different states ($N = 5, 10, \ldots, 100$), providing a pool of size
$|\mathcal{P}| = 20, 40, \ldots, 200$ HMMs. Numbers above points are the N values for BBW, OBW
and IBW that achieved the highest AUCH accuracy on each block. Error bars in (a)
are lower and upper quartiles over ten replications.

higher than that of the reference BBW, most notably when provided with limited training data in the first few blocks. The *incrBC* effectively exploits the
complementary information provided from the pool of HMMs trained with different number of states and different initializations, and using newly-acquired
data. Not surprisingly, OBW leads to the lowest level of accuracy, as one pass
over limited data is insufficient to capture the underlying data structure. IBW
outperforms OBW since it re-estimates HMM parameters iteratively over each
block using a fixed learning rate [2].

The MED and VOTE fusion functions do not improve accuracy with respect
to the Boolean functions produced with *incrBC*, which reflects their inabilities
to exploit the complementary information. The MED function directly combines
HMM likelihood values for each sub-sequence during operations, while VOTE
considers the crisp decisions from HMMs at optimal operating thresholds (equal
error rates). In contrast, *incrBC* applies ten Boolean functions to the crisp decisions provided by the thresholds of each HMM. Then, it selects the decision
thresholds and Boolean functions that improve the overall ROCCH on the validation set $\mathcal{V}$. Exploring all decision thresholds before selection may increase
ensemble diversity, and improve overall system accuracy.

Model selection. Figure 2 presents the impact on accuracy of using the
BC_{search} algorithm proposed for selection of ensembles in the *incrBC* technique (compared to that of *incrBC* from Figure 1). The results in Figure 2a
(using synthetic data) correspond to the first replication of Figure 1a. As shown
in Figure 2, BC_{search} maintains a AUCH accuracy that is comparable to that of

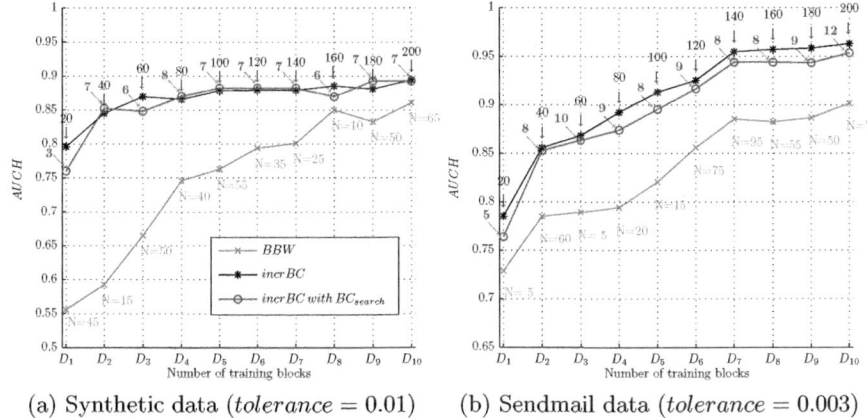

(a) Synthetic data (*tolerance* = 0.01)   (b) Sendmail data (*tolerance* = 0.003)

Fig. 2. Average AUCH accuracy achieved when the *incrBC* technique employs BC_{search} for model selection, over 10 successive blocks of data D_k. For each block, values on the arrows correspond to the number of models (EoHMM size) selected by each technique from the overall pool of size $|\mathcal{P}| = 20, 40, \ldots, 200$ HMMs. The performance of the BBW used to train a single best HMM, where the best state value is selected at each block, is also shown for reference.

incrBC. For each block however, the size of selected EoHMMs, $|E|$, is reduced significantly compared to the original pool of size $|\mathcal{P}| = 20, 40, \ldots, 200$ HMMs. For instance, at the 10th data block in Figure 2a, BC_{search} selects an ensemble of size $|E| = 7$ from the generated pool of size $|\mathcal{P}| = 200$ HMMs (selected with *incrBC*). BC_{search} provides compact and accurate ensembles by exploiting the order in which HMMs are combined and the benefit achieved by cumulative EoHMMs before selecting a new ensemble member.

Model pruning. Figure 3 presents the AUCH accuracy of the full *incrBC* technique, employing BC_{search} and the pruning strategy according to various life time (LT) expectancy values. An HMM is pruned if it is not selected during a *LT* corresponding to 1, 3 or 5 data blocks. Figure 3a illustrates the impact on AUCH accuracy of pruning the pool of HMMs in Figure 2a (synthetic data), while Figure 3b illustrates the impact of pruning the pool of HMMs in Figure 2b (sendmail data). As shown in Figure 3a, the level of AUCH accuracy achieved with *incrBC* decreases when LT varies from ∞ to 1. Early punning (*LT* = 1) of HMMs that have not improved ensemble performance during a given learning stage, may lead to knowledge corruption, and hence a decline in system performance. These HMMs may provide complementary information to newly-generated HMMs, depending on the data. This is illustrated in Figure 3b, where the decline in AUCH for *LT* = 1 is relatively smaller for sendmail data, which incorporates more redundancy than the synthetically-generated data. As shown in Figures 3a and 3b, the performance achieved with a delayed pruning of HMMs (e.g., *LT* = 3 and 5) compares to that of retaining all generated HMMs in the pool (*LT* = ∞). For fixed tolerance and *LT* values, *incrBC* is capable of selecting small EoHMMs and further reducing the

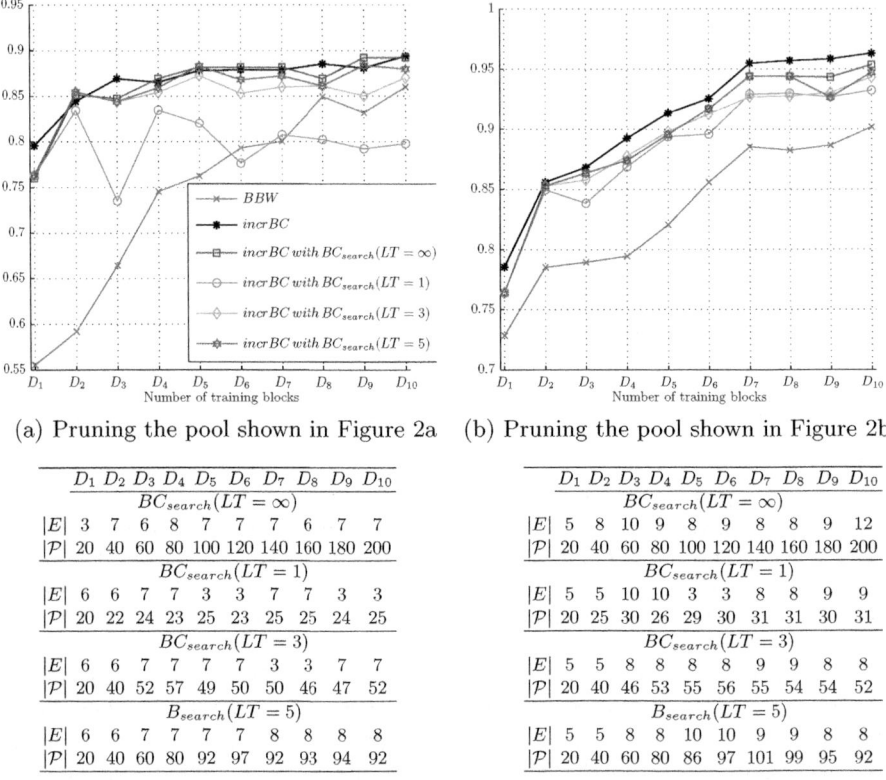

(a) Pruning the pool shown in Figure 2a   (b) Pruning the pool shown in Figure 2b

D_1	D_2	D_3	D_4	D_5	D_6	D_7	D_8	D_9	D_{10}
$BC_{search}(LT = \infty)$									
$\|E\|$ 3	7	6	8	7	7	6	7	7	
$\|\mathcal{P}\|$ 20	40	60	80	100	120	140	160	180	200
$BC_{search}(LT = 1)$									
$\|E\|$ 6	6	7	7	3	3	7	7	3	3
$\|\mathcal{P}\|$ 20	22	24	23	25	23	25	25	24	25
$BC_{search}(LT = 3)$									
$\|E\|$ 6	6	7	7	7	3	3	7	7	
$\|\mathcal{P}\|$ 20	40	52	57	49	50	50	46	47	52
$B_{search}(LT = 5)$									
$\|E\|$ 6	6	7	7	7	7	8	8	8	8
$\|\mathcal{P}\|$ 20	40	60	80	92	97	92	93	94	92

D_1	D_2	D_3	D_4	D_5	D_6	D_7	D_8	D_9	D_{10}
$BC_{search}(LT = \infty)$									
$\|E\|$ 5	8	10	9	8	9	8	8	9	12
$\|\mathcal{P}\|$ 20	40	60	80	100	120	140	160	180	200
$BC_{search}(LT = 1)$									
$\|E\|$ 5	5	10	10	3	3	8	8	9	9
$\|\mathcal{P}\|$ 20	25	30	26	29	30	31	31	30	31
$BC_{search}(LT = 3)$									
$\|E\|$ 5	5	8	8	8	9	9	8	8	
$\|\mathcal{P}\|$ 20	40	46	53	55	56	55	54	54	52
$B_{search}(LT = 5)$									
$\|E\|$ 5	5	8	8	10	10	9	9	8	8
$\|\mathcal{P}\|$ 20	40	60	80	86	97	101	99	95	92

Fig. 3. Average AUCH accuracy achieved when the *incrBC* technique uses BC_{search} and pruning of HMM pools on (a) synthetic and (b) UNM sendmail data. The size of selected EoHMMs and that of the pool are presented for each data block below the graphs. The performance of the BBW algorithm, of *incrBC* without model management (Figure 1), and of *incrBC* when HMMs are combined without any pruning, $LT = \infty$ (Figure 2), are also shown for reference.

size of the pool. A fixed-size pool may be maintained by tuning the tolerance and LT values upon receiving new blocks of data.

4   Conclusions

This paper presents a new ensemble-based technique called incremental Boolean combination (*incrBC*) for incremental learning of new training data according to a learn-and-combine approach. Given a new block of training data, a diversified pool of base classifiers is generated from the data, and their responses are combined with those of previously-trained classifiers in the ROC space. A BC_{search} algorithm selects accurate ensemble of classifiers for operations. This technique allows to adapt Boolean fusion functions and decision thresholds over time, while punning redundant base classifiers. The proposed system is capable of changing its desired operating point during operations, and hence allow to account for changes in prior probabilities and costs of errors.

During simulations conducted on both synthetic and real-world host-based intrusion detection data using HMMs, the proposed system has been shown to achieve a higher level of accuracy than when parameters of a single best HMM are estimated, using reference batch and incremental learning techniques. It also outperforms ensemble techniques that use the median and majority vote fusion functions to combine new and previously-trained HMMs. The system has shown to form compact ensembles for operations, while maintaining or improving the overall system accuracy. Pruning has been shown to limit the pool size from increasing over time, without negatively affecting the overall ensemble accuracy.

The robustness of the proposed learn-and-combine approach depends on maintaining a representative validation set over time, for selection of base classifiers, and for adaptation of decision thresholds and Boolean functions. Future work involves applying *incrBC* to real-world problems that feature heavily imbalanced data sampled from dynamically-changing environments.

References

1. Connolly, J.F., Granger, E., Sabourin, R.: An adaptive classification system for video-based face recognition. Information Sciences (2010) (in Press)
2. Khreich, W., Granger, E., Miri, A., Sabourin, R.: A comparison of techniques for on-line incremental learning of HMM parameters in anomaly detection. In: Proc. 2nd IEEE Int'l Conf. on Computational Intelligence for Security and Defense Applications, Ottawa, Canada, July 2009, pp. 1–8 (2009)
3. Khreich, W., Granger, E., Miri, A., Sabourin, R.: Iterative Boolean combination of classifiers in the ROC space: An application to anomaly detection with HMMs. Pattern Recognition 43(8), 2732–2752 (2010)
4. Kuncheva, L.I.: Combining Pattern Classifiers: Methods and Algorithms. Wiley, Chichester (2004)
5. Mizuno, J., Watanabe, T., Ueki, K., Amano, K., Takimoto, E., Maruoka, A.: On-line estimation of hidden Markov model parameters. In: Proc. 3rd Int'l Conf. on Discovery Science, vol. 1967, pp. 155–169 (2000)
6. Polikar, R., Upda, L., Upda, S., Honavar, V.: Learn++: An incremental learning algorithm for supervised neural networks. IEEE Transactions on Systems, Man and Cybernetics, Part C 31(4), 497–508 (2001)
7. Tan, K., Maxion, R.: Determining the operational limits of an anomaly-based intrusion detector. IEEE Journal on Selected Areas in Communications 21(1), 96–110 (2003)
8. Tao, Q., Veldhuis, R.: Threshold-optimized decision-level fusion and its application to biometrics. Pattern Recognition 41(5), 852–867 (2008)
9. Tsoumakas, G., Partalas, I., Vlahavas, I.: An ensemble pruning primer. Applications of Supervised and Unsupervised Ensemble Methods 245, 1–13 (2009)
10. Tulyakov, S., Jaeger, S., Govindaraju, V., Doermann, D.: Review of classifier combination methods. In: Marinai, S., H.F (eds.) Studies in Comp. Intelligence: ML in Document Analysis and Recognition, pp. 361–386. Springer, Heidelberg (2008)
11. Warrender, C., Forrest, S., Pearlmutter, B.: Detecting intrusions using system calls: Alternative data models. In: Proc. IEEE Computer Society Symposium on Research in Security and Privacy, Oakland, CA, pp. 133–145 (1999)

Bagging Classifiers for Fighting Poisoning Attacks in Adversarial Classification Tasks

Battista Biggio, Igino Corona, Giorgio Fumera,
Giorgio Giacinto, and Fabio Roli

Dept. of Electrical and Electronic Engineering, University of Cagliari
Piazza d'Armi, 09123 Cagliari, Italy
{battista.biggio,igino.corona,fumera,giacinto,roli}@diee.unica.it
http://prag.diee.unica.it

Abstract. Pattern recognition systems have been widely used in *adversarial classification* tasks like spam filtering and intrusion detection in computer networks. In these applications a malicious adversary may successfully mislead a classifier by "poisoning" its training data with carefully designed attacks. Bagging is a well-known ensemble construction method, where each classifier in the ensemble is trained on a different bootstrap replicate of the training set. Recent work has shown that bagging can reduce the influence of outliers in training data, especially if the most outlying observations are resampled with a lower probability. In this work we argue that poisoning attacks can be viewed as a particular category of outliers, and, thus, bagging ensembles may be effectively exploited against them. We experimentally assess the effectiveness of bagging on a real, widely used spam filter, and on a web-based intrusion detection system. Our preliminary results suggest that bagging ensembles can be a very promising defence strategy against poisoning attacks, and give us valuable insights for future research work.

1 Introduction

Security applications like spam filtering, intrusion detection in computer networks, and biometric authentication have been typically faced as two-class classification problems, in which the goal of a classifier is to discriminate between "malicious" and "legitimate" samples, e.g., spam and legitimate emails. However, these tasks are quite different from traditional classification problems, as intelligent, malicious, and adaptive adversaries can manipulate their samples to mislead a classifier or a learning algorithm. In particular, one of the main issues to be faced in a so-called *adversarial classification* task [6] is the design of a *robust* classifier, namely, a classifier whose performance degrades as gracefully as possible under attack [6,12]. Adversarial classification is attracting a growing interest from the pattern recognition and machine learning communities, as witnessed by a recent workshop held in the context of the NIPS 2007 conference [1], and by a special issue of *Machine Learning* [13] entirely dedicated to this topic.

---

[1] http://mls-nips07.first.fraunhofer.de

C. Sansone, J. Kittler, and F. Roli (Eds.): MCS 2011, LNCS 6713, pp. 350–359, 2011.

In this work we consider a specific class of attacks named *causative attacks* in [2,1], and *poisoning* attacks in [11], in which the adversary is assumed to control a subset of samples that will be used to train or update the classifier, and he carefully designs these samples to mislead the learning algorithm. For instance, in intrusion detection skilled adversaries may inject poisoning patterns to mislead the learning algorithm which infers the profile of legitimate activities [11] or intrusions [15,4]; while in spam filtering adversaries may modify spam emails by adding a number of "good words", i.e., words which are likely to appear in legitimate emails and not in spam, so that when a user reports them as spam to the filter, it becomes more prone to misclassify legitimate emails as spam [14].

We argue that the problem of designing robust classifiers against poisoning attacks can be formulated as a problem in which one aims to reduce the influence of outlier samples in training data. The main motivation is that the adversary aims to "deviate" the classification algorithm from learning a correct model or probability distribution of training data, and typically can control only a small percentage of training samples. If this were not true, namely, poisoning samples were similar to other samples within the same class (or even to novel samples which represent the normal evolution of the system), their effect would be negligible. Since bagging, and in particular weighted bagging [19], can effectively reduce the influence of outlying observations in training data [9], in this work we experimentally investigate whether bagging ensembles can be exploited to build robust classifiers against poisoning attacks. We consider two relevant application scenarios: a widely deployed text classifier in spam filtering [16], and a basic version of HMM-Web, an Intrusion Detection System (IDS) for web applications [5].

The paper is structured as follows: in Sect. 2 we review related works, in Sect. 3 we highlight the motivations of this work, in Sect. 4 we describe the problem formulation and the considered applications, in Sect. 5 we present our experiments, and, eventually, in Sect. 6 we draw conclusions and sketch possible future work.

2 Background

The aim of this section is twofold. We first discuss works which investigated the effectiveness of bagging ensembles in the presence of outliers (Sect. 2.1). Then, we shortly review works related to poisoning attacks (Sect. 2.2).

2.1 Bagging in the Presence of Outliers

Bagging, short for *bootstrap aggregating*, was originally proposed in [3] to improve the classification accuracy over an individual classifier, or the approximation error in regression problems. The underlying idea is to perturb the training data by creating a number of bootstrap replicates of the training set, train a classifier on each bootstrap replicate, and aggregate their predictions. This allows to reduce the variance component of the classification or estimation error (in regression) (e.g., [3,7]). Indeed, bagging has shown to be particularly success-ful when applied to "unstable" classifiers (i.e., classifiers whose predictions vary significantly when small variations in training data occur) like decision trees and neural networks. Other explanations to the effectiveness of bagging were also proposed; in particular, in [9] it was argued that bagging equalizes the influ-ence of training samples, namely, it reduces the influence of outlier samples in training data. This was also experimentally verified on a simple task in [10], and exploited in [8] to develop outlier resistant PCA ensembles.

To further reduce the influence of the most outlying observations in training data, *weighted* bagging was proposed in [19,18]. The rationale behind this ap-proach is to resample the training set by assigning a probability distribution over training samples, in particular, lower probability weights to the most outlying observations. The method can be summarised as follows. Given a training set $T_n = \{\mathbf{x}_i, y_i\}_{i=1}^n$, and a set of probability weights $w_1, \ldots, w_n$, for which it holds $\sum_{i=1}^n w_i = 1$:

1. create m bootstrap replicates of T_n by sampling $(\mathbf{x}_i, y_i)$ with probability w_i, $i = 1, \ldots, n$;
2. train a set of m classifiers, one on each bootstrap replicate of T_n;
3. combine their predictions, e.g., by majority voting, or averaging.

Note that this corresponds to the standard bagging algorithm [3] when $w_i = 1/n$, $i = 1, \ldots, n$, and the majority voting is used as combining rule.

The set of weights $w_1, \ldots, w_n$ was estimated in [19,18] using a kernel density estimator. Since kernel density estimation can be unreliable in highly dimensional feature spaces, the authors exploited a *boosted* kernel density estimate, given by

$$f(\mathbf{x}_i) = \sum_{j=1}^n \frac{w_j}{(2\pi)^{d/2}\sigma^d} k(\mathbf{x}_i, \mathbf{x}_j), \tag{1}$$

where $k(\mathbf{x}_i, \mathbf{x}_j) = \exp\left(-\gamma ||\mathbf{x}_i - \mathbf{x}_j||^2\right)$ is a Gaussian kernel, and the set of weights $w_1, \ldots, w_n$ is iteratively estimated as follows. Initially, all samples are equally weighted, i.e., $w_i = 1/n$, $i = 1, \ldots, n$. Each weight is then iteratively updated according to $w_i^{(k+1)} = w_i^{(k)} + \log\left(f^{(k)}(\mathbf{x}_i)/g^{(k)}(\mathbf{x}_i)\right)$, where k represents current iteration, and $g(\mathbf{x}_i)$ is the "leave-one-out" estimate of $f(\mathbf{x}_i)$, given by

$$g(\mathbf{x}_i) = \sum_{j=1}^n \frac{w_j}{(2\pi)^{d/2}\sigma^d} k(\mathbf{x}_i, \mathbf{x}_j) I(j \neq i), \tag{2}$$

where $I(j \neq i)$ equals 0 (1) only when $j = i$ ($j \neq i$). Once convergence or a maximum number of iterations is reached, the final weights are inverted and normalized as

$$w_i = \frac{1}{w_i^{(k)}} / \sum_{j=1}^n \frac{1}{w_j^{(k)}}, \tag{3}$$

so that weights assigned to outlying observations exhibit lower values.

2.2   Works on Poisoning Attacks

Poisoning attacks were investigated in the field of adversarial classification, mainly considering specific applications. According to a taxonomy of potential attacks against machine learning algorithms proposed in [2,1], they can be more generally referred to as *causative* attacks, and can be exploited either to increase the false positive rate (i.e., the percentage of misclassified legitimate samples) or the false negative rate at operation phase. They were thus further categorised as *availability* or *integrity* attacks.

Poisoning attacks were devised against spam filters [14] (based on adding "good words" to spam emails, as described in Sect. 1) and simple online IDSs [2,11]. Instead, in [17] a countermeasure against them was proposed; in particular, the framework of Robust Statistics was exploited to reduce the influence of poisoning attacks in training data (which were implicitly considered "outliers").

3   Motivations of This Work

The aim of this section is to further clarify the scope of this work. As mentioned in Sect. 1, we argue that poisoning attacks can be regarded as outlying observations with respect to other samples in training data. The reason for that is twofold:

1. since the goal of a poisoning attack is to "deviate" the classification algorithm from learning a correct model or probability distribution of one of the two classes (or both), poisoning attack samples have to be different from other samples within the same class;
2. since the adversary is likely to control only a small percentage of training data in real applications, each poisoning sample should be able to largely deviate the learning process.

In addition, we also point out that several defence strategies implicitly deal with poisoning attacks as they were outliers, e.g., [17]. Besides this, as discussed in Sect. 2, a number of works highlighted that bagging (and in particular weighted bagging) can reduce the influence of outliers in training data. Thus, in this work we experimentally investigate whether bagging and weighted bagging can be successfully exploited to fight poisoning attacks in two different adversarial classification tasks, namely, spam filtering and intrusion detection. We also point out that comparing bagging and weighted bagging with the defence strategies mentioned in Sect. 2 is out of the scope of this work, as we are only considering a preliminary investigation.

4   Problem Formulation and Application Scenarios

In this section we briefly describe the problem formulation related to the two case studies considered in this work, namely, spam filtering and web-based intrusion detection. For the spam filtering task, we considered a text classifier proposed in [16], which is currently adopted in several spam filters, including SpamAssassin (http://spamassassin.apache.org), SpamBayes ((http://spambayes.sourceforge.net), and BogoFilter (http://bogofilter.sourceforge.net). It estimates the probability for an email to be spam mainly based on its textual content. We tested the effectiveness of this classifier, as well as that of bagging ensembles, against a poisoning attack proposed in [14], aimed at generating a higher false positive rate at operation phase (i.e., a causative availability attack [1]). The rationale is to modify spam emails by adding "good words" without making them appear as legitimate emails (e.g., using white text on a white background [2]), so that users still report them as spam to the filter, increasing the probability for legitimate emails including those words to be classified as spam.

For the intrusion detection task, we focused on *web applications*. Web applications are largely employed in simple websites, and in security-critical environments like medical, financial, military and administrative systems. A web application is a software program which generates informative content in real time, e.g., an HTML page, based on user inputs (queries). Cyber-criminals may divert the expected behaviour of a web application by submitting malicious queries, either to access confidential information, or to cause a denial of service (DoS). In our experiments we considered a simplified version of HMM-Web, a state-of-the-art IDS for web applications based on Hidden Markov Models (HMMs) [5]. We devised a poisoning attack against this classifier, aimed at allowing more intrusions at operation phase (i.e., a causative integrity attack [1]), as inspired by [2,11]. To this aim, we generate attack queries with (1) a different structure with respect to legitimate queries, and (2) portions of structures similar to intrusive sequences. This attack turned out to be very effective in practice, as it degrades the classifier's performance when very few poisoning attacks are injected into the training set. On the contrary, we noted that the same classifier was very robust to the injection of random sequences or of the intrusive ones.

5   Experiments

We start describing the experimental setup for the spam filtering and web-based intrusion detection tasks. In both experiments performance was evaluated as proposed in [12]. In particular, we computed the area under the ROC curve (AUC) in the region corresponding to FP rates in $[0, K]$: $\mathrm{AUC_K} = 1/\mathrm{K} \int_0^{\mathrm{K}} \mathrm{TP(FP)dFP} \in [0, 1]$, where K denotes the maximum allowed FP rate. This measure may be considered as more informative than the AUC since it focuses on the performance around practical operating points (corresponding to low FP rates).

Spam filtering. Our experiments in spam filtering were carried out on the publicly available TREC 2007 email corpus [3], which is made up of 25,220 legitimate and 50,199 spam emails. The first 10,000 emails (in chronological order) were used as training set, while the remaining 65,419 were used as testing set.

---

[2] http://www.virusbtn.com/resources/spammerscompendium/index
[3] http://plg.uwaterloo.ca/~gvcormac/treccorpus07

The SpamAssassin filter was used to extract a set of distinct tokens from training emails, which turned out to be 366,709. To keep a low computational complexity, we selected a subset of 20,000 tokens (with the information gain criterion), and used them as feature set. In particular, each email was represented by a Boolean feature vector, whose values denoted either the absence (0) or presence (1) of the corresponding tokens in the given email. We exploited the text classifier proposed in [16] to build bagging and weighted bagging ensembles. We considered 3, 5, 10, 20, and 50 as the number of base classifiers, and the simple average to combine their outputs. For weighted bagging, we set the γ parameter of the Gaussian kernel as the inverse of the number of features (i.e., $0.5E^{-4}$), since the latter corresponds to the maximum value of the distance between two samples. Moreover, we performed a further experiment by varying $\gamma \in \{1E^{-3}, 1E^{-5}, 1E^{-6}\}$ besides the default value, to study the effect of this parameter on the robustness of weighted bagging. To keep the kernel density estimation computationally negligible, we computed an estimate of $f(\mathbf{x})$ and $g(\mathbf{x})$ (see Sect. 2.1) by considering a randomly chosen subset of 50 training spam emails (instead of the whole set). We point out that this did not significantly affect the estimation of the probability weights. Poisoning attack samples were created as follows. First, a set of spam emails S was randomly sampled (with replacement) from the training set; then, a number of randomly chosen "good words" (chosen among an available set of good words) was added to each spam in S; and, finally, all spam emails in S were added to the training set. As in [14], we investigated the worst case scenario in which the adversary is assumed to know the whole set of "good words" used by the classifier (which includes all tokens extracted from legitimate emails). In our experiments we noted that the adversary is required to add up to about $5,000$ randomly chosen "good words" to each spam to make the classifier completely useless at 20% poisoning (i.e., using 2,500 poisoning emails). We thus evaluated the performance of the considered classifiers by varying the fraction of poisoning attacks in $[0, 0.2]$ with steps of 2%.

Web-based intrusion detection. We experimented with a dataset which reflected real traffic on a production web server employed by our academic institution. We collected 69,001 queries towards the principal web application, in a time interval of 8 months. We detected 296 intrusive attempts among them. The first 10,000 legitimate queries (in chronological order) were used as training set, while the remaining 58,705 legitimate queries and the intrusive queries were used as test set. Each web application query q has the form $a_1 = v_1 \& a_2 = v_2 \& \ldots \& a_n = v_n$, where a_i is the i-th attribute, v_i is its corresponding value, and n is the number of attributes of q. We encoded each query as the sequence of attributes and their values. [4] The HMM was trained using the Baum-Welch algorithm, to exploit the underlying structure of legitimate sequences, and consequently detect intrusions by assigning them a lower likelihood. To build a simple and effective model, we initialized the HMM with two states: one associated to the emission of symbols in even positions, and the other

---

[4] The whole data set is available at
http://prag.diee.unica.it/pra/system/files/dataset_hmm_mcs2011.zip

associated to the emission of symbols in odd positions. The emission probability of each symbol was initialized as its relative frequency in even or odd positions, depending on the state. The state transition matrix was randomly initialized. Similarly to spam filtering, we carried out experiments also considering bagging and weighted bagging, with 3, 5, 10, 20 HMMs per ensemble, and the simple average as combining rule. In order to apply the kernel density estimator used in weighted bagging (remind that we deal with sequences of non-fixed length), we first extracted all possible bigrams (i.e. contiguous subsequences of length two) from legitimate and intrusive sequences. Then, we represented each sequence as a Boolean feature vector, in which each value denotes either the absence (0) or presence (1) of the corresponding bigram in the given sequence. The length of each feature vector (total number of bigrams) turned out to be $N = 205$. As in spam filtering, the default value of γ has been computed as the inverse of the cardinality of the feature space, i.e., $1/N \approx 5E^{-3}$, and $f(\mathbf{x})$ and $g(\mathbf{x})$ were estimated using a subset of 50 training samples. The sensitivity of weighted bagging to γ was further studied by varying $\gamma \in \{5E^{-4}, 2.5E^{-3}\}$. The poisoning attacks against the HMM were built exploiting the rationale described in the Sect. 4. In particular, poisoning sequences contained only bigrams which were not present in legitimate sequences, but which might have been present in intrusive sequences. As this attack turned out to be very effective, we evaluated the performance of the considered classifiers by varying the fraction of poisoning attacks in $[0, 0.02]$ with steps of 0.2%.

5.1   Experimental Results

In this section we report the results for spam filtering (Fig. 1) and web-based intrusion detection (Fig. 2). We assessed performance using the $AUC_{10\%}$ measure in the spam filtering task (as in [12]), and $AUC_{1\%}$ in the intrusion detection task (since FP rates higher than 1% are unacceptable in this application). Results are averaged over 5 repetitions, as poisoning samples were randomly generated. We do not report standard deviation values as they turned out to be negligible.

First, note that AUC values decreased for increasing percentage of poisoning, as expected. When no poisoning attack is performed (0%), all classifiers behaved similarly, and, in particular, bagging and weighted bagging only slightly outperformed the corresponding single classifiers. Under attack, instead, bagging and weighted bagging significantly outperformed the single classifiers. In particular, the performance improvement was marked when the injected amount of poisoning attacks significantly affected the single classifier's performance (see, for instance, 8-10% of poisoning for spam classifiers).

Increasing the ensemble size of bagging classifiers turned out to significantly improve the performance under attack only in the spam filtering task (Fig. 1, left). The underlying reason could be that bagging can effectively drop the variance of the classification error by increasing the ensemble size (as mentioned in Sect. 2.1, and shown in [7]); thus, increasing the ensemble size may be effective only when poisoning attacks introduce a substantial variance in the classification

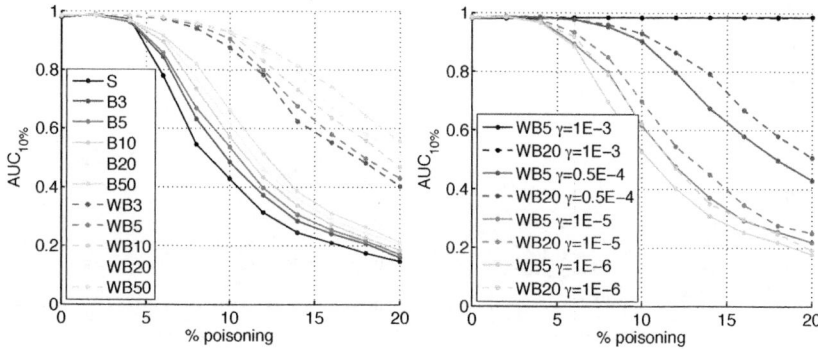

Fig. 1. *Left:* performance of the spam classifier (S), bagging (B), and weighted bagging with default γ (WB) against percentage of poisoning attacks in training data, for different ensemble sizes (3,5,10,20,50). *Right:* performance of WB with ensemble sizes of 5 and 20 against percentage of poisoning attacks in training data, for different γ.

error (whereas this may be not true when the error is highly biased). This aspect can be a promising research direction to investigate.

We focus now on weighted bagging, which significantly improved the performance over standard bagging in both experiments, as expected. This is clearly due to the use of a kernel density estimator, which basically imputes outliers in training data, and reduces their influence. To investigate the effectiveness of weighted bagging more in depth, we considered different values of the γ parameter, as explained in the previous section. The rationale was to alter the performance of the kernel density estimator. From Fig. 1 (right) one can immediately note that a value of $\gamma = 10^{-3}$ in the spam filtering task allowed weighted bagging to completely *remove* poisoning attacks from training data (the performance did not decrease). On the other hand, for $\gamma = 10^{-6}$ performance was very similar to that of standard bagging. Similar results were obtained in the intrusion detection task. Fig. 2 (left) shows that the higher γ, the more gracefully the performance of weighted bagging decreased. It is worth noting that weighted bagging can worsen performance with respect to standard bagging, even in absence of poisoning, if the weights assigned by the kernel density estimator to samples in the same class exhibit a large variance. The reason is that this leads to obtain a set of bootstrap replicates of the training set which do not reflect the correct probability distribution of training samples.

To sum up, standard bagging can provide a significant improvement in performance over an individual classifier, in particular against some kinds of poisoning attacks. The effectiveness of weighted bagging is strictly related to the capability of estimating a reliable set of weights, namely, on the capability of the kernel density estimator to correctly impute the outlying observations. However, when this happens (as in our experiments) weighted bagging can provide a great performance improvement. Besides this, when using a good kernel density estimator the adversary is required to spend more "effort" to build a poisoning

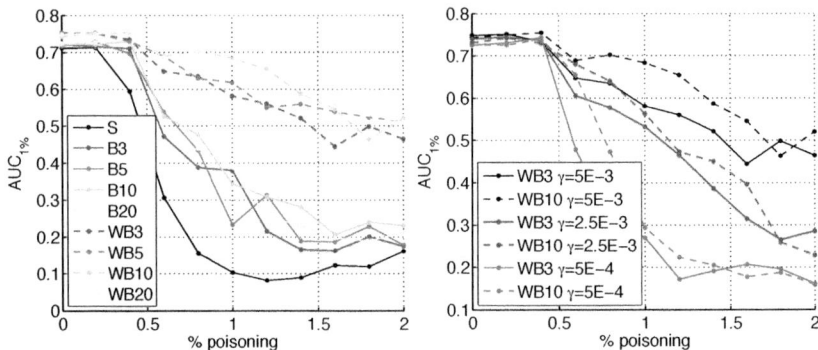

Fig. 2. *Left:* performance of the HMM-web classifier (S), bagging (B), and weighted bagging with default γ (WB) against percentage of poisoning attacks in training data, for different ensemble sizes (3,5,10,20). *Right:* performance of WB with ensemble sizes of 3 and 10 against percentage of poisoning attacks in training data, for different γ.

attack which misleads weighted bagging. For instance, in the case of spam filtering the adversary would be required to add only a few "good words" to each spam email, so that poisoning emails are not easily distinguishable from others. Consequently, he would be required to control a much larger percentage of training data, which may be not feasible in practice.

6   Conclusions and Future Work

In adversarial environments, like spam filtering and intrusion detection in computer networks, classifiers must not only be accurate, but also robust to poisoning attacks, i.e., to the deliberate injection of malicious noise in the training set. In this preliminary study we experimentally showed, for two relevant applications, that bagging ensembles may be a *general*, effective technique to address the problem of poisoning attacks, regardless the base classification algorithm. These results give us valuable insights for future research work. First, we plan to theoretically investigate the effectiveness of bagging against poisoning attacks. Second, we aim to study more general methods for better estimating the set of resampling weights. Lastly, we want to investigate what categories of poisoning attacks can be effectively tackled by increasing the ensemble size (as this emerged as an open problem from our experiments).

Acknowledgements. The authors would like to thank the anonymous reviewers for useful insights and suggestions. This work was partly supported by a grant from Regione Autonoma della Sardegna awarded to B. Biggio and I. Corona, PO Sardegna FSE 2007-2013, L.R.7/2007 "Promotion of the scientific research and technological innovation in Sardinia".

References

1. Barreno, M., Nelson, B., Joseph, A., Tygar, J.: The security of machine learning. Machine Learning 81, 121–148 (2010)
2. Barreno, M., Nelson, B., Sears, R., Joseph, A.D., Tygar, J.D.: Can machine learning be secure? In: Proc. 2006 ACM Symp. Information, Computer and Comm. Sec. (ASIACCS 2006), NY, USA pp. 16–25 (2006)
3. Breiman, L.: Bagging predictors. Machine Learning 24(2), 123–140 (1996)
4. Chung, S.P., Mok, A.K.: Advanced allergy attacks: Does a corpus really help? In: Kruegel, C., Lippmann, R., Clark, A. (eds.) RAID 2007. LNCS, vol. 4637, pp. 236–255. Springer, Heidelberg (2007)
5. Corona, I., Ariu, D., Giacinto, G.: Hmm-web: a framework for the detection of attacks against web applications. In: Proc. 2009 IEEE Int'l Conf. Comm. (ICC 2009), NJ, USA, pp. 747–752 (2009)
6. Dalvi, N., Domingos, P.: Mausam, S. Sanghai, and D. Verma. Adversarial classification. In: Proc. 10th ACM SIGKDD Int'l Conf. Knowledge Disc. and Data Mining (KDD), USA, pp. 99–108 (2004)
7. Fumera, G., Roli, F., Serrau, A.: A theoretical analysis of bagging as a linear combination of classifiers. IEEE TPAMI 30(7), 1293–1299 (2008)
8. Gabrys, B., Baruque, B., Corchado, E.: Outlier resistant PCA ensembles. In: Gabrys, B., Howlett, R.J., Jain, L.C. (eds.) KES 2006. LNCS (LNAI), vol. 4253, pp. 432–440. Springer, Heidelberg (2006)
9. Grandvalet, Y.: Bagging equalizes influence. Machine Learning 55, 251–270 (2004)
10. Hall, P., Turlach, B.: Bagging in the presence of outliers. In: Scott, D. (ed.) Mining and Modeling Massive Data Sets In Science, Engineering, and Business, CSS, vol. 29, pp. 536–539 (1998)
11. Kloft, M., Laskov, P.: Online anomaly detection under adversarial impact. In: Proc. 13th Int'l Conf. Artificial Intell. and Statistics (AISTATS), pp. 405–412 (2010)
12. Kolcz, A., Teo, C.H.: Feature weighting for improved classifier robustness. In: 6th Conf. Email and Anti-Spam (CEAS), CA, USA (2009)
13. Laskov, P., Lippmann, R.: Machine learning in adversarial environments. Machine Learning 81, 115–119 (2010)
14. Nelson, B., Barreno, M., Chi, F.J., Joseph, A.D., Rubinstein, B.I.P., Saini, U., Sutton, C., Tygar, J.D., Xia, K.: Exploiting machine learning to subvert your spam filter. In: Proc. 1st Usenix Workshop on Large-Scale Exploits and Emergent Threats (LEET 2008), CA, USA, pp. 1–9 (2008)
15. Robinson, G.: A statistical approach to the spam problem. Linux J (2001), http://www.linuxjournal.com/article/6467
16. Perdisci, R., Dagon, D., Lee, W., Fogla, P., Sharif, M.: Misleading worm signature generators using deliberate noise injection. In: Proc. 2006 IEEE Symp. Sec. and Privacy (S&P 2006), USA (2006)
17. Rubinstein, B.I., Nelson, B., Huang, L., Joseph, A.D.: Antidote: understanding and defending against poisoning of anomaly detectors. In: Proc. 9th ACM Internet Meas. Conf. IMC 2009, pp. 1–14 (2009)
18. Segui, S., Igual, L., Vitria, J.: Weighted bagging for graph based one-class classifiers. In: Proc. 9th Int. Workshop on MCSs. LNCS, vol. 5997, pp. 1–10. Springer, Heidelberg (2010)
19. Shieh, A.D., Kamm, D.F.: Ensembles of one class support vector machines. In: Benediktsson, J.A., Kittler, J., Roli, F. (eds.) MCS 2009. LNCS, vol. 5519, pp. 181–190. Springer, Heidelberg (2009)

Using a Behaviour Knowledge Space Approach for Detecting Unknown IP Traffic Flows

Alberto Dainotti, Antonio Pescapé, Carlo Sansone, and Antonio Quintavalle

Department of Computer Engineering and Systems, Universitá di Napoli Federico II
{alberto,pescape,carlosan}@unina.it, tonyqui@hotmail.com

Abstract. The assignment of an IP flow to a class, according to the application that generated it, is at the basis of any modern network management platform. In several network scenarios, however, it is quite unrealistic to assume that all the classes an IP flow can belong to are *a priori* known. In these cases, in fact, some network protocols may be known, but novel protocols can appear so giving rise to *unknown* classes. In this paper, we propose to face the problem of classifying IP flows by means of a multiple classifier approach based on the Behaviour Knowledge Space (BKS) combiner. It has been explicitly devised in order to effectively address the problem of the *unknown* traffic too. To demonstrate the effectiveness of the proposed approach we present an experimental evaluation on a real traffic trace.

1 Introduction

In the past decade, network traffic classification – i.e., the association of network traffic flows to the network applications (e.g. *FTP, HTTP, BitTorrent,* etc.) that generate them – has gained large attention from both industry and academy. This is mainly due to two factors: the increasing unreliability of traditional classification approaches, combined with a strong interest in using traffic classification for several practical applications as, for example, the enforcement of Quality of Service and security policies, traffic/user profiling, network provisioning and resource allocation.

Despite the massive contribution of the research community in this field, the analysis of literature shows that there is still no perfect technique achieving 100% accuracy when applied to the entire traffic observed on a network link [15]. Deep Packet Inspection (DPI) is still considered today the most accurate approach, but because of its lack of robustness to the increasing usage of encryption and obfuscation techniques, and also for possible issues related to privacy, is mainly used today as a reference (ground-truth) in order to evaluate the accuracy of new experimental algorithms that should overcome these limitations. Most of these algorithms are based on the application of machine-learning classification techniques to traffic properties that do not need access to packets payload (that is, user-generated content) and, even if their accuracy never reaches 100%, it has been shown that they typically are more resistant to obfuscation attempts and still applicable when encryption is in place [5, 23]. In order to improve the

C. Sansone, J. Kittler, and F. Roli (Eds.): MCS 2011, LNCS 6713, pp. 360–369, 2011.
© Springer-Verlag Berlin Heidelberg 2011

performance of network traffic classifiers based on machine-learning, lately, few approaches based on the combination multiple classifiers have been presented.

In this work, we apply the Behaviour Knowledge Space (BKS) combiner [14] to the problem of traffic classification, attempting the combination of classifiers based on traditional and more recent traffic classification techniques using either packet content or statistical properties of flows. We show on a real traffic trace that it is possible to improve the overall classification accuracy over that of the best-performing classifier.

In several network scenarios, however, it is quite unrealistic to assume that all the classes an IP flow are *a priori* known. In these cases, in fact, some network protocols may be known, but novel protocols can appear, giving rise to *unknown* classes. Another advantage of the proposed approach is that it is straightforward to use it in order to cope with the presence of the *unknown* classes. We also report the BKS combiner performance in detecting *unknown* classes within real traffic traces.

The rest of the paper is organized as follows: related works on traffic classification are presented in Section 2. The BKS combiner is reviewed in Section 3. The tools used for classifying network traffic traces are illustrated in Section 4, while data and base classifiers are presented in Section 5.1. Then, an experimental evaluation of the BKS combiner on real traffic traces is reported in Section 5.2. Finally, some conclusions are drawn.

2   Related Work

In the past years, a large amount of research work has been devoted to traffic classification. Several surveys and papers making comparisons among different techniques [15] [22] [7] [18] have been published. These papers show *pros* and *cons* of different methods, techniques and approaches (DPI- vs statistical- vs port- based) as well as their inability to completely classify network traffic (i.e., reach 100% classification accuracy). On the other side, during the last years, researchers of the machine-learning and pattern recognition communities have developed combination algorithms and approaches for classification problems that allow several improvements, included an increase in overall classification accuracy [16]. A first simple combination approach to traffic classification, for the network traffic classification, was proposed in [20]: three different classification techniques are run in parallel (DPI, well-known ports and heuristic analysis), and a decision on the final classification response is taken only when there is a match between the results of two of them (otherwise the combiner reports "unknown"). Inspired by research in the machine learning and pattern recognition communities related to multiple classifier systems [16], we proposed – in [10] and [9] – the idea of combining multiple traffic classifiers using advanced combination strategies. As for traffic classification, concepts like *En-semble Learning* and *Co-training* have been introduced in [13], where a set of similar classifiers co-participate to learning, while an advanced combination of different traffic classification techniques has been shown in [6]. Finally, it is worth noticing that the approach of combining multiple classification techniques through specific algorithms to build a more accurate multiple classifier system, has been already used with success in other networking research areas as network intrusion and anomaly detection [8].

3   The BKS Combiner

Since some traffic classifiers can be only seen as a *Type 1* classifier (i.e. a classifier that outputs only the most likely class), only criteria that can be applied to classifiers that provide a crisp label as output can be considered. It is worth noting, in fact, that some well-known combination schemes (such as the *Decision Templates* proposed in [17]) cannot be applied to *Type 1* classifiers, since they require class probability outputs (i.e., the so-called *Type 3* classifiers).

Among combiners that can be used for fusing label outputs, we proposed to consider the Behavior-Knowledge Space (*BKS*) approach. This choice is due to the fact that it already demonstrated very good performance in combining traffic classifiers [11] and that it is quite straightforward to use it in order to cope with the problem of *unknown* classes.

BKS derives the information needed to combine the classifiers from a knowledge space, which can concurrently record the decision of all the classifiers on a suitable set of samples. This means that this space records the behavior of all the classifiers on this set, and it is therefore called the *Behavior-Knowledge Space* [14]. So, a BKS is a N-dimensional space where each dimension corresponds to the decision of a classifier. Given a sample to be assigned to one of m possible classes, the ensemble of the classifiers can in theory provide m^N different decisions. Each one of these decisions constitutes one *unit* of the BKS. In the learning phase each BKS *unit* can record m different values c_i, one for each class. Given a suitably chosen data set, each sample x of this set is classified by all the classifiers and the *unit* that corresponds to the particular classifiers' decision (called *focal unit*) is activated. It records the actual class of x, say j, by adding one to the value of c_j. At the end of this phase, each *unit* can calculate the best representative class associated to it, defined as the class that exhibits the highest value of c_i. It corresponds to the most likely class, given a classifiers' decision that activates that *unit*. In the operating mode, the BKS acts as a look-up table. For each sample x to be classified, the N decisions of the classifiers are collected and the corresponding *focal unit* is selected. Then x is assigned to the best representative class associated to its *focal unit*.

In order to detect *unknown* classes we can consider the use of the following decision rule:

$$C(x) = i$$

when $c_i > 0$ and $\frac{c_i}{T} \geq \lambda$, otherwise x is rejected. T is the total number of samples belonging to that focal unit (i.e. $T = \sum_{k=1}^{m} c_k$), while λ is a suitably chosen threshold ($0 \leq \lambda \leq 1$) which controls the reliability of the final decision. By increasing λ we can detect an higher number of *unknown* flows, but this will also correspond to the rejection of a number of known traffic flows whose reliability is quite low. This can even have a positive impact on the overall performance, since most of these flows should be the misclassified ones.

4   Software Tools: TIE and WEKA

To perform experiments on real network traffic traces, and to combine different traffic classifiers, in this work we used TIE[1], TIE a software platform for experimenting with and comparing traffic classification techniques. Algorithms implementing different classification techniques are implemented as *classification plugins* which are plugged into a unified framework supporting their comparison and combination. We refer the reader to [9] as regards the TIE platform as well as the TIE-L7 classification plugin, which implements a DPI classifier using the techniques and signatures from the Linux L7-filter project [1] and that we used in this work to produce the ground truth, whereas, in the following, we describe the new features we introduced in TIE in order to develop this work. Different combination strategies, and in particular one implementing the BKS approach, have been implemented in a TIE's module called *decision combiner*, while a set of support scripts have been developed in order to extract from the ground-truth (generated by TIE-L7) the confusion matrix and the BKS matrix needed for training the combiner. This information is written into configuration files that are read at run time by the combination algorithm selected.

Furthermore, we used the WEKA[2] tool, which implements a large number of machine-learning classification techniques, to be able to rapidly test different machine-learning approaches to traffic classification. We plan to implement few of these techniques as TIE classification plugins, but in order to first study and test a relevant number of machine-learning approaches we implemented a "bypass" mechanism in TIE which is structured in three phases: (i) for each flow, the corresponding classification features extracted by TIE (e.g. first ten packet sizes, flow duration, etc.) along with the ground-truth label assigned by TIE-L7, are dumped in a file in *arff* format (the format used by WEKA); (ii) this file is split in the *training* and *test sets* used to train and test various WEKA classifiers, which generate their classification output in arff format too. (iii) a TIE classification plugin (developed for this purpose) reads the output of a WEKA classifier and use it to take the same classification decision for each flow. Multiple instances of such plugin can be loaded in order to support the output of multiple "WEKA" classifiers at the same time.

Following this approach, TIE has a common view of both WEKA classifiers and TIE classification plugins: all the classifiers are seen as TIE plugins. This allowed us to easily test several classification approaches and to combine several of them plus pre-existing TIE classification plugins not based on machine-learning techniques (e.g. port-based and a novel lightweight payload inspection technique we called Portload). In addition, based on the results of our studies on multi-classification we can later implement in TIE only the best performing classifiers.

---

[1] http://tie.comics.unina.it

[2] http://www.cs.waikato.ac.nz/ml/weka

5   Experimental Analysis

5.1   Data Set and Base Classifiers

For the experimental results shown in this paper we used the traffic trace described in Table 1, in which we considered flows bidirectionally (*biflows* in the following) [9]. To build the ground truth, each biflow has been labeled by running TIE with the TIE-L7 plugin in its default configuration, i.e. for each biflow a maximum of 10 packets and of 4096 bytes are examined.

Table 1. Details of the observed traffic trace

Site	Date	Size	Pkts	Biflows
Campus Network of the University of Napoli	Oct 3rd 2009	59 GB	80M	1M

Table 2. Traffic breakdown of the observed trace (after filtering out unknown biflows and applications with less than 500 biflows)

Application	Percentage of biflows
BITTORRENT	12.76
SMTP	0.78
SKYPE2SKYPE	43.86
POP	0.24
HTTP	16.3
SOULSEEK	1.06
NBNS	0.14
QQ	0.2
DNS	4.08
SSL	0.21
RTP	1.16
EDONKEY	19.21

From such dataset we then removed all the biflows labeled as UNKNOWN (about 167,000) and all the biflows that summed to less than 500 for their corresponding application label. Table 2 shows the traffic breakdown obtained[3]. This set was then split in three subsets in the following percentages: (i) 20% classifiers *training set*; (ii) 40% classifiers & BKS *validation set*; (iii) 40% classifiers & BKS *test set*.

As base classifier, we have considered eight different traffic classifiers, which are summarized in Table 3. The first six are based on Machine-Learning approaches which have been commonly used in the open literature on traffic classification, both in terms of learning algorithms (Decision Tree - J48, K-Nearest Neighbor - K-NN, Random Tree - R-TR, RIPPER - RIP, Multilayer Perceptron - MLP, and Naive Bayes - NBAY) and features [21, 3, 19, 4]. As regards the features, in Table 3 *PS* and *IPT* stands for Payload Size and Inter-Packet Time [12], respectively. J48 and K-NN use the first 10 PS and IPT as feature

---

[3] QQ is an instant messaging application.

Table 3. Single classifiers

Label	Technique	Category	Features
J48	J48 Decision Tree	Machine Learning	PS, IPT
K-NN	K-Nearest Neighbor	Machine Learning	PS, IPT
R-TR	Random Tree	Machine Learning	L4 Protocol, Biflow duration & size, PS & IPT statistics
RIP	Ripper	Machine Learning	L4 Protocol, Biflow duration & size, PS & IPT statistics
MLP	Multi Layer Perceptron	Machine Learning	PS
NBAY	Naive Bayes	Machine Learning	PS
PL	PortLoad	Payload Inspection	Payload
PORT	Port	Port	Ports

vector, MLP and NBAY only consider the PS values, while R-TR and RIP take into account PS and IPT statistics, as their average and standard deviation, as well as the transport-level protocol of the biflow and the biflow duration (in milliseconds) and size (in bytes). The *PortLoad* classifier (*PL*), instead, is a lightweight payload inspection approach [2] that overcomes some of the problems of DPI, as the computational complexity and the invasiveness, at the expense of a reduced accuracy. PL uses the first 32 bytes of transport-level payload from the first packet (carrying payload) seen in each direction. Finally, we have also considered a standard traffic classifier (*PORT*) which is simply based on the knowledge of transport-level protocol ports.

Table 4 shows the biflows accuracy of each base classifier on the test set, for each considered application. Different performance for each application implies that the classifiers are quite complementary each other. Note that the *PORT* classifier has a very low overall accuracy, which in general would suggest to avoid its use in a multiple classifier system. This notwithstanding we decided to consider it since it reaches a very high accuracy on some specific applications. Finally, the last column contains the accuracies that would be obtained by the oracle, that is, by selecting for each biflow the correct response when this is given by at least one of the base classifiers. The overall accuracy obtainable by the oracle (98.8%) demonstrates that the combination of the chosen base classifiers is able to improve the results achieved by the best base classifier (97.2%).

5.2 Evaluation of the BKS Approach

We experimented the combination of the base classifiers from the previous section using the BKS algorithm. When combining the classifiers we experimented with different pools of them, as shown in Table 5, where the overall accuracies for each pool and combiner are reported. The values show that in general it is indeed possible to gather an improvement through combination, as suggested theoretically by the oracle. As it can be expected, this improvement depends on the choice of the classifiers. The port-based classifier has in general a negative impact on the performance of the multiple classifier system, the same happens for the Naive Bayes classifier. This behavior can be easily explained by looking at their rather low performance as base classifiers (Table 4).

Table 4. Classification accuracy – per-application and overall – of base classifiers (best values are in bold font) and oracle

Class	J48	K-NN	R-TR	RIP	MLP	NBAY	PL	PORT	ORACLE
Bittorrent	98.8	97.4	**98.9**	98.6	55.1	79.9	7.7	21.0	99.9
SMTP	95.1	92.9	93.8	**96.0**	90.6	69.2	8.2	96.3	99.4
Skype2Skype	98.8	97.2	96.5	**99.2**	94.6	31.8	98.7	0	99.7
POP	96.0	95.0	98.7	93.9	0	79.6	29.2	**100**	100
HTTP	99.5	98.9	**99.6**	99.3	94.3	63.3	99.1	47.7	100
Soulseek	**98.6**	96.8	98.3	98.1	93	97.7	0	0	99.9
NBNS	78.4	75.9	79.9	**80.4**	9	0	0	0	85.4
QQ	0	0.7	**2.5**	0	0	0	0	0	3.2
DNS	93.6	92.6	95.3	94.4	51.1	86.2	**100**	99.7	100
SSL	96.1	93.1	95.2	93.7	69.5	68.2	**99.1**	0	99.6
RTP	**84.0**	74.1	64.5	77.3	0	41.5	0	0	92.2
EDonkey	93.0	91.7	**93.3**	91.5	72	16.1	92.9	0.1	95.7
overall	**97.2**	95.9	96.3	97.0	82.3	43.7	83.7	15.6	98.8

Table 5. Classification accuracy for different pools of classifiers combined

			Pool of classifiers					Combiner
J48	K-NN	R-TR	RIP	MLP	NBAY	PL	PORT	BKS
X		X	X					97.7
X		X	X			X		97.8
X	X	X	X	X				96.0
X	X	X	X	X	X			97.3
X	X	X	X	X		X		97.9
X	X	X	X	X	X	X		97.7
X	X	X	X	X		X	X	97.7
X	X	X	X	X	X	X	X	97.4

The pool of classifiers achieving the best results is reported in Table 5, using 6 classifiers out of the 8 tested, and closely followed by the second pool in the table that includes only 4 classifiers. The best BKS combiner attained a 97.9% overall accuracy. This value should be interpreted by considering the highest overall accuracy achieved by a base classifier (97.2%) and the maximum theoretically possible combination improvement set by the oracle (98.8%): an improvement equal to 43% of the maximum achievable.

Since the difference between the performance of the best BKS combiner and of the runner-up is not significant, we used the pool reported in the second row of Table 5 for the successive tests. This pool, in fact, does not include the K-NN classifier which is quite time-consuming in the operating phase (an undesirable feature for traffic classification systems). In particular, in Figure 1 the behavior of the chosen BKS combiner as the λ threshold varies is reported. As it can be noted, the proposed approach is able to detect over the 90% of the unknown traffic with a decreasing of the accuracy on known flows of only a 2.5%.

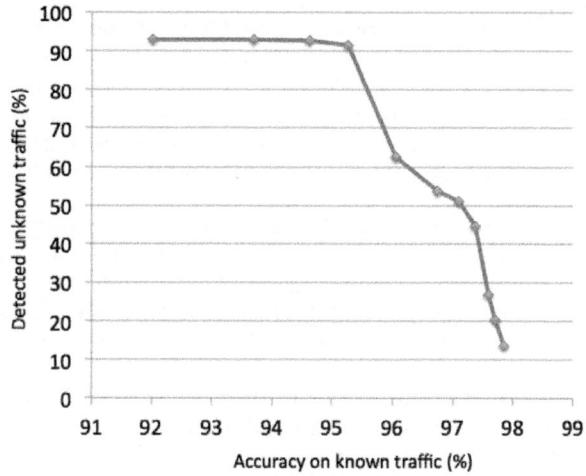

Fig. 1. Trade-off between accuracy and ability to detect the *unknown* classes

6   Conclusion

In this work we have presented and evaluated the Behaviour Knowledge Space combiner for use in network traffic classification. This combiner has been explicitly designed in order to effectively address the problem of detecting *unknown* traffic flows. An experimental evaluation of our proposal on a real traffic trace demonstrated its effectiveness. As future work, we plan to define a methodology for automatically selecting the λ threshold as a function of the considered network scenario (e.g. QoS, security, etc.) as well as to test our approach on other real traffic traces.

Acknowledgments

The research activities presented in this paper have been partially funded by Accanto Systems and by LATINO project of the FARO programme jointly financed by the Compagnia di San Paolo and by the Polo delle Scienze e delle Tecnologie of the University of Napoli Federico II.

References

1. L7-filter, Application Layer Packet Classifier for Linux,
   http://l7-filter.sourceforge.net
2. Aceto, G., Dainotti, A., de Donato, W., Pescapé, A.: PortLoad: taking the best of two worlds in traffic classification. In: IEEE INFOCOM 2010 - WiP Track (March 2010)

3. Alshammari, R., Zincir-Heywood, A.N.: Machine learning based encrypted traffic classification: identifying ssh and skype. In: CISDA 2009: Proceedings of the Second IEEE international conference on Computational intelligence for security and defense applications, USA, pp. 289–296. IEEE Press, Piscataway (2009)
4. Auld, T., Moore, A.W., Gull, S.F.: Bayesian neural networks for internet traffic classification. IEEE Transactions on Neural Networks 18(1), 223–239 (2007)
5. Bernaille, L., Teixeira, R.: Early recognition of encrypted applications. In: PAM, pp. 165–175 (2007)
6. Callado, A., Kelner, J., Sadok, D., Kamienski, C.A., Fernandes, S.: Better network traffic identification through the independent combination of techniques. Journal of Network and Computer Applications 33(4), 433–446 (2010)
7. Callado, A., Szabó, C.K.G., Gero, B.P., Kelner, J., Fernandes, S., Sadok, D.: A Survey on Internet Traffic Identification. IEEE Communications Surveys & Tutorials 11(3) (July 2009)
8. Corona, I., Giacinto, G., Mazzariello, C., Roli, F., Sansone, C.: Information fusion for computer security: State of the art and open issues. Information Fusion 10(4), 274–284 (2009)
9. Dainotti, A., de Donato, W., Pescapè, A.: Tie: A community-oriented traffic classification platform. In: TMA pp. 64–74 (2009)
10. Dainotti, A., de Donato, W., Pescapè, A., Ventre, G.: Tie: A community-oriented traffic classification platform. Technical Report TR-DIS-102008-TIE, Dipartimento di Informatica e Sistemistica, Universitá degli Studi di Napoli Federico II (October 2008)
11. Dainotti, A., Pescapè, A., Sansone, C.: Early classification of network traffic through multi-classification. In: TMA - Traffic Monitoring and Analysis Workshop (in Press 2011)
12. Dainotti, A., Pescapè, A., Ventre, G.: A packet-level characterization of network traffic. In: CAMAD, pp. 38–45. IEEE, Los Alamitos (2006)
13. He, H., Che, C., Ma, F., Zhang, J., Luo, X.: Traffic classification using en-semble learning and co-training. In: AIC 2008: Proceedings of the 8th conference on Applied informatics and communications, pp. 458–463. World Scientific and Engineering Academy and Society (WSEAS), Wisconsin (2008)
14. Huang, Y.S., Suen, C.Y.: A method of combining multiple experts for the recognition of unconstrained handwritten numerals. IEEE Trans. Pattern Analysis and Machine Intelligence 17(1), 90–94 (1995)
15. Kim, H., Claffy, K., Fomenkov, M., Barman, D., Faloutsos, M., Lee, K.: Internet traffic classification demystified: myths, caveats, and the best practices. In: CoNEXT 2008: Proceedings of the 2008 ACM CoNEXT Conference, pp. 1–12. ACM Press, New York (2008)
16. Kuncheva, L.I.: Combining Pattern Classifiers: Methods and Algorithms. Wiley Interscience, Hoboken (2004)
17. Kuncheva, L.I., Bezdek, J.C., Duin, R.P.W.: Decision templates for multiple classifier fusion: an experimental comparison. Pattern Recognition 34(2), 299–314 (2001)
18. Nguyen, T.T., Armitage, G.: A Survey of Techniques for Internet Traffic Classification using Machine Learning. IEEE Communications Surveys and Tutorials (2008)
19. Park, J., Tyan, H.R., Kuo, C.C.J.: Ga-based internet traffic classification technique for qos provisioning. Intelligent Information Hiding and Multimedia Signal Processing, International Conference on 0, 251–254 (2006)
20. Szabo, G., Szabo, I., Orincsay, D.: Accurate traffic classification, jun. 2007, pp. 1–8 (2007)

21. Williams, N., Zander, S., Armitage, G.: Evaluating machine learning algorithms for automated network application identification. In: Tech. Rep. 060401B, CAIA, April 2006, Swinburne Univ. (2006)
22. Williams, N., Zander, S., Armitage, G.: A preliminary performance comparison of five machine learning algorithms for practical ip traffic flow classification. ACM SIGCOMM CCR 36(5), 7–15 (2006)
23. Wright, C.V., Monrose, F., Masson, G.M.: On inferring application protocol behaviors in encrypted network traffic. Journal of Machine Learning Research 7, 2745–2769 (December 2006)

Author Index